OPTICAL AND ACOUSTICAL
HOLOGRAPHY

OPTICAL AND ACOUSTICAL
HOLOGRAPHY

*Proceedings of the NATO Advanced Study Institute
on Optical and Acoustical Holography
Milan, Italy, May 24-June 4, 1971*

Edited by Ezio Camatini

*Technology and Industrial Plant Division
Department of Mechanical and Machine Engineering
Polytechnic Institute of Milan, Milan, Italy*

ℙ PLENUM PRESS · NEW YORK - LONDON · 1972

Library of Congress Catalog Card Number 74-188923

ISBN-13: 978-1-4684-1982-5   e-ISBN-13: 978-1-4684-1980-1
DOI: 10.1007/978-1-4684-1980-1
© 1972 Plenum Press, New York
Softcover reprint of the hardcover 1st edition 1972
A Division of Plenum Publishing Corporation
227 West 17th Street, New York, N.Y. 10011

United Kingdom edition published by Plenum Press, London
A Division of Plenum Publishing Company, Ltd.
Davis House (4th Floor), 8 Scrubs Lane, Harlesden, London, NW10 6SE, England

PREFACE

When Faraday performed the very first experiments on electromagnetic effects, which form the very foundation of modern civilization since virtually everything electrical utilizes them, he was asked of what use his experiments were. He is said to have replied: "What is the use of a baby?" That reply might also be given to those who asked what was the use of basic research which resulted in the development of the laser and holography.

Brigadier General Leo A. Kiley, Commander, Office of Aerospace Research, USAF, in his remarks before the Society of Photo-Optical Instrumentation Engineers at the opening ceremony of the Seminar on Holography held at San Francisco, California, in May 1968, declared, "It has been said that in the twenty-first century at least half of the jobs that will exist then do not exist today. But be that as it may, I am sure that holography will be commonplace."

And later "Even in its embryonic state today, the concept of holography is providing real world applications to the scientific community and I feel that continued rapid growth can be predicted."

Now, only three years after these statements, the development of holography is extraordinary on the scientific side and very promising from the point of view of proven and potential applications. Holography looks like an emerging technique, a valuable tool for engineers.

In this connection, the Advanced Study Institute on Optical and Acoustical Holography held in Milan (Italy) from May 24 to June 4, 1971 has proved to be very fruitful as a stimulating meeting between physicists and engineers for a free discussion, in the most informal way, of the present state-of-art-and-technology of holography as a science and as a technique.

Further progress is linked to two efforts. One is the investigation of the fundamentals of holography. The other main effort is to extend the range of situations in which holography can be employed by introducing variations in the holographic techniques. Of course, each effort enjoys a feedback from the other, so that only the combined work of scientists and engineers will solve the many technological problems connected with holography and its application.

In this line, the proceedings of the Institute appear to be a basic text and a powerful tool for all people working in the field of holography and a valuable contribution to a greater knowledge and to the development of this new technology.

The material in these proceedings covers four main topics: the general theory of holography, optical holography, microwave holography, and acoustical holography. The first topic is introduced by the contribution of Professor Gabor, some historical remarks concerning holography. Optical holography deals with the following subjects: the fundamentals of optical holographic techniques, experience with optical holography, holographic interferometry and mechanical analysis, optical data processing, and such specific problems as the correction of optical aberrations, recording media, information storage, and the application of holography to NDT (nondestructive testing).

As Scientfic Director of the Advanced Study Institute on Optical and Acoustical Holography and editor of these proceedings, my gratitude goes to the Scientific Affairs Division of NATO, which sponsored the seminar and made possible the publication of this volume.

I also wish to thank the lecturers for their papers, which made this book a valuable document of the Institute.

My best congratulations are conveyed to Professor Dennis Gabor, C.B.E., F.R.S., Prix Holweck, "the Father of Holography," who has been very recently honored with the Nobel Prize in recognition of his contribution in the field of holography.

Finally, I am greatly indebted to Mrs. Valeria Sarchi, without whose enthusiastic and extensive help in the organization of the seminar and in the publication of this volume the editor could not have accomplished his task.

Ezio Camatini
Politecnico di Milano, Italy

LIST OF CONTRIBUTORS

Nils H. Abramson, Laser Research Group Division of Production
    Engineering, Royal Institute of Technology, Stockholm,
    Sweden

Byron B. Brenden, Holosonics, Incorporated, Richland, Washington

H. T. Buschmann, AGFA-Gevaert AG, Leverkusen - Bayerwerk, West
    Germany

E. Camatini, Politecnico di Milano, Italy. Scientific Director
    of the NATO Advanced Study Institute on Holography

V. Russo Checcacci, Istituto di Ricerca sulle Onde Elettro-
    magnetiche del C. N. R., Firence, Italy

Jean J. Clair, Faculté des Sciences de Paris, Laboratoire
    d'Optique, Paris, France

A. Consortini, Istituto di Ricerca sulle Onde Elettro-
    magnetiche del C. N. R., Firenze, Italy

A. E. Ennos, Division of Optical Metrology, National Physical
    Laboratory, Teddington, Middlesex, United Kingdom

Dennis Gabor, Imperial College of Science and Technology, London,
    United Kingdom

Emilio Gatti, Centro Informazioni Studi Esperienze, Segrate,
    Milano, Italy

G. Groh, Philips Forschungs Laboratorium Hamburg GmbH, Hamburg,
    West Germany

Leonard A. Kersch, G C O, Incorporated, Ann Arbor, Michigan

H. Kiemle, Siemens AG, Research Laboratories, Munich, West Germany

Emmett N. Leith, University of Michigan, Ann Arbor, Michigan

C. K. Megla, Director, Electronic Research, Technical Staff
    Division, Corning Glass Works, Raleigh, North Carolina

George W. Stroke, State University of New York, Stony Brook,
    New York, and Harvard University, Cambridge, Massachusetts

CONTENTS

ACOUSTICAL HOLOGRAPHY

CLOSING ADDRESS

CONTRIBUTION TO THE VOLUME

INTRODUCTORY SPEECH

Emilio Gatti

CISE (Centro Informazioni Studi Esperienze), Segrate

Milano (Italy)

The holography was born long before than quantum electronics (in 1948) by the pioneering efforts of the British physicist Gabor, whom we have the pleasure and the honour to welcome at this conference.

He and the other experts, whom we equally heartily welcome, will deal with the subject of holography from the very beginning up to the most modern developments which are rapidly growing.

So, I felt much embarassed in choosing the subject of this introductory talk, which is due only to the courtesy and strong pressure of the organizers.

My major choice is to make the talk short in order to leave most of the time to the experts and to try to give a very general outline of the subject.

Visual perception of an object is due to a series of cascaded processes. The object is first illuminated by a light source. The wave scattered by the object carries on information about the object itself: the pupil of the eye cuts a small portion of the wavefront which is focused by the cristalline in order to form an image, which is brought to the brain as the result of photochemical reaction of the retina and a complex signal processing of association type locally done.

The light propagation from the object to the eye is a complete deterministic process described by Maxwell equation or more simply by Huygens-Fresnel principle, according to the classical theory.

1

It is peculiar that in order to observe objects our retina does not need wavefronts or rays but needs what we call an image, that is a particular cross section of a wavefront where the uncertainty in directions of wave vectors is maximum and, on the contrary, the spread in a bundle of rays coming from a point of the object is minimum. As our eyes, so our photographic plates need such particular cross section in order to record an image of an object which can be in turn reproduced on our retina observing the photography.

At this particular section of a bundle of quasi monochromatic rays we have a resolution of the order of the wavelength which allows immediately the evaluation of the information capabilities of a given area of the image. (One should also take into consideration the number of brilliance steps distinguishable in any independent area λ^2).

The same information is present at any other cross section of the bundle of rays considered but, in order to reconstruct the bundle of rays, one needs the double of information elements than needed in the focal plane; furthermore, he needs to know, not only amplitudes but phases also.

This is just what holography does by mixing techniques with carryer beams using as nonlinear elements, photographic plates or photocromic materials.

In the perception process:

Object – Light wave – Image

holography breaks the process in two processes by freezing first the light wave in a recorded hologram:

Object – Light wave, Carrier wave – Hologram.

The reconstructed light wave can be seen or photographed as if it were the original one.

It is interesting to note that the most general monochromatic wave has at the most, without any quantum mechanics consideration, ∞^2 degrees of freedom: i.e. amplitude and phases at a given plane; and, therefore, it is able to carry on information about surfaces and not about volumes. We generally look at surfaces of objects and, when we think to see them three dimensionally, we think to surfaces in a three dimensional manifold. If we have to look at volumes, we must put a lot of a priori information in our evolution or try to introduce other variables: for instance, different wavelengths or colours in order to distinguish different depths.

The only purpose of the above review of some basic concepts is to introduce the main subject of this Seminar that will engage you for two weeks in a very stimulating meeting and for which I wish a fruitful work.

WELCOME TO MILAN!

E. Camatini

Politecnico di Milano, Italy

Scientific Director of the NATO Advanced Study
Institute on Holography

It is my pleasure to welcome you to Milan and to express in
advance the deepest gratitude for your valuable contribution, as
lecturers or participants, to this Advanced Study Institute on:
DEVELOPMENT OF LASER: OPTICAL AND ACOUSTIC HOLOGRAPHY

Holography is something which sounds like a magic name to
most people and when you try to explain to them that hologram
means "entire recording", from the Greek word "holos", they still
do not understand, nor show any gratitude to Prof. Gabor, the
Father of Holography, who invented that name.

This situation is typical of any other when we speak about a
new science or technique.

Some years ago, on the door of a laboratory in California I
saw a picture from a popular magazine of a light beam blasting a
missile from the sky, boldly titled "The incredible Laser".
Few years later, on the same door a second legend has been added:
"For credible Lasers see inside".
That defaced picture neatly summarizes the brief history of the
Laser, one of the most exciting discoveries of modern physics.

Today, many people are familiar with Laser. It might be that
within few years holography become so popular that we shall see the
tourist to use the holographic equipment instead of the tradition-
al camera to take shots of Scala Theater or something else.

Holography with Laser is a very recent development and much
research is still currently in progress. It is therefore hard to

judge the relative importance of its application, but many of those already under development promise to have widespread use in the future.

Applications of holography include information storage, record ing of images in depth, the use of holograms as optical elements, and as a means of performing precise interferometric measurements on three-dimensional objects of any shape and surface finish.

However, of the many fascinating possibilities for holograms none seems to hold greater promise than that as a tool for scientific research.
The computer has given the scientist a remarkably powerful tool for analysing immense quantities of data, and to match its abilities and exploit its power to the full, the scientist needs single systems for gathering very large amounts of data at minimum inconvenience to himself.

Holography promises to provide one such system: a single means for gathering data that can then be analysed by a complex system.

Although still in the experimental stage of development, acoustic holography appears destined to play an important future role in such areas as medical diagnosis, non distructive testing, underwater viewing and underground exploration. Its full utilization must await development of certain key components for the imaging system and sophistication of the over-all technology involved.

However, we are not here to deal with the proven and potential applications of holography only; aim of the Seminar is a free discussion, in the most informal way, on the present state-of-the art-and-technology of holography, that means as a science and as a technique.
In other words, the Institute will be very successful if only a stimulating meeting be promoted between plysicists and engineers and the scientific and technical aspects of the subject be dealt with in a balanced way.
In this line, your cooperation, both of lecturers and participants, will be highly appreciated; personally I do not want to interfere at all with your work, therefore the success of the Seminar will mean your success and the merit of this success will be yours, and yours only.

Before the Seminar is declared officially open, please let me express the deepest gratitude to the Scientific Affairs Division of NATO, from which this Institute has been sponsored, and to the FAST, which hosts the meeting and puts at our disposal its very efficient organization: in particular, to the President of FAST, ing. Rossari, and to the President of the Cultural Committee of FAST, prof. Dadda.

I am also indebted to Professors Caldirola and Gatti for their
encouragement during the organization of the Institute; to Professors
Vienot and Sona, whose contribution to the finalization of the pro-
gramme was extremely precious; to the lecturers, who bring into
the Seminar the high contribution of their science and experience;
to the participants, who came here with the scope of promoting a
constructive discussion with the lecturers.
To all of you I cordially wish a pleasant stay in Milan and a
fruitful work.

THE PRINCIPLE OF WAVEFRONT RECONSTRUCTION, 1948

Dennis Gabor, FRS

Imperial College of Science and Technology

London (U.K.)

24 years ago, in 1947, I was very interested in improving the
electron microscope. The electrons used in microscopy have wavelengths of the order of 0.05 Angstroms, hence one might have thought
that they not only allow us to see atoms, but even explore their
details. This of course is a naive view; there are some very fundamental reasons which will forever prevent us to explore the inside
of individual atoms. But at least, one would think, it ought to be
possible to resolve atomic lattices, which have usually atom spacings
of the order of 2 - 3 Angstroms. Unfortunately, it turned out that
electron lenses, electric or magnetic, cannot be made perfect, they
have a spherical aberration which at best would allow us to go down
to 3 - 4 Angstroms. This was the theoretical limit in 1947, now it
is practically reached, while in 1947 the practical limit was about
12 Å.

I put it in my head that I had to think out a method for overcoming this barrier. By Abbe's formula the smallest distance which
can be resolved by a microscope is

$$d = \frac{0,6\,\lambda}{\sin\theta}$$

where θ is the half-angle of the cone of radiation which enters the
lens. This is the condition for at least one diffracted beam to go

into the lens aperture, because it is the diffracted beams which
carry the information on the object. It means that we must double
the aperture if we want to go from 4 Å to 2Å. But if we double the
aperture in the microscope, the spherical aberration disc increases
8 times in diameter, as it goes with the cube of θ . This means
that we get a hopelessly diffused picture.

Could we not sharpen up this picture somehow? Some trick is
needed, because, as a little analysis shows, the overlapping patches
of atoms merge into the noise level. I realised that the trick is
the use of coherence. Coherent waves interfere, they produce interference fringes. Would it not be possible by adding a known, simple
wave, to the unknown, complex wave which comes from the object to
produce a fringe system which contains all the information, and from
which the object could be sharply reconstructed by some optical
method?

I was delighted to find, that the problem has a solution and a
very simple, general and direct solution. Let me give a simple explanation. It is not quite the one which occurred to me first, but
soon after the first flash of insight. Imagine that we have a bad
electron lens, so that the spherical wave emanating from a point
object is distorted in such a way that the actual wavefront deviates
from the original, spherical wave, by a distance which is proportional to the fourth power of the angle θ . Now let this wave interfere with a plane wave, parallel to the image plane. We then get on
the plate a system of circular interference fringes, which will not
be the usual Fresnel-zone pattern, because the outside fringes will
be increasingly closely spaced. But now let us illuminate this
fringe-pattern with a plane, monochromatic light wave. As the phases
of the plane wave and of the distorted spherical wave coincided in
the fringes, and of course they will be right for the illuminating
wave, they must be right also for the object wave. Consequently the
diffraction at the diffraction pattern must reconstruct, as a component, the original wave. We now have a light wave, and in light
optics we are free to correct the spherical aberration of this wave.
Hence putting behind it a lens which has a spherical aberration
opposite to that of the electron lens, we can reconstruct the o-
riginal, undistorted waveform, and a true image of the object.

This was the principle of Wavefront Reconstruction, and you
will note that I had rightaway conceived it in a form a little more
sophisticated than as it is usually explained nowadays; with e-
lectrons and bad electron lenses for the taking of the photograph,
with correcting lenses in the reconstruction. The reason was of
course that I had a special object in mind; the improvement of the
electron microscope. Why should one bother in light optics with
such a complicated two-stage process, when coherent light was so
weak and uncomfortable to use, when we had such perfect lenses,even
achromatic ones? Little did I think at that time that after 24

years the application of holography in electron microscopy would
still be in a primitive stage, while the simple optical experiments,
which I considered only as model or feasibility experiments, would
give rise to a new branch of optics, with some 2000 papers and a
dozen books!

Though some experts gave immediate recognition and appreciation
of the principle of wavefront reconstruction, practical applications
remained few, until the advent of the laser. The early workers,
Hussein El-Sum and Alberto Baez, seized on the feature of holography
that it was lensless, and applied it to X-rays, for which we have
no good lenses. With my collaborators Michael Haine and James Dyson
I carried on for a few years attempts to improve the electron microscope with holography, but with only a modest success. We had started too early. Twenty years ago the electron microscope was still
far from the theoretical limit. The resolving power was limited by
such apparently trivial effects as the vibration of the column,
stray magnetic fields, creep of the object stage, and the growth of
"contamination" on the objects. Now when the patient work of e-
lectron microscopists has at last reduced the practical resolution
limit to the theoretical limit, now would be the time for somebody
to undertake this work, but it does not appear that candidates are
forthcoming.

Unknown to me, a most interesting branch of holography was developing from 1956 onwards in the Willow Run Laboratory, attached
to the University of Michigan. It was holography with electromagnetic waves and reconstruction by light, which was called "Side
Looking Radar" or "Synthetic Aerials". It was classified work; the
first publication by Cutrona, Leith, Palermo and Porcello appeared
in 1960. An aeroplane flies in a straight line and emits pulse
trains, in a wide, horizontal fan. These return from the various
objects, and the return times are marked as the y-coordinate on a
film which moves steadily and slowly in the x-direction. If now one
analyses the pulses which return from a single point in the plane,
one finds that these will be recorded on a stretched-out hyperbola.
Sometimes the pulse train arrived in phase with the local oscillat-
or, this will give maxima, sometimes in counterphase, this gives
minima along the hyperbola. The distribution of these maxima and
minima is the same as of those along a radius in a Fresnel-zone pattern, that is to say they are one-dimensional holograms. Cutrona,
Leith et al stretched the hyperbolas straight by means of an ingenious optical system, and then reconstructed the object plane by illumination with a monochromatic mercury lamp. The reconstructions
were of impressive perfection. So, curiously, in the first 12 years,
the aim of holography was the reconstruction with light of electron
or X-ray records, with wavelengths about 100,000 times shorter than
light, and reconstruction from electromagnetic holograms, with wavelengths about 100,000 times longer.

Optical holography came into its own only in 1962, when Leith
and Upatnieks first applied the laser to it. In 1963 they published their first results, which were already much superior to any-
thing I could obtain in 1948 with a mercury lamp. The helium-neon
laser, right from the start, could supply light with about 3000
times the coherence, and a few million times the intensity of the
mercury lamp. (The coherence length of a line of a high-pressure
lamp is about 0.1 mm, that of a laser at least 30 cm, but nowadays
many meters have been achieved, exceptionally even kilometers.)
This made it possible to produce large holograms on fine-grain, low-
speed photographic material. Moreover, it made it possible to
separate the illuminating beam from the coherent background or
"reference beam". In 1948 I was forced to arrange everything in
one line, moreover I had to use very small objects, and to make
small holograms on high-speed emulsions. It could not therefore be
avoided that the two images which are formed in the reconstruction,
the "virtual" and the "real" one, appeared in one line, and disturb
ed each other. This dis not bother me much at that time, because
in electron microscopy the coherence was just as insufficient, and
at that time there were no electron-beam splitters. (It was only
later, in 1956, that Möllenstedt produced his ingenious beam
splitter, the "biprism".) I had a good method for getting rid of
the second image. An optical system which corrects the spherical
aberration in one of the two images, <u>doubles</u> the aberration in the
order, and thereby washes out the unwanted image so evenly that it
hardly disturbs the sharp one. (An example of this method is reproduced as the last plate on G.W. Stroke's book. <u>Introduction to
Coherent Optics and Holography</u>, Academic Press, 2d ed. 1969, p.324.)

The high coherence of the laser enabled Leith and Upatnieks to
eliminate the second image by a very simple method; simply by putting the reference beam at an agle to the object beam. In the reconstruction the two images are now <u>angularly</u> separated, by twice
this angle, and if their angular extension is not more than this,
they are now completely separated, not only by depth, but also by
direction. Moreover, the separation of the reference beam from the
illumination made it possible to image also opaque objects, not
only almost transparent ones, and one could adjust the ratio of the
reference beam and object beam intensities to an optimum. There was
even a third advantage: one could use photographic material with al
most any H & D (Hurter & Driffield) characteristic. In fact the
fine-grain plates, such as the Eastman Kodak 649F emulsion have a
gamma of about 4 - 5; they are much too hard for most purposes. Yet
with such a hard process, one can make a good holographic portrait
on an emulsion which is almost black-or-white. The reason is that
the error-terms produce harmonic components in the hologram, which
diffract the light outside the angle used for the reconstructed
image.

However, the simple in-line method which I used in 1948, proved itself also remarkably efficient when combined with the superior
coherence and intensity of the laser. All the remarkable pulse-
holograms which Brooks, Heflinger Wuerker and their collaborators
produced in the TRW Laboratories from 1965 onwards, are made with
the in-line method.

From 1963 onwards progress was rapid. In 1964 Leith & Upatnieks published the first "diffused" holograms, which were taken
with diffuse, wide-angle illumination of the object. This is not
important in the case of rough objects, which diffuse the light any
way in a wide angle, but it is vital for objects which diffract
only, without diffusing, such as transparencies.The diffuse illumin
ation had two very important consequences. One was that, while in
regular illumination the information on the object or any part of
it was contained only within its own diffraction pattern, it was
now diffused over the whole hologram. Any point of the hologram
now "saw" any point of the object, and looking through the hologram
with two unaided eyes, one could see the object in three dimensions.
Holography was of course three-dimensional from the start, but previously one could observe this only by racking a microscope or short
focus eyepiece through the field, and seeing one layer after the
other coming out sharp.

The diffuse hologram is the superposition of a very great
number of "regular" holograms thrown over each other, and distribut
ed over the whole area, at random. In consequence, any small area
of the hologram now contains the information on the whole object.
If one damages 10% of the hologram by dust or scratches, this will
reduce the quality of the reconstruction by 10%, instead of erasing
10% of the object. This is a valuable property of diffuse holograms, which has found application in the production of printed
circuits, in the SELECTAVISION system of the RCA, and, most im-
portantly, in holographic stores for computers, where it introduces
an enormous factor of safety. Unfortunately, diffuse illumination
produces also the phenomenon of "speckle noise". I will have more
to say on this phenomenon and its avoidance in the General Theory.
I want to say now only that one cannot blame the hologram for it .
The hologram records faithfully all information, but it cannot
separate the wanted information on the object, from the unwanted
information on the roughness of the diffuser.

From 1965 progress in holography became so rapid, that I cannot
give a full report on it, I will pick out only a few highlights.
One was that in 1965 several workers independently discovered the
basis of holographic interferometry. (Brooks, Heflinger & Wuerker, Powell & Stetson, J.M. Burch, Hillebrand & Haines.) One can
call it also "non-simultaneous interferometry". The principle was
hidden in my old equations of 1948. If one takes two holograms in
succession on the same plate, with the same reference beam, in the

reconstruction the two waves, which were "frozen in" in succession, will be "revived" simultaneously, and they will interfere with one another. One can thus obtain a direct interferometric record of the small deformation of an object. I need not enlarge on this; holographic interferometry has become the chief engineering application of holography, and it will be very thoroughly treated in this Seminar.

Another very important development, which must not remain unmentioned even in a short survey, was started by Yu.N.Denisyuk and P.J. Van Heerden in 1962, just before the introduction of the laser into holography. This was "deep" or "volume" holography. Van Heerden initiated holographic volume stores, which in fact have not become as important to this day as could have been expected. Denisyuk combined the idea of holography, with the old (1891) method of photography in natural colours by Gabriel Lippmann.

One can interpret the transmission holograms, of which I have talked so far, as records of standing waves. If two coherent beams intersect, standing waves will be produced which bisect the angle between the two directions. If now, as Denisyuk has proposed one arranges the object and the reference source at the two opposite sides of the photographic emulsion, the standing waves will form small, acute angles with the emulsion surface. Their spacing will be approximately half a wavelength, so that there will be something like 20 of them even in the usual small emulsion thickness of 5 microns. Where there were maxima of the electric field, silver will be precipitated. In very fine grain "Lippmann emulsions" these will be small globules of 20-50 Å diameter. Such globules absorb almost nothing, and they scatter light almost isotropically. But these spherical wavelets are very weak, they will have an appreciable effect only in directions in which the amplitudes add up in phase. Now, if the system of Lippmann-layers is illuminated from the position of the reference source, this will happen just in the direction of the object ray, and only for a certain wavelength. If the emulsion has not shrunk, it will be the original wavelength. Therefore, by Denisyuk's idea, one obtains reflecting holograms, in natural colours. Denisyuk had no laser at his disposal, and he could carry out only a "feasibility proof"; a two-colour reflecting hologram was first produced by G.W. Stroke and A. Labeyrie in 1965.

This is a brief and incomplete historical review of holography. I have left out such important fields as pattern and character recognition, the holographic deblurring of images, and many others. You will be able to hear all about these in this Seminar.

WAVEFRONT RECONSTRUCTION

Dennis Gabor, FRS

Imperial College of Science and Technology

London (U.K.)

A reference beam with complex amplitude A (x,y,z) $e^{-i\omega t}$ and an object beam coherent with it, with amplitude $B(x,y,z)e^{-i\omega t}$ fall simultaneously on a thin photographic emulsion in the plane $z = 0$. We can operate with "scalar light" so long as the polarizations are the same, i.e. if the electric vectors in the two beams are parallel or nearly parallel. If the angle of the two electric vectors is considerable, a correction must be made, because the amplitudes are added vectorially. The photographic plate (or any other energy detector,) completely ignores the phase factor and records only the resultant energy, which is proportional to the <u>joint intensity</u>

$$I = (A + B)(A^* + B^*) = AA^* + BB^* + (AB^* + A^*B) \qquad 1.$$

The first two terms are the intensities of the beams taken singly, the last is the interference term. If there is a phase difference ϕ between the complex vectors A and B, the interference term is

$$(AB^* + A^*B) = |A||B| \cos \phi \qquad 2.$$

This is assuming perfect coherence, otherwise we must add to this
the Van Cittert-Zernike coherence factor

$$\gamma = \frac{I_{max} - I_{min}}{I_{max} + I_{min}} \qquad\qquad 3.$$

I_{max}, I_{min} are the maximum and the minimum of the intensity in an
interference pattern when A and B are equal. γ is zero for incoherent light, unity for full coherence. Under the conditions in
which laser holography is carried out, one can usually assume full
coherence. A correction is needed only when the optical paths for
A and B from the laser to the plate differ by not much less than
the coherence length.

Let us now first assume that we carry out the photographic
process with a gamma $\Gamma = -2$. We choose this, because this makes
the amplitude transmission of the record proportional to the origin
al intensity I. (The intensity transmission is proportional to the
square of I.) Let us now illuminate the hologram with the reference beam A. We obtain a transmitted amplitude

$$(AA^* + BB^*)\ A + (AA^*)\ B + A^2B^* \qquad\qquad 4.$$

where the first term is essentially the illuminating beam, unmodifi
ed; the second term is the reconstructed wave; the third term is
the twin image.

Under the usual conditions of holography, the factor of A in
the first term is almost uniform. (Even in "focused holography"
it is almost uniform, because the uniform term AA^* is usually much
larger than BB^*). Hence this term is essentially the illuminating
beam unmodified.

The second term, (AA^*) B is what we are after: the reconstruct
ed wave. AA^* is exactly uniform in a plane reference beam, very
nearly so with a spherical reference beam from a distant source.
However, the reconstruction is somewhat spoilt by the last term,
A^2B^* which represents the "twin object", and has exactly the same
energy as the reconstructed wave. Its significance is most easily

understood if A = const. i.e. in the case of a plane reference wave, parallel to the plane of the emulsion. In this case $A^2 = AA^*$ and the twin wave B^* differs from the reconstructed wave B only by all its phases changed into the opposite; delay changed into advance. This means that B^* appears to come from an object which is the mirror image with respect to the hologram plane of the object which emitted B. It is easy to show that if the illuminating wave A is not parallel to the plate, each ray will be skewed by twice the angle, and the "twin" objects will no longer be in line. It may be noted that the "twin" image will be perfect only in the case of plane reference beams. In the case of spherical reference beams it will have the type of aberrations which one obtains with a spherical mirror. These have been analysed in detail by R.W. Meier. It can be shown that the twin image and its aberrations will be very nearly the same as if the object were viewed in a spherical mirror having the shape of the reference beam at the hologram.

There is no need to go in detail into the peculiarities of the twin object, as it has been thoroughly eliminated first by the skew reference beam, (Leith & Upatnieks,) later also even more completely in reflecting holograms. Let the reference beam be a plane wave, but at an angle θ to the plate normal

$$A = |A| \, e^{ikx}$$

where

$$K = (2\pi/\lambda) \sin \theta$$

We now have, as before $AA^* = |A|^2 = $ const. but $A^2 = |A|^2 e^{2iKx}$. This means that the factor B^* now modulates a wave which is turned by twice the angle θ. The twin object has suffered an "affine transformation"; every ray leading to it from the hologram is twisted by the same angle 2θ. If the object, seen from the hologram, subtends an angle less than 2θ, the two reconstructed waves are completely separated.

For simplicity of explanation, I have previously assumed Γ = − 2, but now we can rid ourselves of this condition. Γ = − 2 meant using a positive process, or taking a print of the original hologram, which always spoils it. There is no need at all to adhere to this rule if we make the reference beam strong relative to the object beam. Let us call the ratio of the interference term to the background term

$$X = \frac{2|A||B| \cos \phi}{AA^* + BB^*}$$

and let this be small against unity. We can then use the binominal
expansion for the amplitude transmitted by the original, (negative)
hologram, processed with an arbitrary gamma

$$I^{-\frac{1}{2}\Gamma} = (AA_+^* \cdot BB^*)^{-\frac{1}{2}\Gamma} \left[1 - \frac{1}{2}\Gamma X + \frac{\frac{1}{2}\Gamma\left(\frac{1}{2}\Gamma + 1\right)}{2} X^2 + \dots \right] \qquad 5.$$

The first term is large, but in the case of a skew reference
beam it is harmless, because it is essentially the illuminating
beam, which is at an angle θ to be object beam. The object beam,
(with its twin) is represented by the second term, and the twin
part of it is harmless, as explained before. The third term, which
is an error term, is small, but not always very small. It is made
almost harmless by the fact that the interference term is now squared. Now the central ray of the object forms an angle θ with the
reference beam, so the interference fringes will be grouped around
a fringe system cos Kx = cos $\left[(2\pi/\lambda)x.\sin\theta \right]$ and this squared
gives

$$\cos^2 Kx = \frac{1}{2}(1 + \cos 2Kx)$$

The first term of this falls into the illuminating beam, the
second gives diffracted beams at angles ± 2θ , which (if the object is not too wide,) fall well outside the object beam, and
forms "secondary objects". Thanks to these fortunate circumstances the holographer has a wide choice in his emulsions and photographic processes. The photographic process in holography has
been investigated in great detail by A. Kozma, who has shown that
under certain circumstances even a black-or-white hologram can
give satisfactory reconstructions. The chief reason for this is
that in holograms the phase is so much more important than the in
tensity. This has been demostrated also by A. Metherell in acous
tical holography.

VOLUME HOLOGRAMS

So far we have assumed recording in one plane, that is to say with very thin emulsions, but it is clear that the equations hold true also for every thin layer of a thick emulsion – in the recording. There will be a difference however in the reconstruction, because the holograms in the various layers will interfere with one another. Two new effects arise: colour selectivity and directional selectivity.

A thin hologram which has been taken with one wavelength, can be illuminated with any other wavelength, and it will give a reconstruction. There will be, however, changes in the size and the position of the reconstructed object, also certain aberrations. If one illuminates with a wavelength X times longer than the original, the diffraction angles will be increased in the ratio X. (More exactly, the sines of the diffraction angles will be increased in this ratio, and this gives rise to aberrations.) The distances from the hologram will be reduced in the ratio 1/X, hence the shape of the object will be distorted. This effect is rather troublesome in acoustical holography, (where X is much smaller than unity,) because it prevents the correct, proportional reproduction of depths, except if the hologram is reduced in the same ratio, which is technically difficult.

A deep or volume hologram on the other hand will give good reflectances only for a narrow range of wavelengths. This is of course the reason why these are suitable for holography in natural colours. I will quote only the formula for the amplitude selectivity if the illuminating light with $k' = 2\pi/\lambda'$ is different from the original $k = 2\pi/\lambda$, assuming that the emulsion thickness d has not changed in the process. λ and λ' are the wavelengths in the emulsion, usually 2/3 of the vacuum wavelengths.

$$S_c = \frac{\sin\left[\frac{1}{2}(k'-k)(\cos\theta + \cos\theta_n)d\right]}{\frac{1}{2}(k'-k)(\cos\theta + \cos\theta_n)d} \qquad 6.$$

θ and θ_n are the polar angles, (angles with the plate normal,) of the reference ray and the object ray, both in the medium of the emulsion. (The cosines are usually not far from unity.) This formula has been derived by several authors. For small incidence angles the first zeros of the reflectance are at

$$\frac{\lambda' - \lambda}{\lambda'} = \pm\frac{\lambda}{2d}$$

This means that the total waveband transmitted is a fraction of the wavelength equal to twice the reciprocal of the number of Lippmann layers in the emulsion. For a 5 micron emulsion and green light this is about ± 1/25.

This formula reveals a difficulty of holography in natural colours. If one illuminates the hologram with laser light, it pays to make the thickness large, because the reflection coefficient for the centre of the band increases with the square of the number of Lippmann layers – but any small shrinkage of the emulsion will be fatal. On the other hand, if one uses natural light, (white or line spectra from high pressure arcs,) one gains nothing beyond a certain thickness, because the admitted spectral width decreases in the ratio as the peak intensity increases. This practical difficulty is not very serious for monochrome reflecting holograms, but it becomes very troublesome with two or three colours. As it does not pay to increase the depth beyond about 10 microns, the two or three colours must share the same gray scale in a limited number of Lippmann layers. One can then allow for one only 1/2 or 1/3 of the amplitude range, and this, theoretically, reduces their intensities to 1/4 or 1/9. Practically the reduction is even worse. One can help this to some extent by putting a structure like a Dufay-filter on top of the plate during the exposure, but other means of improvement appear desirable.

DIRECTIONAL SELECTIVITY

Volume holograms differ from thin holograms also in their feature that one cannot freely displace the illuminating source from the position of the reference source. At certain positions the reflectance drops to zero. The phenomenon is rather complicated and I must refer to D. Gabor & G.W. Stroke, The Theory of Deep Holograms, Proc. Roy. Soc. A, 304, 275-89, 1968.

I have proposed to make use of this phenomenon in my scheme of stereoscopic projection of three dimensional pictures on holographic screens. (U.S. Patent No.3,479,111.) The two projectors are so displaced that the pictures projected by them are visible only from alternate viewing zones in the theatre.

HOLOGRAPHIC MIRRORS

For completeness a third variety of hologram may be mentioned, first realized by N.K. Sheridon of Xerox Corporation in 1968. The hologram is taken either like a transmission hologram, or like a reflection hologram, but on a thin photoresist, coated on top of glass. When this is etched, a reflection hologram gives a sawtooth-profile, just like the "diffraction mirrors" proposed by Toraldo di Francia, (and realised by him for electromagnetic waves.) They are ideally "blazed". If they are silvered, they can have reflection coefficients of 90%, and can put all the energy into one diffrac-

tion order. Alternatively, the hologram is taken as for transmission, with object and source at one side. This gives a sinusoidal profile, but by a happy chance, these sinusoids, if they are
a little skewed, also give a "blazing" effect, and allow to put
more than 50% into one diffraction order, instead of the maximum of
33.9% for sinusoidal gratings. Spherical diffraction gratings of
surprising perfection have been produced recently by this method by
Antoine Labeyrie. By a further happy chance, they can be made
stigmatic not only for one wavelength but for a spectrum, and in
their line are already superior to the best ruled spherical gratings; the results of just a hundred years of development.

INFORMATION THEORY IN HOLOGRAPHY

Dennis Gabor, FRS

Imperial College of Science and Technology

London (U.K.)

HUYGENS' PRINCIPLE

Huygens' Principle, expressed in modern form, states that the information in a light beam is an invariant. The same information can be extracted from it at any cross section.

The classical formulation of Huygens' Principle is somewhat non-physical. It states that if one knows all the data on a closed surface which contains all the sources, one can calculate the light effects everywhere. But it is physically impossible to measure the light vector as a function of space and time at any surface. A reasonable formulation must mean: From the measurable data at a first surface one can deduce the measurable data at another surface.

Such a physical description of natural, polychromatic, incoherent light is very complicated. White light is a stochastic phenomenon. Its "spectrum" is not the Fourier transform of the amplitude as functions of time, but a periodogram; it contains only the powers, that is to say the squared amplitudes, integrated over time. It was known to Boltzmann, Lord Rayleigh and Michelson that the power spectrum is the (cosine) Fourier transform of the autocorrelo gram ϕ of the amplitudes $A(t)$

$$\phi(x) = \lim_{T=\infty} \frac{1}{2T} \int_{-T}^{+T} A(t)\, A(t+\tau)\, dt$$

It was only much later, (1928-30) that Norbert Wiener (and independently Khintchine) formulated this theorem in a mathematically rigorous form. Nowadays it is used very efficiently in Fourier
Spectroscopy. The autocorrelogram is directly recorded by delay-
ing one-half of the light by a movable Michelson-mirror, and measuring the joint intensity. From this the power spectrum is calculated by computers, using the Cooley-Tukey algorithm for the Fourier
transformation.

Norbert Wiener was also the first to think of a full spatial-
temporal representation of natural light in terms of measurable
data. In two papers (1,2) he proposed describing it by a "co-
herency matrix". These papers were completely disregarded by
physicists until the coherence matrix was re-invented in 1954 by
the author, and in 1955 independently by H. Gamo. (3,4) Wiener's
theory was incomplete and contained an error. (The Hilbertian
coherence matrix applies only to nearly monochromatic beams, the
general representation requires two Hilbertian matrices, as pointed out by Gamo). Even the description of almost monochromatic but
not fully coherent beams is rather complicated. In order to $_2$ describe a beam which carries with itself N invariant data N^2
measurements are needed for full specification. Thus the real
physical implementation of Huygens' Principle is practically very
inconvenient.

Fortunately all difficulties vanish when we consider fully
coherent light. Only two data are required to describe it fully
at any point: amplitude and phase. But in order to put physical
sense into Huygens' Principle, we must specify more clearly what
we mean by a "point". The beam does not carry an infinite number
of independent data with it, but only a finite number. Let us
count the degrees of freedom of a light beam.

This was done first by Max von Laue, in 1907. For simplicity
let us consider rather narrow bundles of light, almost at right
angles to a plane x,y. We now divide up the plane into rectangles,
 Δx,   Δy, and the angular space into small solid angles ΔΩ .
Let Δv be the frequency width in a wavelet, and T the observation
time. Then Laue's result is, that the degree of freedom, that is
to say the number of independent data in such a beamlet is

$$F = (2 \times 2) \, \Delta x . \Delta y . \Delta \Omega . \Delta v . T / \lambda^2 \qquad\qquad 1.$$

The first factor 2 stands for the two directions of polarization, the second factor 2 for the phase. In experiments with coherent light the factor $\Delta\nu T$ is always of the order unity, because we have <u>temporal</u> coherence only in these limits. In principle the data could be extracted from the beam in the time which the coherence length takes to pass a point. (Of the order of 10^{-9} with helium-neon lasers.) If we continue the experiment beyond this point we get no new data, only repetition. (This is of course the <u>classical</u> view, in reality we get an improvement of the data by reducing the photon noise, which the classical theory neglects.)

So for practical purposes we can say that the number of independent data in a coherent beam is the invariant

$$F = \frac{\text{Beam Area x Solid Angle of Beam}}{\text{Wavelength squared}} \qquad\qquad 2.$$

This formulation is not quite precise, it applies well only to <u>compact</u> areas. It would give the result zero for an aperture in the form of an infinitely thin annulus, though in reality the degree of freedom in this case is finite. (The number of Fourier components around the ring, up to a certain maximum spatial frequency, defined by the beam angle.) But it is good enough for almost all practical applications.

GAUSSIAN BEAMLETS

It will be useful now to give a physical interpretation to all this mathematics. I will do this in two ways. The first is the representation of a complex but coherent beam by dividing it up into Gaussian beamlets. The second will be the representation by "eigenfunctions".

In optics (as also in communication theory,) two descriptions have equal importance: spatial distribution and Fourier description. The Fourier description is equivalent to the expansion of the distribution in terms of plane waves, with spatial frequencies k_x, k_y

$$k_x = \left(2\pi/\lambda\right)\sin\theta_x \qquad\qquad\qquad k_y = \left(2\pi/\lambda\right)\sin\theta_y$$

The Fourier expansion of the distribution at some plane x,y gives directly the amplitude distribution at infinity, or in the focal plane of a lens.

A well known mathematical theorem of Fourier Theory states that a beamlet cannot be limited both in x and in the corresponding spatial frequency k_x so as to infringe the Uncertainty Relation

$$\langle x^2 \rangle \langle k_x^2 \rangle \geq 1 \qquad\qquad 3.$$

Both x and k_x are measured from the centroids of their distribution in terms of energy. So the nearest we can get to a beam sharply limited both in spatial distribution and in angular spread is to realise the case when the inequality changes into an equality. This is the case for a Gaussian Beam

$$a\,(x,y,0) = \exp -\left[\frac{x^2}{2\langle x^2 \rangle} + \frac{y^2}{2\langle y^2 \rangle} \right] \qquad\qquad 4.$$

The Fourier Transform of this is, (apart from an unimportant factor,)

$$A\,(k_x,k_y) = \exp -\left[\frac{1}{2}\langle x^2 \rangle\, k_x^2 + \frac{1}{2}\langle y^2 \rangle\, k_y^2 \right] \qquad\qquad 5.$$

hence

$$\langle x^2 \rangle\,\langle k_x^2 \rangle = 1 \qquad\qquad \langle y^2 \rangle\,\langle k_y^2 \rangle = 1 \qquad\qquad 6.$$

The more one restricts the beam in the x-direction, the more it spreads angularly in the x,z plane.

The Fourier representation makes it easy to extend the beam into space. One has only to add a factor exp $(ik_z z)$ to the factor exp $i(k_x x + k_y y)$. k_z follows from the wave equation

$$k_x^2 + k_y^2 + k_z^2 = k^2 = (2\pi/\lambda)^2 \qquad\qquad 7.$$

At the small incidence angles used in holography one can use the approximation

$$k_z = k - \frac{1}{2k}\left(k_x^2 + k_y^2\right) \qquad\qquad 8.$$

which gives only a 1% error for wave normals 25° off the axis. This gives, (writing out only the x-part,)

$$a(x,y,z) = e^{ikz}\int\!\!\!\int_{-\infty}^{\infty}\exp\left[-\frac{1}{2}\left(\langle x \rangle^2 - iz/k\right)k_x^2 + \cdots\right]e^{i(k_x x + k_y y)}dk_x\,dk_y \qquad 9.$$

which is, apart from a constant factor

$$e^{ikz}\exp\left[\frac{1}{2}i\frac{z}{k}\left(\frac{x^2}{\langle x^2 \rangle^2 + z^2/k^2} + \cdots\right)\right]\exp\left[-\frac{1}{2}\left(\frac{x^2}{\langle x^2 \rangle^2 + z^2/k^2\langle x^2 \rangle} + \cdots\right)\right] \qquad 10.$$

From this we see:

1.) A beam which is Gaussian in one plane remains Gaussian in all other planes. It spreads out from the plane z = 0 by the law for the mean square width

$$\langle x^2 \rangle_z = \langle x^2 \rangle_0 + z^2 / k^2 \langle x^2 \rangle_0 \qquad \text{11.}$$

Ultimately, at a large distance, the r.m.s. width propagates in straight lines with angles

$$\theta_x = \lambda / 2\pi \langle x^2 \rangle^{\frac{1}{2}} \qquad\qquad \theta_y = \lambda / 2\pi \langle y^2 \rangle^{\frac{1}{2}} \qquad \text{12.}$$

as an elliptical cone.

2.) The curvature of the wave, as given by the unitary complex factor in eq. 10, first goes through a maximum, and ultimately the factor becomes

$$\exp\left[ik \left(x^2 + y^2 \right) / 2z \right]$$

that is to say the Gaussian beam spreads, at great distances, like a spherical wave centering on z = 0.

Gaussian beams are rather exceptional, because they have no side lobes, and they have no phase jump by π in the focus, like the beams studied by E. Wolf. (Born-Wolf, Principles of Optics.) They give a simple illustration of the theorem 2.:- As many degrees of freedom (complex) as one can fit Gaussian beamlets into the object area and the aperture.

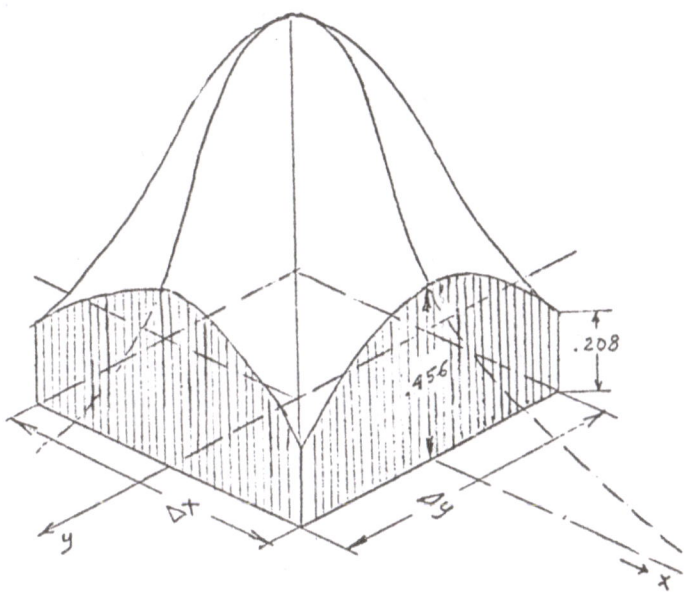

Fig. 1. Gaussian elementary signal exp $\left[-(x^2 + y^2)\right]$ in its
cell $\Delta x \cdot \Delta y = 2\pi$.

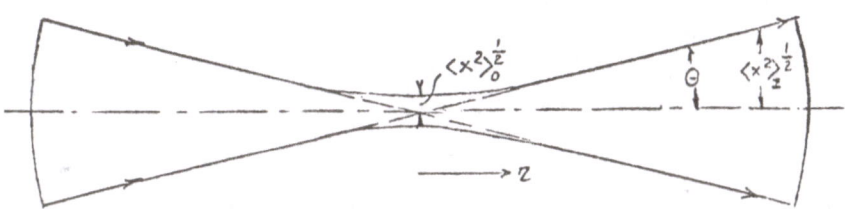

Fig. 2. The spreading of the r.m.s. radius of a Gaussian beam.

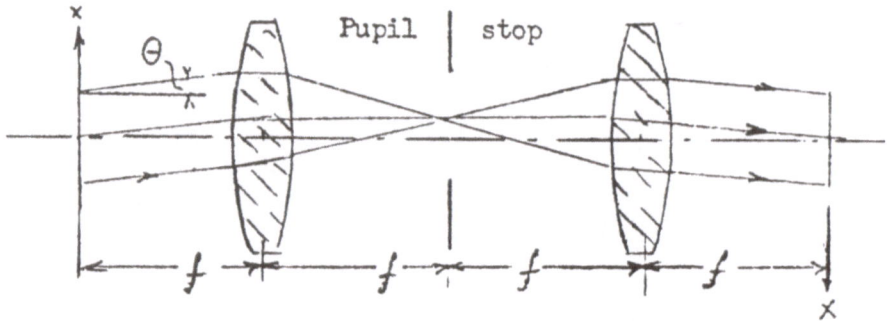

Fig. 3.   Illustrating information transmission by eigen solution.

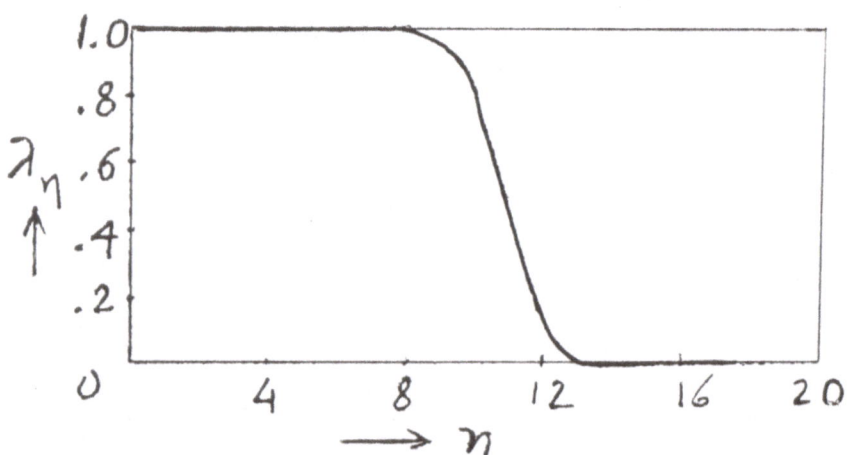

Fig. 4.   The eigenvalues λ_n as a function of n for 11.5 degrees
          of freedom, after G. Toraldo di Francia (1969).

TRANSMISSION OF OPTICAL INFORMATION BY EIGENSOLUTIONS

The representation by Gaussian beams is somewhat unsatisfact-
ory only because they are not quite orthogonal. Another representation first pioneered by Slepian and Pollack, 1961, (5) is free
from this.

Consider, for simplicity a one-dimensional case of a one-to-
one optical transformation. The coordinate is x, and as the
Fourier conjoint it is practical to use the Fourier-coordinate
ξ = u/ λ f, where u is the coordinate in the Fourier-plane of the
lens, with focal length f. Let the aperture in the Fourier (or
Fraunhofer) plane of the lens be limited by $\pm$ ξ_m. It can then be
shown that if s(x) is the original image, the final image $\bar{s}$(x) will
be

$$\bar{s}(x) = \int_{-\frac{1}{2}x}^{\frac{1}{2}x} \frac{\sin \pi \xi_m (x - z)}{\pi (x - z)} \, s(z) dz \qquad 13.$$

$\pm$ $\frac{1}{2}$X are the limits of the object. The sinc-function under the
integral is the "point-figure" in the optical transmission; the
image of a point will be spread out in this shape because of the
aperture limitation.

Consider eq. 13 now as an integral equation, and ask for its
eigensolutions, that is to say for those shapes ψ_n(x) which are
transmitted without distortion, just reduced by a factor λ_n, the
"eigenvalue".

$$\lambda_n \, \psi_n (x) = \int_{-\frac{1}{2}x}^{\frac{1}{2}x} \frac{\sin \pi \xi_m (x - z)}{\pi (x - z)} \, \psi_n(z) dz \qquad 14.$$

The eigensolutions of this Laplacian integral equations are, as
Slepian and Pollock have shown, the prolate spheroidal functions.
(Tabulated for instance in M. Abramowitz & I.A. Stegun, Handbook
of Mathematical Functions, U.S. Dept. of Commerce, Washington 1964,
p. 752-69.)

These eigenvalues have a remarkable property, pointed out by
Toraldo di Francia. They are practically unity up to an

$n_{max}= X \xi_m = X \; \theta/\lambda$ that is to say up to the <u>degree of freedom</u>, and then they drop suddenly to zero. This is the sharpest form-ulation of the finite information in an optical system. The drop is so sharp, that it is entirely hopeless to try to get more than this number of data out of the optical system. As Toraldo di Francia has shown, it would require an impossible degree of pre-cision, (soon to be drowned by noise,) to get something out of the quickly vanishing tail of the independent data.

PHOTON NOISE

By counting the degrees of freedom, we have introduced some physical sense into Huygens' Principle, but we cannot be satisfied with it, because the information is still infinite. Each degree of freedom in the classical theory is a continuous variable, which could transmit an infinite number of data. Physical intuition tells us, that the information must be finite. At some point the information must become uncertain by noise effects.

This additional "physical sense" is supplied by the photon theory of light, pioneered by Einstein in 1906. It is interesting to introduce this theory by an imaginary experiment.

Consider a holographic set-up. A photographic plate or some other detector collects coherent energy $\underline{E}$ from the reference source, and $\underline{e}$ from an object. The joint exposure is then

$$E + e + 2(E.e)^{\frac{1}{2}}\cos\phi$$

The last term represents the interference fringes. It is seen that however small the signal energy, and however insensitive the detector, this term can be large enough to be detectable by increas-ing $\underline{E}$. So in the limit we could transmit information without energy – if $\underline{E}$ were uniform and noiseless, as is assumed in the classical theory.

As soon as we recognize this as absurd, we get a hint at the nature of fundamental noise. The mean square amplitude of the signal term is $\underline{2E}.e$ Let us postulate that the signal becomes un-recognizable at some minimum energy ε at which the mean square fluctuation exceeds this by a factor k, that is to say when

$$\langle \delta E^2 \rangle = 2kE\epsilon \qquad\qquad 15.$$

Dividing both sides by ϵ^2 this becomes

$$\langle \delta (E/\epsilon)^2 \rangle = 2k.(E/\epsilon) \qquad\qquad 16.$$

We have now a pure number $N = E/\epsilon$ at both sides, and information most stop when

$$\langle \delta N^2 \rangle = 2kn \qquad\qquad 17a.$$

Now give k the value $k = \frac{1}{2}$ and we obtain Poisson's law for the fluctuation of rare events

$$\langle \delta N^2 \rangle = N \qquad\qquad 17b.$$

which indicates that the fluctuation of the energy is of the nature of shot noise, and the arrival of at least one photon is needed for information. (Einstein was led by somewhat similar considerations of energy fluctuation in a cavity to the photon hypothesis, but without using the concept of information.)

    We can now even go a step further, and find the connection of photon energy with wavelength. Information cannot be increased by any a <u>posteriori</u> operation on the message. But we can change the photon energy by the Doppler effect, by reflecting the photons on a mirror which moves towards it with velocity <u>v</u>. This changes the energy into

$$\varepsilon' = \varepsilon \, (1 + 2v/c) \qquad \qquad 18.$$

and the wavelength into

$$\lambda' = \lambda / (1 + 2 \, v/c) \qquad \qquad 19.$$

The only invariant in this process is $\varepsilon \cdot \lambda = \varepsilon' \cdot \lambda'$ or ε/v so this must be a natural constant. Calling this <u>h</u>, we have the Planck-Einstein Law

$$\varepsilon = h\nu \qquad \qquad 20.$$

(That such heuristic considerations hit the truth at times does not mean that we can guess the laws of Nature a <u>priori</u>. It is reasonable to assume that we cannot get information without energy, but why should Nature be reasonable?)

The photon theory at once makes the information in a light beam finite. One could ask of course naively, "has it not made it fundamentally uncertain?". If there is shot noise in a light beam, can we ever be sure to get the right data? The answer to this is given by Claude E. Shannon's Statistical Theory of Communication. Any small message, such as the number of photons arriving in an elementary cell of an image can be uncertain. But if we take a <u>large complex</u> of such elementary messages, we can determine with a probability as near to certainty as we like which complex message was meant. In electrical communication this is achieved by long

codes.   In optics we have a great advantage over electrical communications, where a code of 128 bits is the largest which has been
ever realised.   In optics we code whole pictures, with many millions
or even billions of bits. We can then with practical certainty
distinguish, by the Shannon-Tuller formula just

$$F. \log_2 \left(\frac{P + N}{N} \right) \quad \text{bits of information}$$

where F is the number of degrees of freedom, previously defined.
P is the signal energy and N is the noise energy, assuming for both
a Gaussian distribution.

    I will now state without proof, because it is already intuitive
ly clear, that the photon noise makes Huygens'Principle physically
fully meaningful. The information in a light beam, in the Shannon-
sense and taking only photon noise into consideration, is finite
and invariant in every cross section of a light beam so long as no
energy is lost. It is the same in the object plane and in the
Fourier (or Fraunhofer) plane of a lens.   It even remains the same
if we put a diffuser into the light beam, (unless it diffuses some
energy backwards,) and we can extract the information from it any-
where by holography. But holography brings into the subject a new
source of noise; photographic grain noise, and this requires a
little discussion.

PHOTON NOISE AND PHOTOGRAPHIC NOISE IN HOLOGRAPHY

    By the Ornstein-Selwyn Law, the probability of a small density
fluctuation δ in an area A, around the mean density level corresponding to the exposure E has the Gaussian law

$$p(\delta)d\delta = P. \exp \left(- \frac{A}{G(E)} \delta^2 \right) . d\delta \qquad\qquad 21.$$

where G is the characteristic grain area, which is a slow function
of the mean exposure E. This means that the mean square density
fluctuation $\langle \delta^2 \rangle$ is inversely proportional to the area A, just
as in the case of shot scattered at random. As for a given number
n of photons the signal density is proportional to n/A and the

r.m.s. grain noise is proportional to $1/A^{\frac{1}{2}}$, the best signal-to-noise ratio is obtained by making the area as small as possible. But the Ornstein-Selwyn law breaks down at areas of the order of the grain size, and the best conditions are obtained at a small multiple of the grain size.

In holography, that is to say with the superposition of a coherent background, the conditions are quite different. Let A again be the elementary area, n the number of signal photons, N that of background photons. The signal density is then proportional to

$$S(N/A)(N.n)^{\frac{1}{2}}$$

where S is the slope of the density vs exposure curve. (In terms of photons.) We can drop this factor, because it affects signal and noise in the same way. (We use it only to put the mean density to the point of maximum slope.)

The mean square noise, measured in units of energy density squared, is the sum of photon noise and of grain noise, because the two are independent, and is, in the limits of the Ornstein-Selwyn law

$$(N^{\frac{1}{2}}/A)^2 + C/A$$

where C is some constant, (neglecting the small dependence of the grain size on exposure.) Hence the r.m.s. signal-to-noise density ratio has the form

$$n^{\frac{1}{2}}/(1 + CA/N)^{\frac{1}{2}}$$

But as we are free to choose the background (reference beam)
intensity as we like, we can make N proportional to the area A,
which means that in reception with a coherent background we are
free to choose the scale of the image. We are only limited by the
consideration that the exposure must not be too far from the one
with optimum slope. In fact reference beam to signal intensity
ratios of 10-15 are currently used, and for contrasty objects even
50 : 1 has advantages.

In 1955 I have carried out a lengthy investigation on the
photon detection by a photographic emulsion, (EK XXX,) with and
without coherent background, which was never published (6). I will
only quote a few results from it, always in terms of bits/elemen-
tary area for a certain number of photons. The coherent background has the greatest advantage in the case of binary, black-or-
white pictures. An ideal photon detector gives ½bit for one photon,
direct recording requires about 400, the same emulsion with optimum
coherent background only 8. Practical certainty; 1 bit per elementary area, is achieved in the case of the ideal detector with 4
photons, without coherent background about 700, with it about 20.
In this case it is advisable to use a background about 25 - 50 times
signal intensity. This I have experimentally confirmed at about
the same time with my collaborator W.P. Goss.(JOSA, 56, 849, 1966.)
An object, so poorly illuminated that it was invisible became well
visible when a background 10-50 times stronger was added at the
photographic plate, or at the eye. It was also confirmed, theoretic
ally and experimentally by J.W. Goodman, R.B. Miles & R.B. Kimball.
(JOSA, 58, 609, 1968.)

SPECKLE NOISE

Speckle noise stands apart from other noise phenomena, in that
it is not a random error but precise but unwanted information on
the roughness of objects or of diffusers. It is not at all an
effect peculiar to holography, it is rather a direct consequence
of coherence. Optical holographers know it as "laser speckle" but
it appears in a particularly damaging form in acoustical holography.

If a microscopically rough object, such as a sheet of paper,
is illuminated uniformly with incoherent light, it appears uniform.
In laser light it appears speckled. It becomes much worse if one
puts as small pinhole before the eye, it becomes almost invisible
if it is photographed with a large power objective. I have called this speckle "subjective", because it is produced by the viewing device, though the first reason is of course the roughness of
the paper. Following G.W. Stroke one can explain it as follows.
If the optical system were perfect, so that it would image a point
on a sharply limited elementary area, there would be no speckle,
because the object is uniformly illuminated. But an instrument

which is good enough for incoherent light, is by far not good
enough for coherent light. The point-spread figure of light amplitudes, caused partly by diffraction, partly by geometrical aberrations, falls off only with the square root of the intensity, and
extends with its tail over many elementary areas. In incoherent
light this does not matter, because the small amplitudes are first
squared, and then summed. But in coherent light they are first
vectorially summed and then squared, and this can produce enormous
fluctuations in intensity.

There exists also what I have called (7) "objective speckle".
If one illuminates an object such as a transparency through a
diffuser, the wavelets scattered by the diffuser width add up and
subtract at random, and the transparency will be unevenly illuminated. This is objective speckle, because the non-uniformity is
really there, at the object. It is important to distinguish the
two, because the objective speckle can be eliminated, while the
subjective speckle can be reduced only; by throwing away a great
part of the resolving power, and averaging over many resolution
areas.

Though there is a physical difference between these two cases,
the mathematical result is the same. In both cases one considers
the interference of plane wavelets which fill a certain aperture,
evenly but with random phases. One then calculates the mean
square fluctuation energy (or "noise power",) relative to the
square of the mean energy, ("signal power") in areas $A = X^2$. For
small areas, in which the relative fluctuation is large, the result
is rather complicated, but for larger areas the asymptotic result
is very simple. The ratio of the mean square energy fluctuation
to the square of the mean is, asymptotically

$$\text{Noise}-\text{to}-\text{Signal} = (\lambda/x)^2 \qquad\qquad 22.$$

where λ is the finest fringe spacing, produced by the marginal
rays which pass the aperture. (This result is derived for square
apertures, but it holds true nearly enough for round ones. Note,
however, that a narrow ring aperture gives a somewhat more favourable result.)

In words: Random illumination within a certain solid angle
with coherent light produces a speckle noise ratio which is the
inverse of the resolution elements contained in the sample area.

A tolerably good television picture must have a signal-to-noise ratio of 1000 or 30 decibels. This can be obtained only at the cost of a useful resolution which is about 30 times the resolution determined by diffraction. A good television picture has 40 db signal-to-noise, and this means that the useful resolution is only one hundredths of the diffraction limit.

Holography makes this neither better nor worse. The speckle in a picture taken in a laser light with a lens of a certain area, is the same as for a hologram of the same area. Though the loss of utilizable resolution caused by speckle appears heavy, it is quite tolerable in optical holography. A hologram of, say 10 x 10cm has a diffraction resolution angle of the order of 10^{-5}, which would be one micron for an object point 10 cm from the hologram. Throwing away 99% of this still means a resolution of 100 microns = 0.1 mm, which is still a little better than what the eye can resolve.

Of course looking at a hologram, it is not the hologram, but the pupil of the eye which will be responsible for the speckle. One can do the hologram justice only by taking a picture of the reconstructed object with a reasonably large lens. But one can improve the speckle also in visual observation, by illuminating the hologram not with a laser, but with a sodium lamp, or one line of a mercury lamp. These have coherence lengths of the order of 0.1 mm. As the eye pupil is about 5 mm in diameter, looking at a point behind the hologram the eye uses at most a disc of 5 mm diameter in the hologram. With an angle of the order 0.1 radian the coherent discs in mercury or sodium illumination shrink to about 1 mm. This gives a diffraction angle of the order 10^{-3}, but this is still satisfactory for the eye. It pays therefore to use in the reconstruction for visual observation light with only a minimum of coherence. For the same reason reflecting holograms illuminated with white light are almost free from speckle, because the coherence, selected by the hologram itself, is only just the minimum required.

While in optical holography speckle is a tolerable nuisance, it is almost fatal to acoustical holography. All the objects in acoustical holography are rough, (objects with glare only can hardly be recognized,) there are no "transparencies" in acoustics. Moreover the wavelengths which can be used in acoustical holography are not all that short relative to the apertures used that one could throw away 1 - 2 orders or resolution. Consequently most acoustical holograms have extremely strong speckle.

ELIMINATING THE "OBJECTIVE"SPECKLE IN TRANSPARENCIES

In the case of transparencies one uses diffuse illumination in order to obtain the well known benefits of diffuse holograms, (unaided 3-dimensional vision, insensitivity to scratches etc.) but by this one produces an "objective" speckle, i.e. non-uniform illumination.

I have shown (l.c.) that one can retain these advantages of diffuse holograms, by illuminating the hologram not through a diffuser, with random phases, but by a system of plane waves, with properly adjusted amplitudes and phases. In one dimension these waves form a <u>fan</u>, with equally spaced angles. (More exactly: equally spaced sines of the angles, i.e. spatial frequencies which are integer multiples of one another.) One can then give these plane waves such amplitudes and phases that all interferences drop out, except those between the marginal waves. These, however, if they are sufficiently spaced angularly, will produce only inteference fringes at or beyond the resolution limit, which are harmless.

Such a system of waves can be produced by a special grating (amplitude <u>and</u> phase,) illuminated by a plane wave. By crossing two such gratings one obtains very nearly the equivalent of a diffuse hologram, but without speckle. To my knowledge such gratings have not yet been realised. I recommend them to those who are interested in making holograms of transparencies.

REFERENCES

(1,2)   N. Wiener, J. Math. and Phys. <u>7</u>, 109, 1928, J. Franklin
        Inst.<u>207</u>, 525, 1929
(3)   D. Gabor, Light and Information, lecture 1954, in
        <u>Astronomical Optics</u>
        Z. Kopal, 1956, Optical Transmission, <u>Information Theory</u>,
        C. Cherry, 1956
(4)   H. Gamo. Matrix Treatment of Partial Coherence, Vol.3
        <u>Progress in Optics</u>- Ed. E. Wolf, North Holland Publ. Co.
        <u>Amsterdam, 1964</u>
(5)   D. Slepian & H.O. Pollack, Bell S.T.J. <u>40</u>, 43, 1961
        For a full discussion and other references see
        G. Toraldo di Francia, Rivista del Nuovo Cimento,(1) <u>1</u>,
        460, 1969
(6)   To appear in abridged form in <u>Advanced Holography</u>,
        Ed. El-Sum, Plenum Press
(7)   D. Gabor, <u>Laser Speckle and its Elimination</u>, IBM Journal
        of Research, <u>14</u>, 509 - 14, Sept. 1970

HOLOGRAM INTERFEROMETRY

A.E. Ennos

Division of Optical Metrology National Physical

Laboratory Teddington, Middlesex(U.K.)

Holography is a method whereby optical wave-fields can be stored in a photographic plate and 'played out' again at any later time. The wavefronts so generated can be used to take part in optical interference phenomena. Hologram interferometry can thus be defined as the study of the interference effects taking place in wave fronts reconstructed from holograms in which some optical path difference exists between the wave-fronts. Because of the complex nature of the wave-fields that can be stored holographically, hologram interferometry has a much wider application as a measuring technique than has convenional interferometry.

The characteristic features of hologram interferometers are best illustrated by a direct comparison of a conventional instrument (e.g. a Michelson interferometer or its modifications) with a 'live' transmission hologram interferometer (Fig.1). Whereas in the former the light is split and recombined by a semi-reflecting mirror, the holographic instrument effects the recombination at the hologram itself. The hologram is formed by interference between light in the two 'arms' of the interferometer, and after processing the plate, it is replaced in its original position. Since the hologram will regenerate either of the two beams used to manufacture it, viewing along either beam direction can be used to investigate phase disturbances. Comparing the two instruments:

1) The conventional interferometer must be constructed of accurately worked optical components whose alignment must be critically adjusted before us. The hologram interferometer, on the other hand, can be constructed from poor quality optical elements. The wave aberrations from these are recorded in the

41

hologram and automatically subtracted when the phase disturbance
is to be measured. The interferometer is thus a null-recording
instrument.

2) Complex wave fields occurring at different <u>times</u> may be compared, by recording the first wavefront in the hologram used to construct the instrument. This is impossible with conventional inter
ferometers.

3) Modification of optical parameters e.g. wavelength, refractive
index, etc. can be made between effecting the comparison of two
complex wave fields. This has application in holographic countour-
ing.

4) By inserting a diffuser in one arm of the interferometer before
the initial hologram is recorded, many different directions of view
can be used, giving three-dimensional information about any subsequent phase disturbances. The holographic interferometer is then
equivalent to a whole set of conventional interferometry. Similar
ly, light scattered from an opaque surface can be used to generate
the hologram, when the phase disturbances caused by a displacement
of the surface can be subsequently measured.

5) By making use of the multiple storage capability of holograms,
combination of more than two wave-fields can be effected (multiple
beam interference) or comparison of selected wave-fields made
(multiplexing, using separate reference waves).

BASIC TYPES OF INTERFEROMETRY

'Live' interferometry: Correct replacement of the initially-
recorded hologram allows interference between the reconstructed
waves from the object transmitted by the hologram plate

$$A = \alpha M a \; \exp\left[-i\phi(x)\right] + \tau_0 a \exp\left\{-i\left[\phi(x) + \delta\right]\right\}$$

where τ_0 =   mean amplitude transmittance of hologram

α =   slope of τ , log E characteristic

M =   modulation transfer function

δ =   phase disturbance

$$I = a^2 \left\{ \left(aM - \tau_o \right)^2 + 4aM \ \ \tau_o \cos^2 \frac{\delta}{2} \right\}$$

where the first term is the D.C. term and the second term is the Modulation-term.

The fringes do not have maximum contrast since 100% modulation is not obtained. As in a conventional interferometer, fringe formation can be followed and the zero fringe identified.

Double exposure interferometry: Two exposures of half the normal exposure time are given without disturbing the optical elements or hologram plate.

$$A = \frac{I}{\sqrt{2}} aMa \ \exp\left[-i\phi(x)\right] + \frac{I}{\sqrt{2}} aM \exp\left\{ -i\left[\phi(x) + \delta\right]\right\}$$

$$I = 2 a^2 M^2 a^2 \cos^2 \frac{\delta}{2}$$

Fringes have 100% contrast. Since a is negative, 'live' fringes are in anti-phase with double exposure fringes, for an unbleached hologram.

Multiple-exposure interferometry: N exposures are made with a constant phase difference δ between exposures

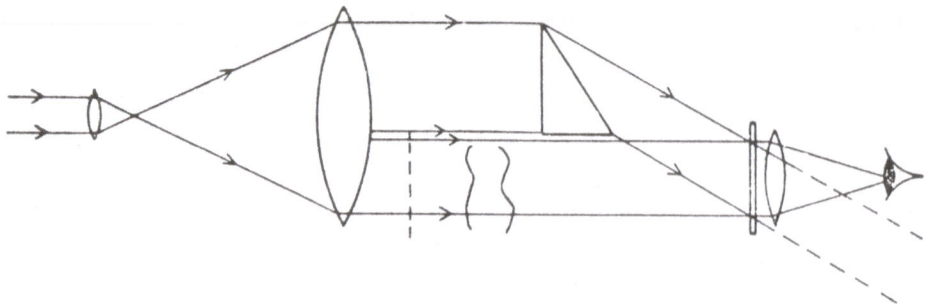

Fig. 1. A "live" transmission hologram interference.

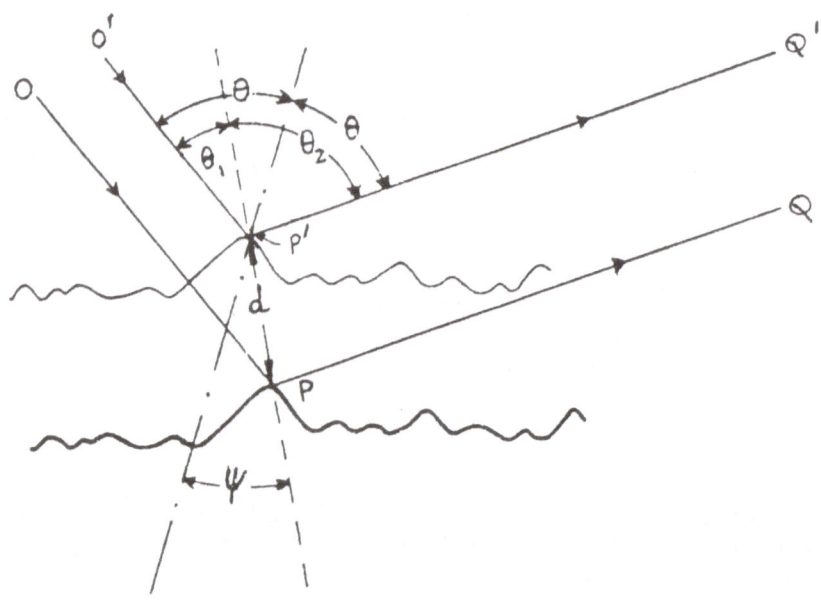

Fig. 2a. Relation between interference fringes and surface dis-
          placement using the concept of homologous rays.

$$A = \frac{a M}{\sqrt{N}} \left\{ \exp\left[-i\phi(x)\right] \right\} \left\{ 1 + \exp\left(-i\delta\right) + \exp\left(-2i\delta\right) + \ldots \right\}$$

$$I = a^2 M^2 a^2 \qquad \frac{\sin^2\left(N \frac{\delta}{2}\right)}{\sin^2\left(\frac{\delta}{2}\right)}$$

This is a light distribution similar to that obtained by diffraction at N equally spaced apertures.

RELATION BETWEEN INTERFERENCE FRINGES AND SURFACE DISPLACEMENT

In two dimensions, using the concept of homologous rays (Fig.2.a)

$$\delta = d \left(\cos\Theta_i + \cos\Theta_t\right) = 2d\cos\left(\frac{\Theta_t - \Theta_i}{2}\right) \cos\left(\frac{\Theta_t + \Theta_i}{2}\right)$$

$$\delta = 2\,d\cos\Theta\,\cos\psi$$

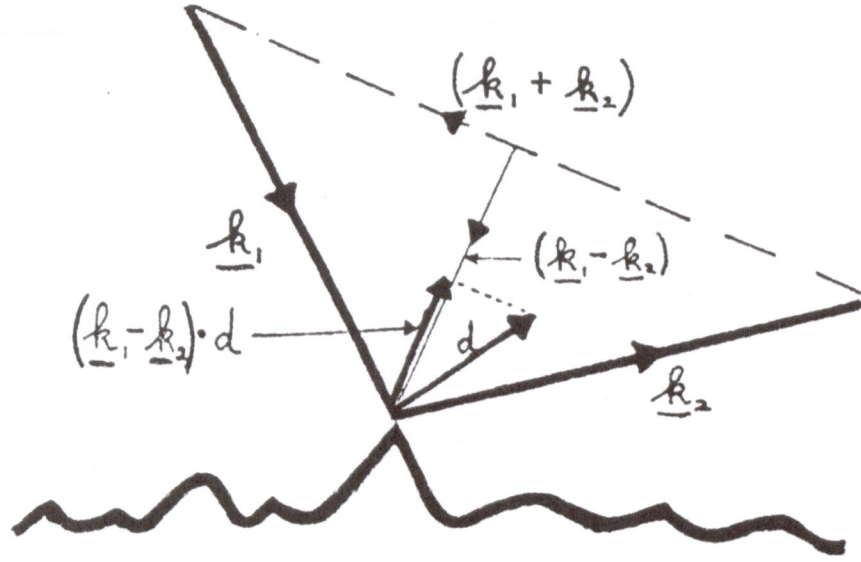

Fig. 2b.   Relation between interference fringes and surface
           displacement in vectorial notation.

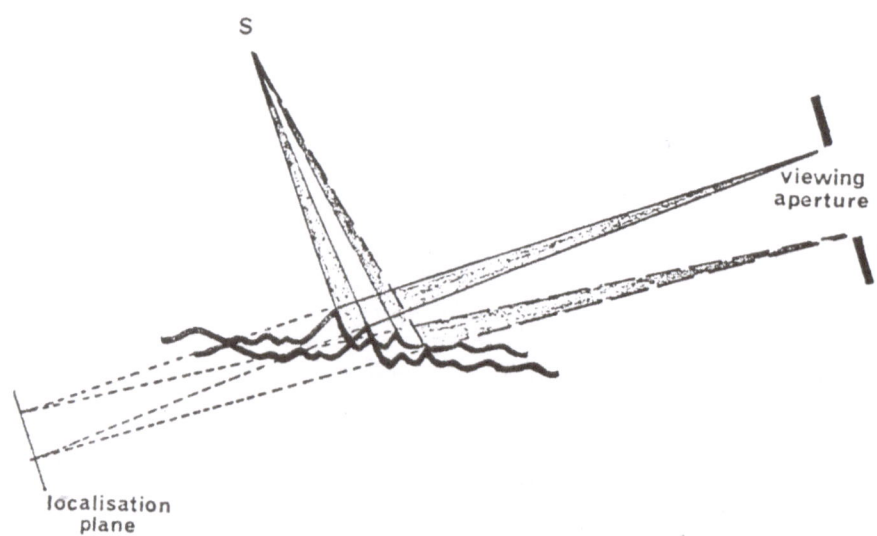

Fig. 3.   Fringe formation: the apparent localization of the
          fringe pattern.

or, in vectorial notation:

$$\delta = \left(\underline{k_1} - \underline{k_2}\right) \cdot d$$

where $\underline{k_1}$ and $\underline{k_2}$ are the illuminating and viewing direction vectors.
(Fig.2.b)

Bright fringes occur for $\delta = n\lambda$, dark fringes for $\delta = \left(n + \frac{1}{2}\right)\lambda$

The sensitivity of measurement is proportional to $\cos\theta$
and the measurement obtained is $d^* = d\cos\psi$, which is the
resolved part of the motion lying in the direction bisecting the
illuminating and viewing directions.

FRINGE LOCALIZATION AND VISIBILITY

Simple fringe formation theory does not take into account the
necessity for a finite viewing aperture. Over the small range of
viewing angles that are encompassed by this aperture, the conditions for interference may vary. This gives rise to an apparent
localization of the fringe pattern in a plane other than that of
the object, depending on the type of surface movement(Fig.3).
Visibility of the fringes is a maximum in this plane. For a
generalized deformation, localization will only occur along a line
in space, but in particular cases, it has distinct characteristics.

Examples

1. Lateral translation of object surface orthogonal to viewing
   direction: a set of parallel straight fringes, normal to the
   direction of movement, is seen localized in front of or behind
   the surface.

2. Tilt about an axis lying in the surface: a set of straight
   fringes, parallel to the tilt axis, are formed, localized
   close to the surface.

3. Rotation of the surface about on axis parallel to the viewing
   direction: straight parallel fringes localized in front of or
   behind the surface: their direction varies according to the
   viewpoint.

4. Translation of the surface in the viewing direction: circular
   fringes of ill-defined localization are formed.

Reducing the viewing aperture will bring fringes into better focus with the surface, but laser speckle sets a limit to this. Optimum results are obtained with a slit aperture oriented in a direction orthogonal to that in which the fringes change position most rapidly with eye movement.

For a full theoretical treatment of the geometry of fringe localization, see the following references: (1),(2),(3),(4),(5).

METHODS OF FRINGE INTERPRETATION

1) 'Static' method. The fringes are labelled according to the fringe order number n and values of d cosψ obtained by applying equation 1. This presupposes a knowledge of the zero order fringe, which can be obtained either by using the 'live' fringe technique in conjuction with a permanent double-exposure record, or by noting which fringe does not move relative to the surface when the viewpoint is changed. The motion of any point may be calculated by using fringe patterns obtained from widely-spaced viewpoints.

Example: a two-viewing system for obtaining in-plane motion d_0 of a stretched metal foil. Viewing from direction $\pm\theta$ to the normal,

$$d_o = \frac{N_2 - N_1}{2 \sin\theta} \cdot \lambda$$

where N_2 and N_1 , are the order numbers of corresponding fringes on the two reconstructions. The method is most sensitive to movements in the line of sight of the viewing direction. See reference (6).

2) 'Dynamic' method: The number of fringes N moving across a point of the surface is counted as the viewing direction is changed. Suppose this direction changes from $\underline{k_2}''$ to $\underline{k_2}'$

Then

$$\left.\begin{aligned}\left(\underline{k_1} - \underline{k_2'}\right) \cdot d &= N_1\,\lambda \\[6pt] \left(\underline{k_1} - \underline{k_2''}\right) \cdot d &= N_2\,\lambda\end{aligned}\right\} \quad \left(N_1 - N_2\right)\lambda = N\lambda = \left(\underline{k_2''} - \underline{k_2'}\right) \cdot d$$

This method is most sensitive to movements orthogonal to the mean viewing direction, given vectorially by $\left(k_2'' - k_2'\right)$
See reference (7).

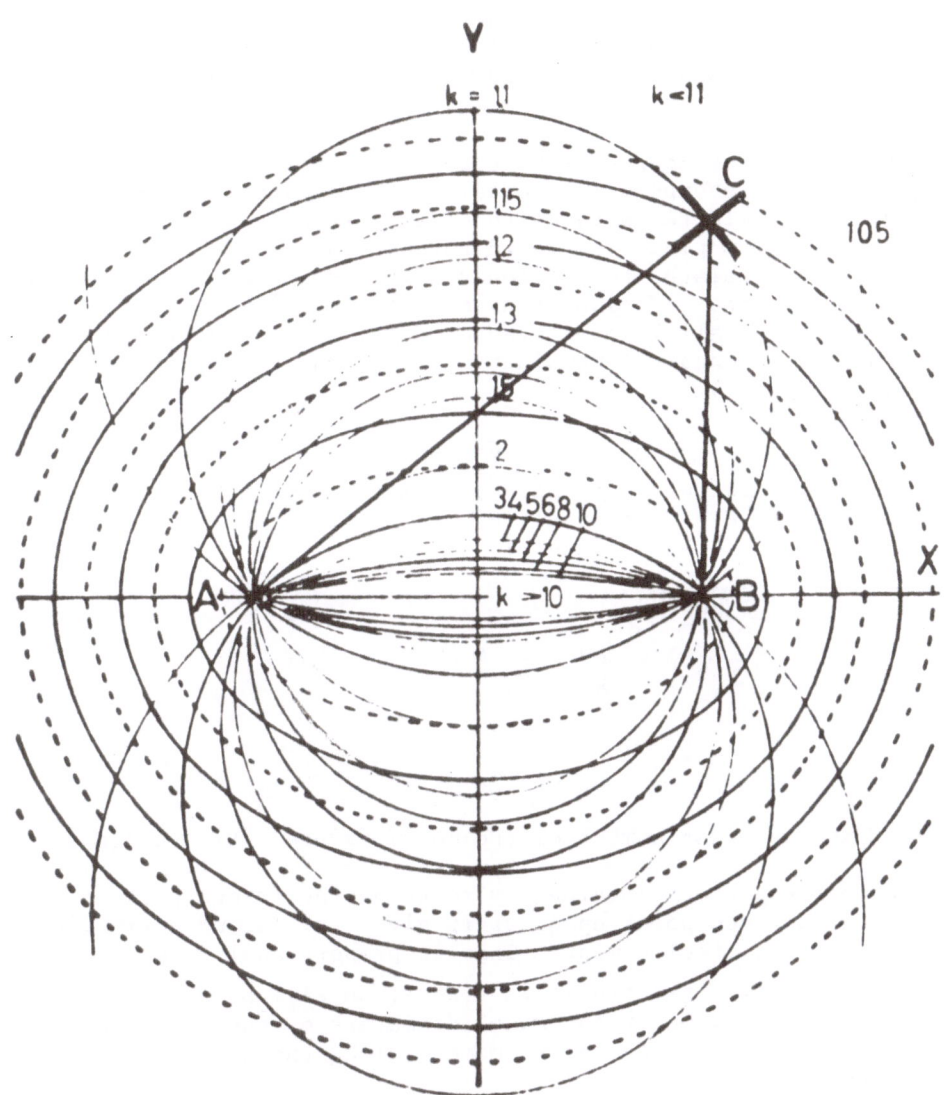

Fig. 4. The holo-diagram, after N. Abramson (1969, 1970).

3) Use of the 'Holo-Diagram': This is a diagram (Fig.4) consisting of ellipses whose foci are the source and observing positions,
and which trace out object points which will give rise to a
constant path length. A series of circles, called k lines, on the
diagram, indicate the relative fringe sensitivity to be expected
for surface movements normal to an ellipse. The diagram is a
geometrical aid to the solving of a generalized displacement using
the 'dynamic' method as described in 2. See reference (8),(9)(10).

4) Observation of Source-Plane Fringe Patterns: If the real
image of a double exposure hologram is formed and an aperture
placed over a selected area of the image, fringes will be observed
when the hologram plate is viewed through this aperture. The
fringe pattern relates to the generalized motion of the surface at
the aperture point. Motion transverse to the viewing direction
give straight parallel interference bands, and motion in the line
of sight gives rings. A combination gives an off-axis ring pattern.
See reference (11).

POST-RECORDING TECHNIQUES

1) The moiré pattern obtained by a small lateral translation δ
of a fringe pattern upon itself gives the differential of the
displacement i.e. the surface slope. Slope:

$$\frac{\partial d}{\partial x} = \frac{N \lambda}{2\delta \cos\theta \cos\psi}$$

where N is the order number of the moiré fringe. See reference (12).

2) The moiré pattern obtained by rotating the fringe pattern by
180° about a given point upon itself gives the second derivative
$\frac{\partial^2 d}{\partial x^2}$ of the displacement. This is proportional to bending
moment. The moiré pattern is in general an ellipse or hyperbola.
The squares of the major and minor diameters of the conic section
are proportional to the bending moment along the axes. See
reference (13).

REFERENCES

(1)   Froehly, Monneret, Pasteur and Vienot 1969, Optica Acta,16,343.
(2)   Tsuruta, Shiotke and Itoh, 1969, Optica Acta, 16, 723.
(3)   Stetson,1969 Optik, 29, 386.
(4)   Walles, 1969, Arkiv for Physik, 40, 299.
(5)   Steel, 1970, Optica Acta, 17, 873.
(6)   Ennos, 1968, J.Phys.E.,1, 731.
(7)   Aleksandrov and Bonch-Bruevich, 1967, Soviet Phys.Tech.
       Phys.12, 258 Sollid, 1969, Appl.Opt.8, 1587.
(8)   N. Abramson, 1969 Appl.Opt. 8, 1235.
(9)   N. Abramson, 1970 Appl.Opt. 9, 97.
(10)   N. Abramson, 1970 Appl.Opt. 9, 2311.
(11)   Gates, 1969, Optics Technology, 1, 247.
(12)   Boone and Verbiest, 1969, Optica Acta, 16, 555.
(13)   Stetson, 1970, Optics Technology, 2, 80.

VIBRATION ANALYSIS BY HOLOGRAPHY

A. E. Ennos

Division of Optical Metrology National Physical

Laboratory Teddington, Middlesex (U.K.)

The principles of holographic interferometry may be used to give information about the modes of a vibrating surface. Interference fringes so obtained are contours of equal amplitudes of vibration. As in the case of the measurement of static deformations, the contours relate to vibrational movement resolved in a direction bisecting the illuminating and viewing directions. For normal incidence, the amplitude of the in-and-out motion is recorded. Live fringe techniques can yield information on relative phases of vibration across the surface.

TIME-AVERAGED RECORDING TECHNIQUE

A hologram is recorded while the object is vibrating at constant frequency and amplitude. The reconstructed image surface appears covered in an interference fringe pattern.

Theory: A hologram recorded while an element of surface is executing a characteristic motion described by a function Ω_t will give rise to a reconstructed amplitude which is the time average of the instantaneous amplitude

$$\bar{A} = \frac{a}{T} \int_0^T \exp\left\{-i\left[\phi(x) + \frac{4\pi}{\lambda}\,\Omega_t\right]\right\}\,dt$$

where $\phi(x)$ and a are the phase and amplitude of the light emitted
by a stationary point described by its coordinate x and T is
the periodic time of the motion. The reconstructed amplitude is
the square of the modulus of $\bar{A}$.

1.   Sinusoidal motion, amplitude d:

$$\Omega_L = d \sin \left(\frac{2 \pi t}{T} \right)$$

$$\bar{A} = \frac{a}{T} \exp \left[-i\phi(x) \right] \int_0^T \exp \left[-\frac{4 \pi i d}{\lambda} \sin \left(\frac{2 \pi t}{T} \right) \right] dt$$

$$= J_0 \left(\frac{4 \pi d}{\lambda} \right) a \exp \left[i \phi(x) \right]$$

Intensity I:

$$I = \left| \bar{A} \right|^2 = J_0^2 \left(\frac{4 \pi d}{\lambda} \right) a^2 = J_0^2 \left(\frac{4 \pi d}{\lambda} \right) I_0$$

where I_0 is the intensity of the non-vibrating areas.

2.   Steady motion.

Let velocity normal to the surface be v and distance travelled $Z = vT$:

$$\bar{A} = \frac{\lambda a}{4 \pi v T} \left[1 - \exp \left(-\frac{4 \pi i}{\lambda} \right) vT \right] \exp \left[-i\phi(x) \right]$$

Intensity I:

$$I = |\bar{A}|^2 = \frac{\sin^2\left(\frac{2\pi v\tau}{\lambda}\right)}{\left(\frac{2\pi v\tau}{\lambda}\right)^2} \, a^2 = \frac{\sin^2\left(\frac{2\pi z}{\lambda}\right)}{\left(\frac{2\pi z}{\lambda}\right)^2} \, I_0$$

The Bessel function and the sinc function are compared with a cosine squared function in Fig.1.

BESSEL FUNCTION

$$J_0^2(x) \simeq \frac{2}{\pi\,x} \cdot \cos^2\left(x - \frac{\pi}{4}\right)$$

Fringes obtained for a sinusoidally vibrating surface are approximately regular in spacing, after the initial two fringes. Their amplitude decreases approximately inversely as the order number. (Successive amplitudes 1.0, 0.162, 0.080, 0.061 ...) and the positions of the maxima are displaced one quarter of a fringe order spacing compared to two-beam fringes.

SINC FUNCTION

$$\frac{\sin^2 x}{x^2}$$

Fringe amplitude falls off very much more rapidly than for the $J_o^2(x)$ function. (Amplitude of first fringe $= \frac{4}{9\pi} = 0.045$)
The first minimum corresponding to $x = \pi$ occurs for a displacement z of $\frac{\lambda}{2}$

COMBINATION OF VIBRATION MODES

For combination of modes of the same frequency, but different phase, the modes must be added vectorially by phasor addition. For combination of irrational frequencies, the characteristic function becomes the product of the two separate $J_o(x)$ functions. See reference (1).

REAL-TIME TECHNIQUES

In these methods, comparison is made between a hologram record of the stationary surface and the vibrating surface itself. The processed hologram has either to be accurately replaced in its original position, or a liquid gate used to process the plate in situ. A number of different methods can be used.

1) <u>Time-averaging of 'live' fringes</u>: the instantaneous reconstruct-ed intensity is given by

$$I_t = 2\, I_o \cos^2 \left[\frac{2\pi d}{\lambda} \sin \left(\frac{2\pi t}{\tau} \right) \right]$$

Averaging gives

$$\bar{I} = I_o \left[1 + J_o \left(\frac{4\pi d}{\lambda} \right) \right]$$

The fringes observed have the same configuration as time-averaged hologram fringes, but have half their number. The fringe contrast is poor, due to non-zero value of the minima.

2) <u>Fringe-Spoiling Technique</u>: the hologram plate is deliberately misaligned on replacement after processing (or the direction of illumination changed) to give a fiduciary band of fringes across the surface. If these are spaced at equal intervals h in the y direction, the intensity distribution before vibration is

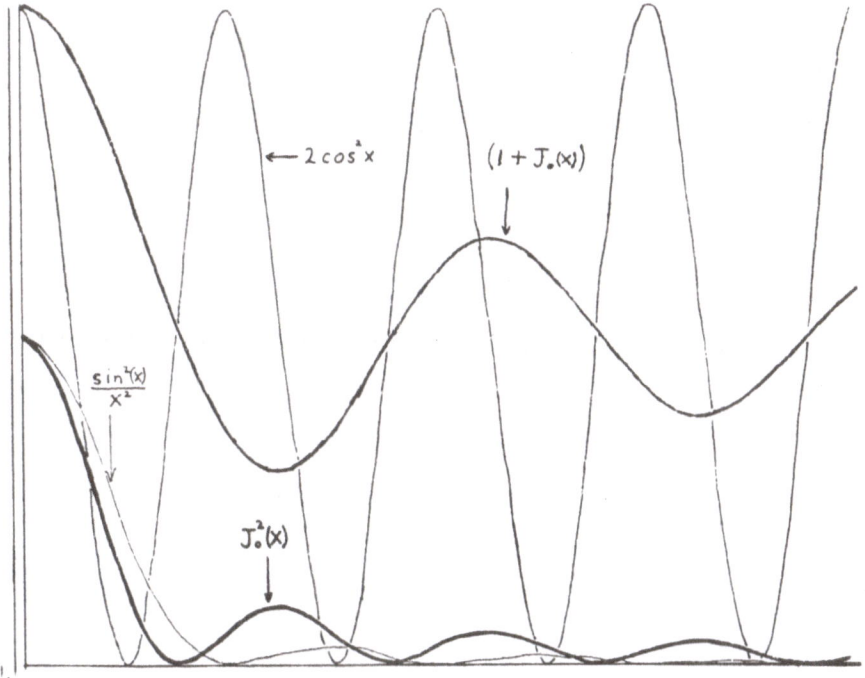

Fig. 1.   Comparison of the Bessel function and the sinc function
with a cosine squared function.

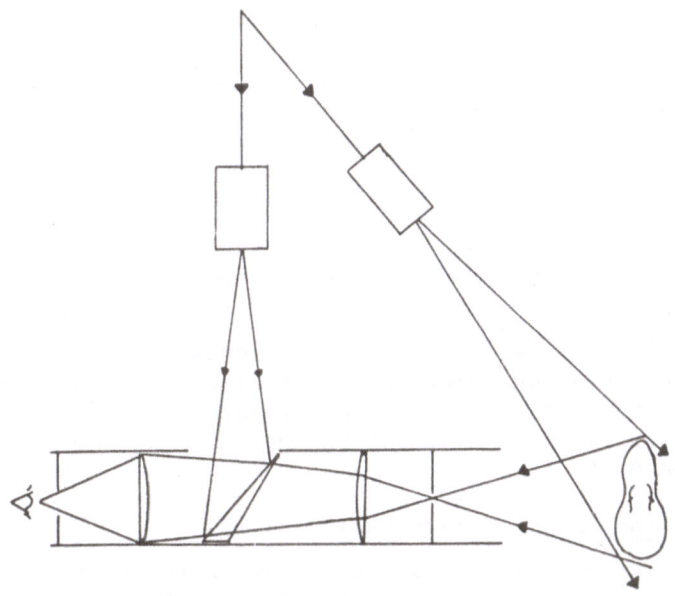

Fig. 2.   The Speckle interferometer.

$$I = I_o \left[1 + \cos \left(\frac{2\pi y}{h} \right) \right]$$

When the surface vibrates with amplitude d, and the time average is taken

$$\bar{I} = \left[1 + I_o \left(\frac{4\pi d}{\lambda} \right) \cos \left(\frac{2\pi y}{h} \right) \right]$$

The original fringes become alternately blurred and sharp, but with diminished contrast. Reversal of phase takes place at every minimum. See reference (2)

3) Stroboscopic Techniques. 'Live' fringes are viewed while pulsing on the laser illumination at the same frequency as that of the vibrating surface. Alternately a double exposure recording can be carried out pulsing on the light at twice the frequency. The pulse width must be short compared to the vibration period $(\frac{1}{10} - \frac{1}{20})$ for fringes to remain sharp. The technique is wasteful of light when using a CW laser. The live technique allows analysis of relative vibration phase when the strobe frequency is varied slightly. (See reference (3)).

EXTENSION OF RANGE OF MEASURABLE AMPLITUDE USING TIME-AVERAGED HOLOGRAPHY

1) Subfringe technique for measurement of very small vibrations:
A small amplitude phase variation is impressed upon the reference beam at the vibration frequency, so that the bias point of the fringe pattern falls on the steep part of the $J_o^2(x)$ characteristic. Sensitivity to a vibration amplitude of 1nm is claimed. See reference (4)

2) Increasing range of measurable amplitude

   (a) A large amplitude phase variation is impressed on the reference beam by reflecting it from a vibrating mirror.

The apparent nodal areas then correspond to the amplitude
of the mirror, so that biasing at any amplitude can be
achieved. See reference (5).

(b) A generalised form of stroboscopy in which more than two
pulses of light per vibration period are directed on to
the object while recording the hologram. Up to 25 wavelengths of amplitude can be recorded. See reference (6).

SPECKLE PATTERN TECHNIQUES

Although not strictly holographic, these provide rapid real-
time method for determining resonant frequencies and optimum siting
of the vibration excitors.

SPECKLE INTERFEROMETER

This instrument (Fig.2) combines the speckle pattern seen on
the surface of the laser-illuminated object with a smooth reference
beam. The resultant speckle pattern is highly sensitive to surface
movement. On vibrating surfaces the speckle pattern is blurred
out. Nodal regions remain speckly and can be distinguished. See
reference (7).

The system has been used in conjunction with television
techniques to give a large visual display - electronic filtering
can then be used to increase speckle contrast. See reference (8).

NODAL ANALYSIS BY DIRECT OBSERVATION OF SPECKLE

Appearance of a speckle pattern does not vary with movement
of the surface towards the observer, if single-direction illumination is employed. However, when illuminated by light from a large
diffuser, the surface itself is immersed in a stationary speckle
field. The observed speckle pattern will then be blurred out only
if the vibration amplitude exceeds the 'depth' of a speckle. This
depth can be made larger by reducing the solid angle subtended by
the diffuser at the object. Nodal areas can thus be distinguished,
with variable sensitivity to vibration amplitude. See reference
(9).

REFERENCES

(1) Stetson, 1969, J.Phys.E.2,609.
(2) Biedermann and Molin 1970 J. Phys.E.3, 669.
(3) Archbold, Burch & Ennos Nature 1968, 217, 942.
(4) Metherell et al. 1969 J. Opt.Soc.Am.59,1534.
(5) Aleksoff. 1969 Appl. Phys. Lett.14, 23.
(6) Mottier 1970 Proceedings of International Symposium on
     Applications of Holography Besançon
(7) Archbold, Burch, Ennos & Taylor, 1969 Nature, 222, 263

(8) Butter & Leendertz, 1971, Optics & Laser Technology,3,26.
(9) Eliasson and Mottier, Brown Boveri Research Report
     KLR-70-18 October 1970 and J.Opt.Soc.Amer, 61,559,1971.

HOLOGRAM INTERFEROMETRY AND LASER SPECKLE METHODS:

FURTHER APPLICATIONS

A.E. Ennos

Division of Optical Metrology National Physical

Laboratory Teddington, Middlesex (U.K.)

COMPARISON OF SHAPES

The wavefront reflected from a surface may be stored in a holo gram and subsequently reconstructed to take part in optical inter- ference with any other wavefront. In applications where surface deformation is to be measured, the two wavefronts, although complex, are very similar, and the concept of homologous rays can be used. If two different surfaces are to be compared, however, the wave- fronts scattered from each will in general be too dissimilar in detail to yield an interference pattern showing the mean difference in shape. This difficulty can be overcome to some extent by illu- minating the surfaces at a high angle of incidence, when an increas ing proportion of the light will be scattered specularly. The near ly smooth waves so generated can then be compared interferometric- ally. The high angle of incidence will of course decrease the sensitivity of measurement.

For example, for a rough surface with a Gaussian distribution of surface profile height, measured from the mean height, reflected intensity I :

$$I = R\exp\left[\frac{-8\pi^2 \, h_o^2 \, \cos^2\phi}{\lambda^2}\right]$$

where ϕ is the angle of incidence,R the reflection coefficient, and h_o the roughness parameter.

61

When

$$h_o \cos\phi = \frac{\lambda}{4}$$ $$I = 10^{-2} R \qquad [\sim 1\%]$$

When

$$h_o \cos\phi = \frac{\lambda}{9}$$ $$I = \frac{1}{e} R \qquad [\sim 35\%]$$

This principle has been used to compare cylinder bores with a master shape. See reference (1)

USE OF COMPUTER-GENERATED HOLOGRAMS

In precise comparison of shapes, a computer-generated hologram can be used to generate the wave from the master shape. The principle use to which this has been applied is the testing of aspheric optical surfaces. In this case the hologram has only to generate a simple wavefront of constant amplitude but varying phase. The hologram can be generated either in binary form or as the calculated off-axis interferogram between a place wave and the aberrated wave. After suitable photographic reduction it is used in one arm of a modified Twyman-Green interferometer, in which the first order diffracted wave from the hologram interferes with the wave generated by the aspheric element. See reference (2).

HOLOGRAPHIC CONTOURING

Holograms of scattering objects record the phase of the light which has travelled from the source via the surface of the object. Without altering hologram or object positions, this phase can be varied by changing one of three recording perameters; the illuminating angle, the wavelength or the refractive index of the surrounding medium. A double exposure hologram recorded when such a change is made between exposures will reconstruct interference fringes giving surface depth contours.

1) Varying angle of illumination. To obtain fringes corresponding to contours of equal height spacing, the illuminating and viewing directions must be orthogonal. Thus this method is not of practic-

al use since some areas of surface may be in shadow. Contour height spacing ∆h:

$$\Delta h = \frac{\lambda}{\sin\left(\dfrac{\Delta \gamma}{2}\right)}$$

where $\Delta \gamma$ is the change in illuminating beam angle. See reference (3).

2) Varying Wavelength.

A simple Fresnel off-axis hologram system is inconvenient since a wavelength change give rise to both a change in the direction and magnification of the reconstructed image. This can be overcome by using focused-image holography.

Contour height spacing ∆h:

$$\Delta h = \frac{\lambda_1 \lambda_2}{2 \Delta \lambda} \qquad \text{where} \quad \Delta \lambda = \lambda_1 - \lambda_2$$

Example : Argon wavelengths formula

$$\lambda_1 = 488.0 \text{ nm} \qquad \lambda_2 = 514.5 \text{ nm}$$

$$\Delta h = 4.75 \text{ μm}$$

See reference (4).

3) Varying Refractive Index of Medium. The object under study must be immersed in a vessel with a flat window, and the object illuminated at normal incidence with collimated light.

Contour height spacing   ∆h:

$$\Delta h = \frac{\lambda}{2 \left(N_1 - N_2\right)}$$

where N1 and N2 are the refractive indices of the two media.

Example: Liquids e.g. water and
           ethylene glycol, $\Delta h = 7.5 \, \mu m$

           Gases a) Air varied in pressure
                      by 1 atm. $\Delta h = 2 \quad \mu m$

                  b) Freon varied in pressure
                      by 1 atm. $\Delta h = 0.6 \, \mu m$

   See reference (5).

SPECKLE PATTERN TECHNIQUES

   Holography records phase and amplitude of a wavefront, which
can be related to the detailed surface structure. In metrology,
one is concerned only with comparison of phases, which relate to
the change in length, so that much of the information in a hologram
is superfluous. It is also very often difficult to extract from
hologram interferograms the motion of a surface resolved in one
particular direction (especially the in-plane directions) since the
fringe pattern relates to the generalized deformation. Both these
drawbacks can be overcome by making use of coherent speckle effects.

   The random speckle observed when a rough surface scatters
light is due to inter-modulation of the scattered light vectors
emanating from individually resolved areas. The size of these is
dependent only upon the numerical aperture of the viewing instrument, or eye.

   Angular speckle size = $1.22 \, \dfrac{\lambda}{a}$ where a is the aperture diameter.

Example: A lens of focal ratio F will record speckles of
           diameter $\delta = 1.22 \, \lambda F$

When $F = \dfrac{1}{4}$ $\delta = 3 \, \mu m$

   The amplitude and phase of the light from each speckle is
randomly distributed, but the light is still coherent. The distribution of brightnesses in a speckle pattern obeys the following exponential relation

$$W(I) = \frac{1}{I_o} \exp\left(-\frac{I}{I_o}\right)$$

where W(I) is the brightness probability density and I the actual brightness of a speckle. (This predicts that the highest probability is that the speckle has zero brightness i.e. is dark).

Two superimposed speckle pattern fields can combine coherently to form a third independent speckle pattern having the same brightness distribution characteristics. However, this speckle field is highly sensitive to movement of the surfaces giving rise to the original speckle patterns. A movement of one surface in the line of sight causes each speckle to vary over a complete bright-to-dark cycle for each δ of displacement, each speckle operating independently. The speckle pattern field thus becomes alternately uncorrelated and correlated again with its original distribution.

Correlation can be effected by imaging the speckle pattern and using the recording as a shadow mask. 'Live' speckle correlation fringes result. The technique has particular importance in measuring in-plane displacements of a surface. The surface is illuminated by two oblique beams, inclined at angle $\pm$ Θ to the normal. These give rise to two independent speckle patterns which interfere in the viewing direction. The interferometer is insensitive to movements in the surface normal direction, but for a lateral movement speckle correlation fringes appear corresponding to increments of displacement:

$$\delta = \frac{\lambda}{2 \sin \Theta}$$

See reference (6)

By increasing the speckle size sufficiently, using a lens of low aperture, television camera techniques may be used to record the fringes.

Correlation may also be carried out by recording a simple double exposure of the image, using a high contrast film and making use of its non-linear characteristic. Regions in which the two speckle patterns remain correlated record with a lower optical density than those where the patterns are uncorrelated. See reference (7)

Double esposure speckle photography with single-direction illumination can also be used to measure relatively large lateral movements of a surface ($\sim$ 1 mm) by examining the optical transform of the recorded image. Doubling of the speckle pattern gives

rise to Young's fringes from which the direction and magnitude of
the displacement at any point can be found. With two-directional
illumination, small relative movements of the surface, as caused
by strain, can be measured in the presence of a large body movement. See reference (7).

REFERENCES

(1)   Archbold, Burch and Ennos 1967 J. Sci. Instrum.44, 489.
(2)   Wyant and McGovern 1971. Appl. Opt. 10, 619.
(3)   Hildebrand and Haines 1967 J. Opt. Soc. Am.57,155.
(4)   Zelenka and Varner 1968. Appl. Opt. 7,2107.
(5)   Shiotke, Tsuruta and Itoh 1968, Jap.J.Appl.Phys.7,904.
(6)   Leendertz 1970, J.Phys.E.3,214.
(7)   Archbold, Burch and Ennos 1970 Optica Acta, 17,833.

Some representative photographs showing various interferometric applications follow:

Fig. 1. Interference fringes on a steel ring subjected to diametral
         loading. Reconstruction from a double exposure hologram.

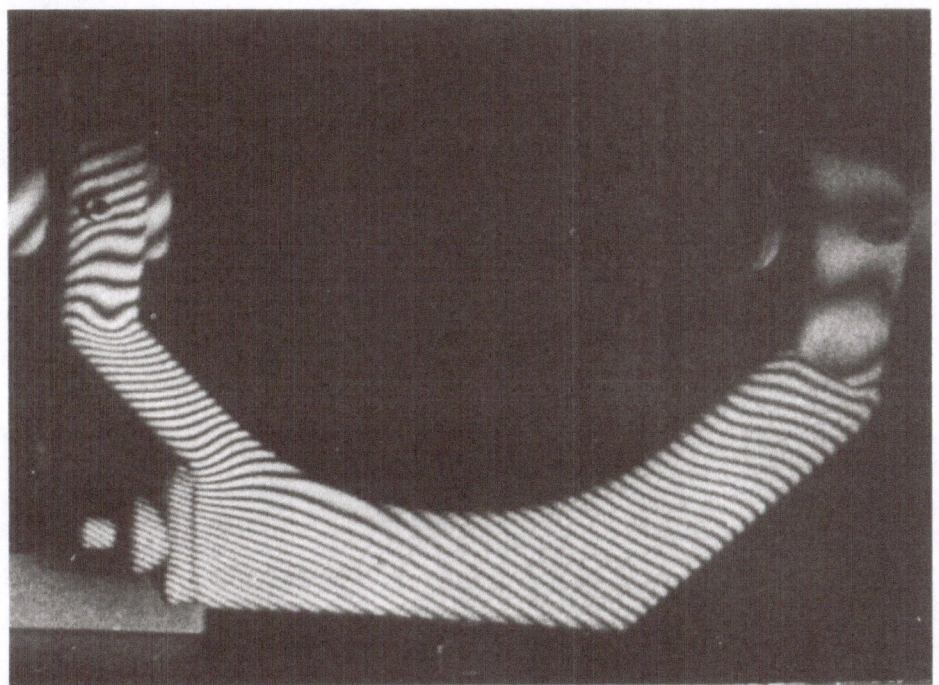

Fig. 2. Fringes on yoke-shaped structure held rigidly at its ends
        and loaded at bottom left hand corner in a direction
        towards the observer. The abrupt change in density of
        fringes indicates a change in degree of bending, showing up
        two weak regions.

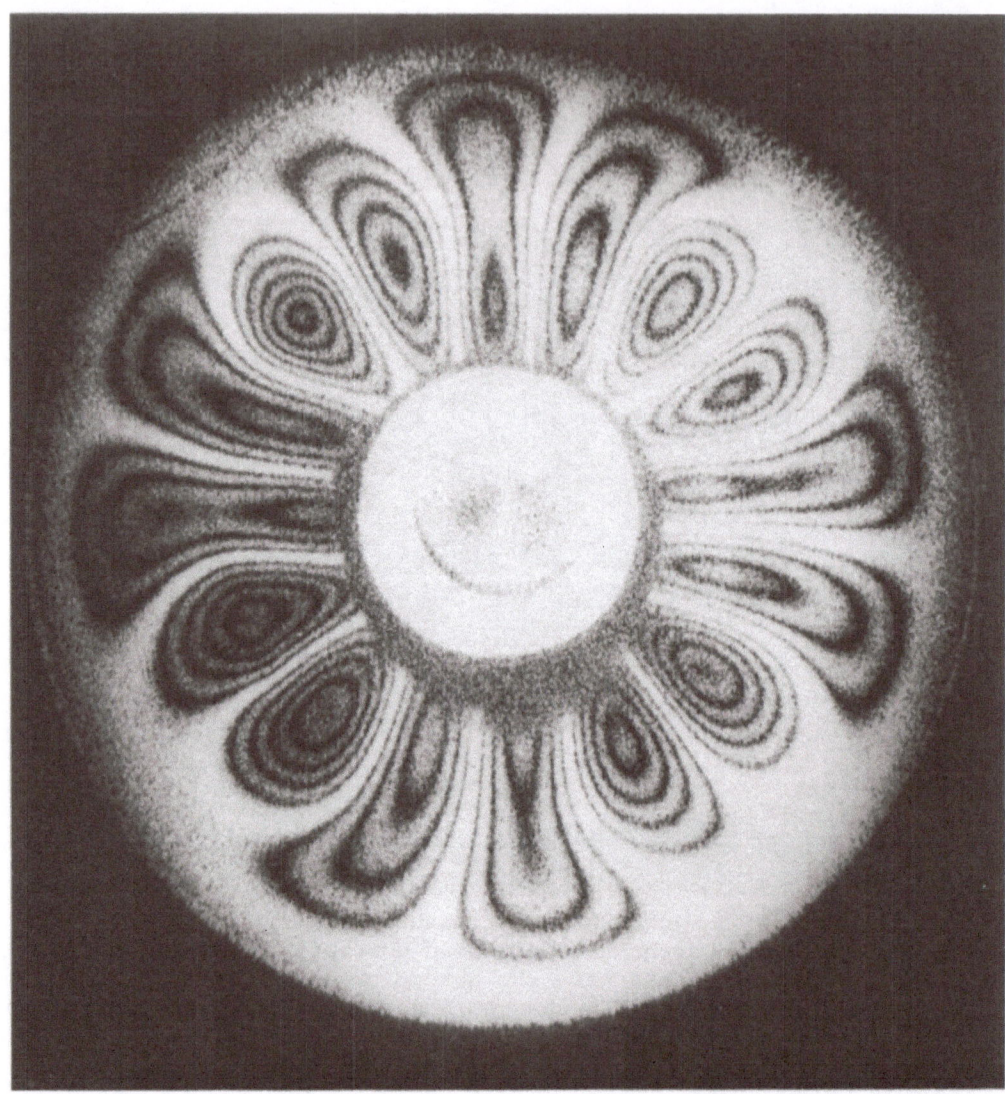

Fig. 3. 'Time-averaged' interference fringes on the surface of a
vibrating hollow cylinder. The pattern is displayed over
the full 360° surface by using a conical mirror illuminating the viewing arrangement. (courtesy K.A. Stetson).

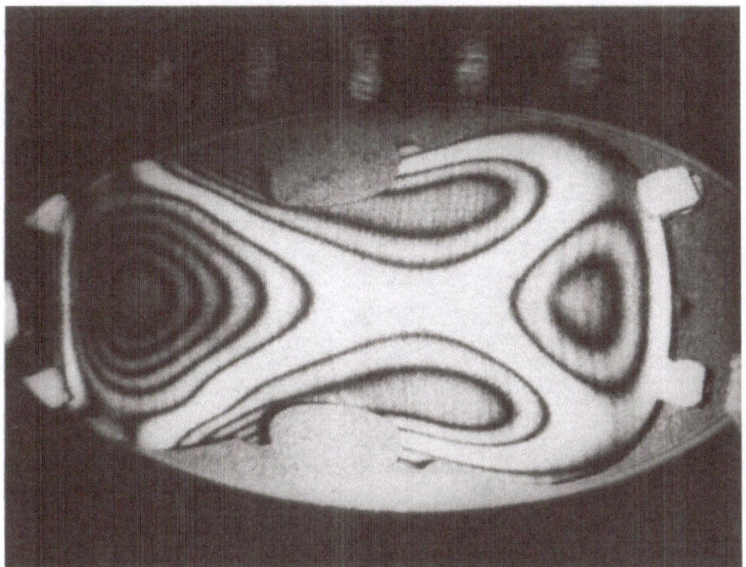

Fig. 4. 'Time averaged' fringes on a freely-held violin plate
(courtesy K.A. Stetson).

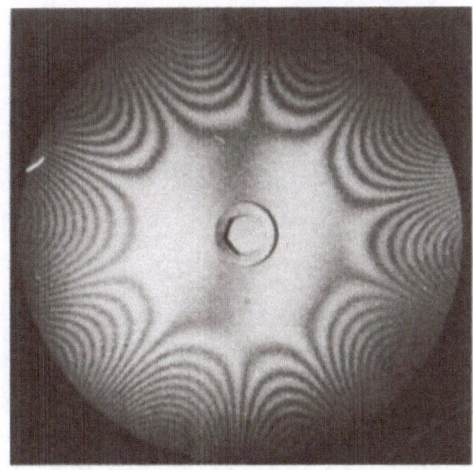

Fig. 5. Vibration fringes on a circular disc, obtained by stroboscopic interference holography.

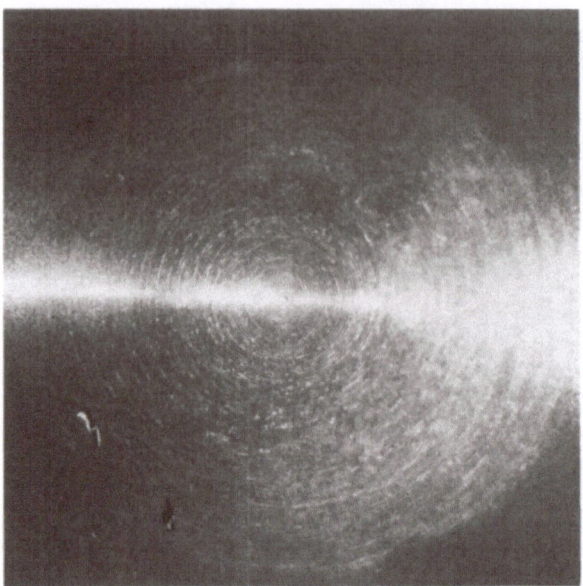

Fig. 6. Contour fringes showing the profile of an electron microscope condenser lens, obtained by double-exposure
holography and change of refractive index of the immersion
medium. Contour spacing 92 micrometres.

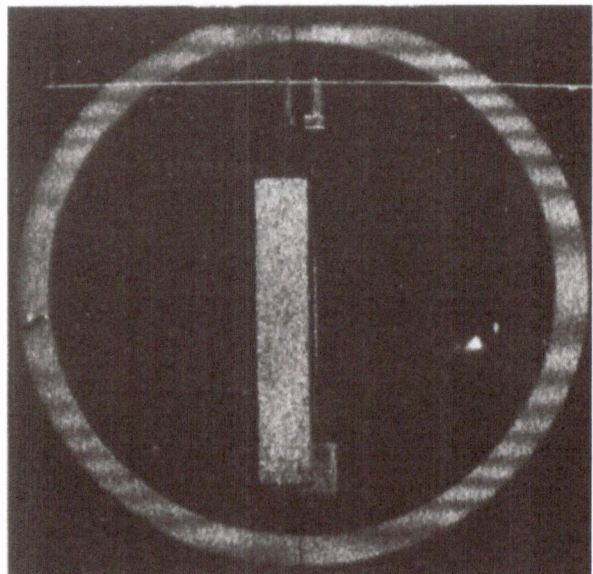

Fig. 7. Speckle pattern correlation fringes on the surface of the
steel ring shown in Fig. 1, obtained by double exposure
photography with two directions of illumination. The fringes
are contours of constant surface displacement, resolved in
the horizontal direction. The contour spacing is approximately one wavelength of the illuminating laser light.

SOME OPTICAL INFORMATION PROCESSING SYSTEMS

G. Groh

Philips Forschungs Laboratorium Hamburg GmbH
Hamburg (West Germany)

INTRODUCTION

Optical information processing systems have met increasing interest during the past 20 years. Potential applications are in all fields in which large quantities of information are to be processed. Best-known are optical filtering techniques which by very simple means, e.g., are capable of differentiating an image in order to emphasize contours and fine structures, of suppressing certain spatial frequencies of the image information in order to suppress or emphasize periodic structures, of introducing a phase shift between light diffracted into the first and zero order, respectively, in order to make visible small phase changes in otherwise indiscernible microscope specimens, etc.

The number of important applications has been considerably increased since with the invention of holography a new principle has been found which offers the possibility to realize even arbitrarily complicated, complex filter functions. Now correlation filters to be applied to the problem of automatic pattern recognition or image enhancement can be produced. Inverse filters for information retrieval from coded images formed by generalized imaging systems have become feasible. Optical multiplexing systems capable of distributing image information on several thousand spatially separated channels could be constructed using special holograms. Important applications are in systems for multiple imaging of etching masks of integrated circuits, in optical mass memories and in multiplex arrangements for optical pattern recognition. More recently the old problem of forming three-dimensional images of X-ray objects could be solved by applying holographic techniques.

This survey is by no means complete. However, it covers part of the field, in which we, at the Philips Research Laboratory in Hamburg, could pick up experimental and theoretical experiences. Being fully aware of the fact that our own experiments in this field are by no means representative for all the beautiful work having been performed all over the world I am going to demonstrate the general considerations of this lecture by our own experimental results, since to my opinion the audience will get a better feeling for the problems and difficulties when the lecturer is talking on experiments he was personally involved.

OPTICAL FILTERING USING COHERENT LIGHT

Most optical information processing systems use coherent light.

Fig. 1 shows a typical arrangement. Light from a monochromatic point like source, usually realized by a small pinhole being illuminated by a focussed laser beam, is collimated by the lens L_c. The plane wave U illuminates the object transparency, e.g. a slide, which modulates the phase and amplitude of the wave according to the image information stored. I.e., the wave immediately behind this transparency is proportional to its amplitude transmission function. This modulated wave is labelled g(x,y).

If the transparency is in the front focal plane of the lens L_1, it is known from diffraction theory that the distribution of the light amplitude in the back focal plane of this lens can be described by the Fourier transform of g(x,y)

$$G(\xi,\eta) = f_1 \left[g(x,y) \right] = \int\int_{-\infty}^{+\infty} g(x,y) \exp\left[-2\pi i(\nu x + \mu y) \right] dx dy \qquad 1.$$

where ν and μ are the spatial frequencies, which are directly proportional to the coordinates in that plane

$$\xi = 2f_1 \nu \qquad\qquad \eta = \lambda f_1 \mu \qquad\qquad 2.$$

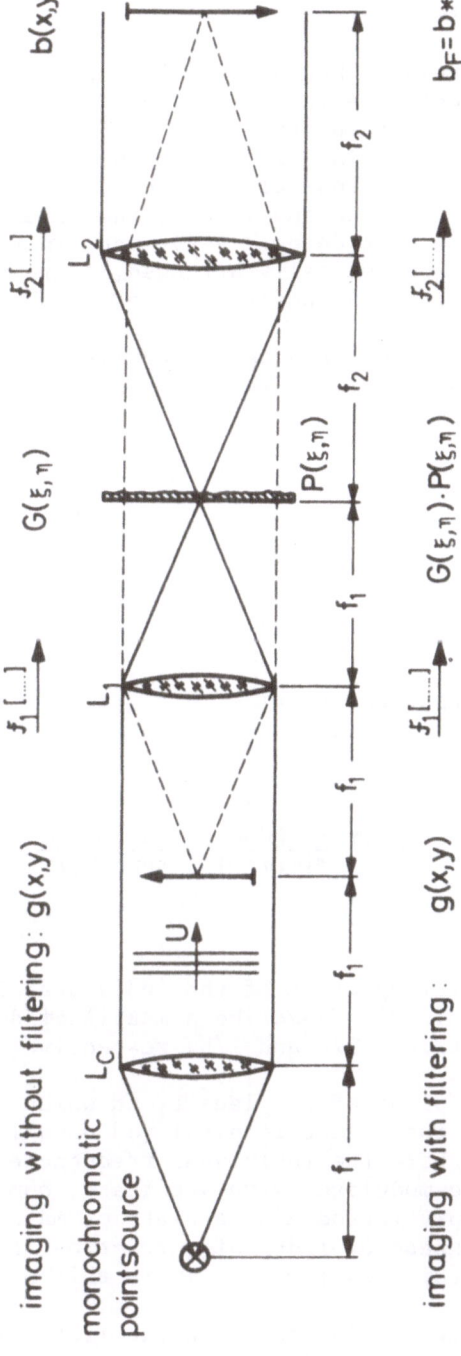

Fig. 1 Coherent optical filtering arrangement

The index at the symbol f, [   ] for the Fourier transform operator is to remind the **reader** that the focal length f_1 is a scale factor in these relations.

To give an impression how a spatial Fourier spectrum looks like in Fig. 2 two characteristic examples are exhibited. Fig. 2a shows the Fourier spectrum of the etching mask of a single integrated circuit (IC). In Fig. 2b the Fourier spectrum of about 100 in both directions periodically arranged patterns of the same circuit is shown. While the envelope of the intensity profile does not seem to have been changed very much, in the second case there is a screen of periodically arranged, sharp points. Its pitch corresponds to the fundamental period in the primary image.

In mathematical terms this can be easily described. If $f(x,y)$ represents the individual image the total periodic arrangement of $N \cdot M$ images is given by:

$$g(x,y) = \sum_{n=1}^{N} \sum_{n=1}^{M} \delta(x-na) \, \delta(y-mb) \, f(x,y) \qquad\qquad 3.$$

Its Fourier transform is (1)

$$G(v,\mu) = F(v,\mu) \; \frac{\sin(\pi Nav)}{\sin(\pi a \, v)} \; \frac{\sin(\pi Mb\mu)}{\sin(\pi b \, \mu)} \qquad\qquad 4.$$

I.e., the Fourier spectrum of the individual IC is modulated by the sin-functions, which describe a matrix of dots of size $2/Na$ and $2/Mb$ and periods $1/a$ and $1/b$, respectively.

The back focal plane of the lens L_1 in which the spatial Fourier spectrum of the object is displayed usually is called the filter plane, since here the individual frequencies are spatially separated and can be modulated with arbitrary, complex functions $P(\xi \quad \eta \quad)$ by simply introducing a plate of material whose absorption and thickness or index of refraction is a corresponding function of the coordinates ($\xi \quad \eta$) in this plane.

Usually, the light modulated with the filter function $P(\xi \, \eta \quad)$ propagates and enters a second Fourier transform lens L_2. Since

two sequentially performed Fourier transformations, except for a
change of the sign of the argument and possibly the scale yield
the original function the filtered or unfiltered image of the
object is displayed in the back focal plane of lens L_2.

As follows from the convolution theorem the filtered image
is given by the convolution

$$b_F = b * p = \int\int\limits_{-\infty}^{+\infty} b(x,y)\, p(x'-x, y'-y)\, dxdy \qquad 5.$$

of the ideal, i.e. unfiltered, image b with the point spread
function p of the coherent imaging system, where the latter and
the filter function P are Fourier transform pairs. Of course,
b_F only describes the light amplitude in the image or, more general,
output plane of the coherent system. If the light is to be detected or recorded the intensity

$$b_F \, b_F' = |b_F|^2$$

is the representative quantity.

When looking at this very simple theory one should bear in mind
that it is only a first order approximation. Not even the simplest
third order aberrations have been taken into account. Especially
the point spread funtion of the system is assumed to be shift-
invariant within the used image field. In general this essential
assumption is only approximately fulfilled in rather small image
fields or in systems with small apertures: f-members of 1/5 to
1/10 and even smaller are typical. The correspondingly poor
resolution is a severe limitation for a number of applications.

A simple and instructive example of image filtering has
already been described by Abbe more than seventy years ago.
Obviously, periodic structures in images can be easily suppressed
by simply blocking the corresponding spatial frequencies in the
filter plane. This principle has recently been applied to suppress
the periodic structure of IC-masks in order to emphasize nonperiodic
faults for inspection (2). Fig. 3 shows an experimental result
obtained in our own laboratory.

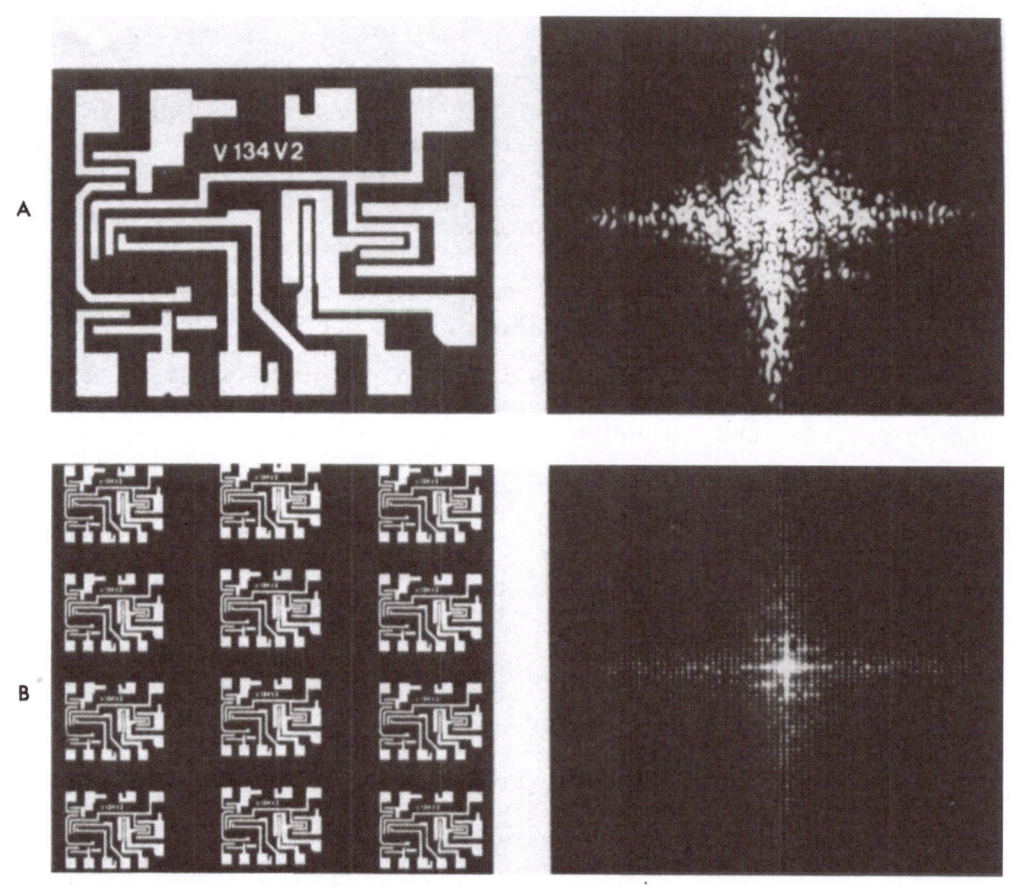

Fig. 2   Spatial Fourier spectrum of a
         a) single IC-mask
         b) periodic arrangements of IC-masks

Obviously, the complimentary filter, which stops all the light except for characteristic spatial frequencies, will remove periodic structures from a periodic image. This principle has recently been applied to restauration of images of faulty IC-masks (3) . Fig. 4 shows an experimental example. In this case the filter simply consists of a screen of small pinholes periodically arranged with pitches of $\lambda f_1/a$ and $\lambda f_1/b$, respectively, where, according to equation (3), a and b are the distances between the individual images in x- and y-direction.

HOLOGRAPHICALLY PRODUCED COMPLEX FILTERS

As Marechal and Croce (4) and O'Neill (5) have shown it is possible to achieve as much flexibility in optical filtering as exists in electrical filtering. An elegant method for synthesizing even complex optical filters has been proposed and demonstrated by Van der Lugt (6) . It utilizes holography.

According to this proposal a mask, whose amplitude transmittance is proportional to the desired impulse response p of the system, is placed in the object plane of the optical filtering arrangement (Fig. 1). In the back focal plane of lens L_1, where the Fourier transform P of p is displayed, an obliquely incident, plane reference wave R is superimposed, which is coherent with respect to the wave illuminating the object transparency. The interference pattern is recorded on a photographic plate situated in the filter plane. The processed plate will be referred to as Fourier transform or filter hologram of p.

Suppose the characteristic of the photographic material is linear, i.e. its amplitude transmission τ is proportional to the intensity I:

$$\tau = a - bI \qquad\qquad 6.$$

which itself is given by

$$I = |R + P|^2 = |R|^2 + |P|^2 + RP^* + R^*P \qquad\qquad 7.$$

P is the desired pupil function P^* is its conjugate complex value.

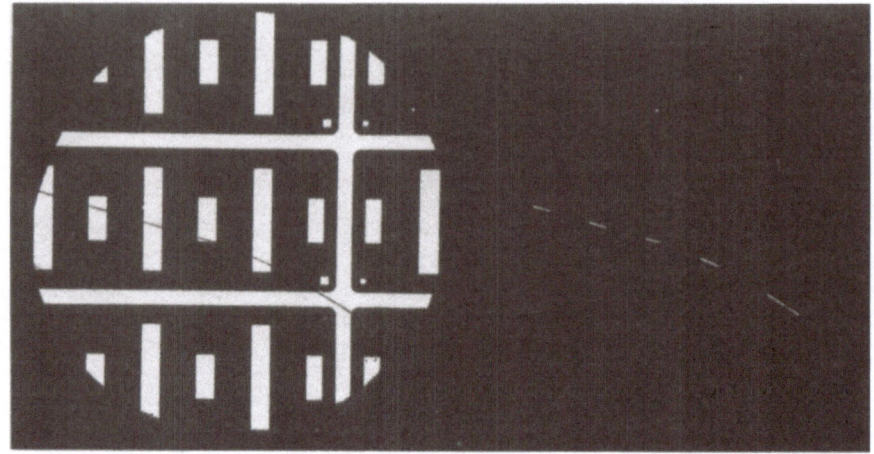

Fig. 3   Faulty IC-mask and the image of the fault emphasized for
         inspection by optical filtering

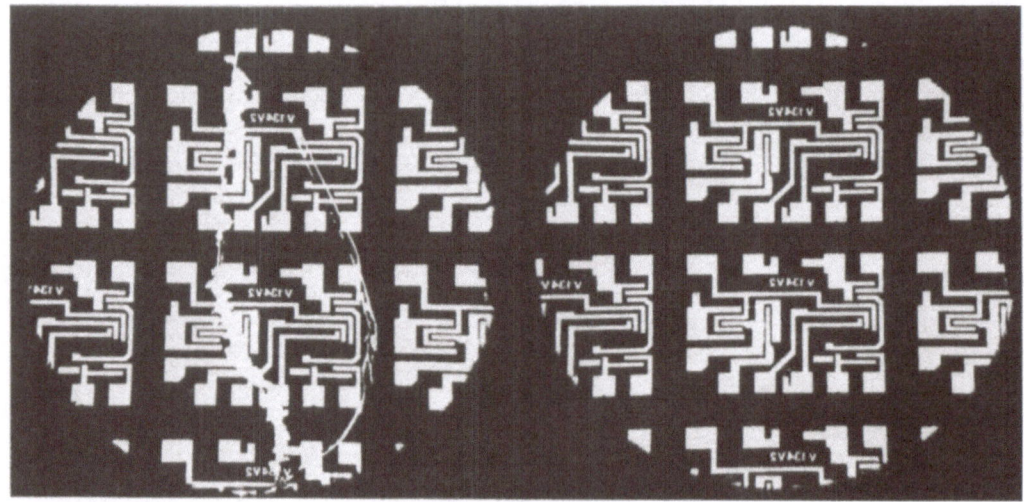

Fig. 4   Faulty IC-mask and its restored image

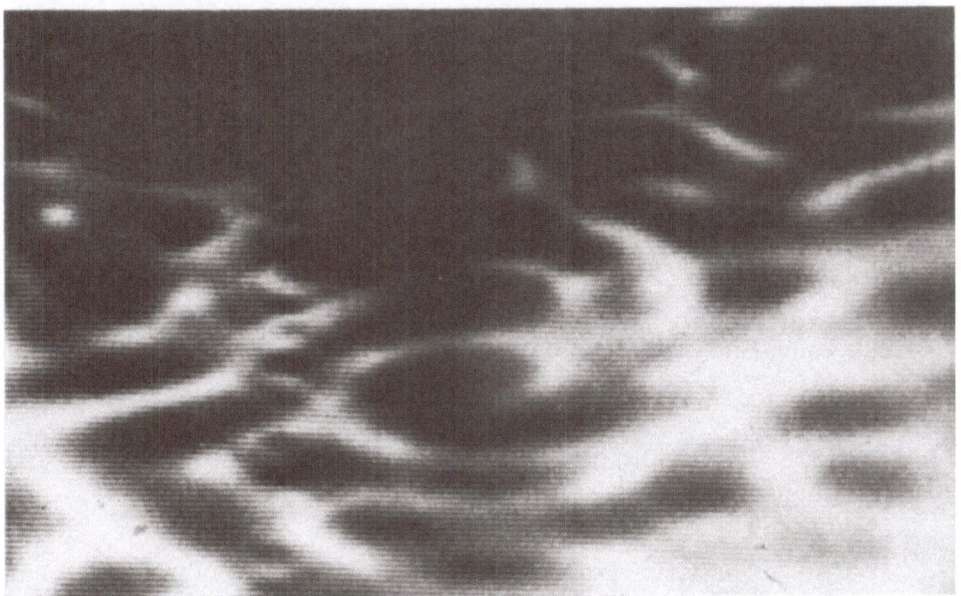

Fig. 5 Picture of a motor-car (Renault R4) and a microscopically
enlarged section of its filter hologram

Upon illumination with the wavefront G being the Fourier transform of a general object transparency g three waves are generated by diffraction at the hologram.

$$G \left(|R|^2 + |P|^2 \right)$$

is the so-called zero order term essentially describing the undiffracted light.

$$G R P^* \quad \text{and} \quad G R^* P$$

are the plus and minus first order terms we are mainly interested in.

By a second Fourier transformation as e.g. performed by lens L_2 in Fig. 1 these wavefronts are transformed in the cross correlation integral

$$g \circledast p = \int\!\!\!\int_{-\infty}^{+\infty} g(x,y) \cdot p^*(x'-x, y-y') \, dx \, dy \qquad 8.$$

and the convolution integral

$$g^* p = \int\!\!\!\int g(x,y) \cdot p(x'-x, y'-y) \, dx \, dy \qquad 9.$$

Respectively due to the oblique incidence of the reference wave these processed images are spatially separated. In most applications, however, only the wavefront corresponding to the correlation integral (8) is utilized.

Before discussing applications it has to be noticed that the theory presented here is only a very rough approximation. The reason is that the intensity of typical Fourier spectra covers a dynamic range of more than 4 orders of magnitude, whereas the linearisation of the characteristic of the photographic emulsion according to equation (6) is only valid for a considerable smaller range. In praxis this leads to the result that holographically produced filters can only be optimized for a comparatively narrow band of spatial frequencies. One has to find out in which range of the Fourier spectrum the most essential information is located and than to choose the exposure time and the intensity of the reference wave according to the average intensity of those most important spatial frequencies.

In general lower frequencies are overexposed and higher frequencies are underexposed. Fig. 5 shows a microscopically enlarged small section of a filter hologram of a motor car. The course and partly overexposed structure corresponds to the intensity of Fourier spectrum. The straight horizontal line are due to interference of the Fourier spectrum with the reference wave. The usable information is holographically stored only in those parts in which the interference lines appear.

One important application of this method for synthesizing complex filters is pattern recognition. A filter hologram of the pattern to be searched is recorded and placed in the filter plane. Then the image to be investigated is put into the input plane of the optical filtering arrangement (Fig. 1). In the output plane according to equation (8) the crosscorrelation function $g * p$ of the pattern p and the image g is displayed. If g contains p in the corresponding part of the output plane the autocorrelation function is formed which in general is well destinguished by a pronounced sharp central peak. For recognition this peak height has to be discriminated.

Fig. 6 shows an experimental example. The image to be inspected contained two motor-cars one of them, the Renault R4, was to be detected. The crosscorrelation function was formed, picked up by a TV-camera and displayed on TV-monitor. Fig. 7 shows an oscillogram of the videosignal.

It should be mentioned in this context that the previously discussed loss of the lower spatial frequencies is extremely advantageous for pattern recognition. The reason is that the undiffracted light contains only information on the average bright-

Fig. 6.   Picture of two motor-cars and the autocorrelation
          peak indicating the recognized car.

ness of the objects which roughly spoken is the same for all parts
of the image. Due to the overexposed central part of the spectrum
this trivial information is suppressed and only the finer structures of the object being represented by higher spatial frequencies
are compared. Though not quite correct this fact often is expressed by saying that the differentiated objects are correlated. This
leads to a considerably smaller central peak of the autocorrelation
function whose peakheight is by far more sensitive to changes of
the shape and the structural information content.

Though as a matter of principal arbitrarily complicated
patterns can be recognized by this method there are some important
difficulties which up to now considerably delayed practical applications.

1) The method is sensitive to trivial changes of the object information as are scale and rotational orientation.

2) The method is extremely sensitive to changes of the phase and
the scattering properties of the object.

3) In many cases the peak height of the autocorrelation function
is an insufficient criterium to distinguish between similar
patterns.

The first difficulty can be overcome by rotating the filter
or better the object and using a variable scale Fourier transform
arrangement. The second difficulty has been solved on a laboratory scale by using film transparencies in a liquid gate. Regarding the third difficulty some proposals exist to overcome it by
utilizing coded reference sources (7) or by post processing of
the correlation functions.

A more recent proposal which hopefully will help to solve
some of the problems mentioned above will be discussed in section:
Multiplex arrangement for pattern recognition using an optical
relay tube.

Before, however, another important application of optical information processing for deblurring or decoding images is to be
described.

GENERALIZED IMAGING SYSTEMS

Unconventional imaging systems such as e.g. arrays of pinholes get increasing importance at wavelength where no ordinary
imaging devices, as lenses or mirrors, exist. Moreover, image
information often is obtained from antenna systems whose collected
data have to be processed in order to retrieve the desired image.
In other cases images are blurred due to motion of the camera or

the object during exposure or since the imaging device was not
focussed. There is no time to discuss all possible cases were
images have to be processed. Rather the purpose of this section
is to describe the general class of images which as a matter of
principle can be decoded or deblurred and to discuss the optical
means and their limitations.

For this reason the following considerations will be restricted to all kind of coded or blurred images which can be described
by the convolution integral of the true image information g with
the point spread function p of the system, i.e. the latter has to
be shift invariant.

In these cases mathematically spoken the problem of image deblurring or decoding can be reduced to deconvolution. This
process can be performed in the Fourier transform plane of an
optical filtering arrangement (Fig. 1). There the product of the
Fourier spectra of the image $G = f[g]$ and the point spread function
$P = f[p]$ is displayed. Thus as a matter of principle the deconvolution process can be performed by placing a filter whose amplitude transmittance is inversely proportional to P in the filter
plane. Such a filter is called the inverse filter. The problem
is its realisation.

The usually used approach is to utilize the identity:

$$\frac{1}{P} = \frac{1}{PP^*} \cdot P \qquad\qquad 10.$$

by forming a sandwich of two filters. The first one has stored
the inverse function $1/|P|^2$ of the intensity of the Fourier
spectrum of the point spread function by utilizing the logarithmic characteristic of photographic materials (9). Obviously,
this is only approximately possible since at the zeros of $|P|^2$
that function becomes infinite. The second filter containing the
conjugate complex P^* of Fourier transform of the point spread
function can be synthesized holographically by using the third
term in equation (7).

Again, due to the rather limited dynamic range of the photo-
grafic emulsions both functions can only be approximately realized.
By carefully controlling the photographic processes involved
Stroke could recently extend the usable dynamic range to $10^3:1$.

Another principle, which avoids the inverse filter, has been recently discussed in the literature (9) , (10) . The essential assumption again is that the intensity distribution of the coded image can be described as the convolution g $*$ p of the object g(x,y) with the point spread function p(x,y) of the system. I.e., except for a translation p(x,y) has to be shift invariant over the object field. Then the object information g(x,y) is retrievable, if the autocorrelation function p $\circledast$ p of the point spread function p(x,y) is a δ-function or similar to it. According to earlier suggestions this data processing can be performed by recording the Fourier transform hologram of the blurred object information g $*$ p using the point spread function p (1. method) or the δ-function (2. method) as a reference source. On suitable reconstruction one obtains g $*$ p $\circledast$ p.

The disadvantage is that an individual hologram has to be recorded from each coded image. In order to avoid this drawback it was proposed (11) to record once the Fourier transform hologram of the point spread function p using a δ-function (i.e. a suitably illuminated pinhole) as the reference source. The essential terms stored in the hologram are given by equation (7) where P and R denote the Fourier-transforms of the point spread function p and the δ - function, respectively. Using this hologram it is quite simple to construct a decoding apparatus for all coded images obtained by the corresponding generalized imaging system. One has just to put that hologram into the filter plane of an ordinary Fourier transform arrangement where the Fourier transform f [g $*$ p] = G.P of the coded image is displayed (G being the Fourier-transform of the decoded image g). Then, according to the third term of equation (7) the virtual image:

$$f^{-1} [G.P.P^* R] = g * p \circledast p * \delta \qquad\qquad 11.$$

is reconstructed in the surrounding of the position of the reference source used in the recording process. This convolution of the object function σ and the autocorrelation function of the point spread function p is equal to the decoded image if

$$p \circledast p \approx \delta \qquad\qquad 12.$$

Another advantage of this method is that the zero order image terms of the hologram are not broadened since they correspond to δ -functions when equation (12) is fulfilled. Whereas for the two methods suggested earlier the zero order image is practically determined by the autocorrelation function of the decoded image.

The feasibility of the method is demonstrated by Fig. 8. The incoherently illuminated object Fig. 8b was imaged by a system of statistical distributed small pinholes, whose point spread function is shown in Fig. 8a. The resulting image (Fig. 8c) was recorded. Fig. 8d shows the retrieved image.

The difficulty with this method is to find a point spread function p whose autocorrelation function is similar to a δ -function. For point arrays distributions have been calculated having a compact nonredundant autocorrelation (12).

In all examples discussed up to here optical information processing has been applied to reduce the information. There are, however, important applications for multiplexing the image information.

Two simple examples will be described in the next section.

MULTIPLE IMAGING

The holographic method for multiple imaging has been indepently proposed by several authors (13). Except for some modifications, they all use socalled point holograms, i.e., holograms which upon illumination with a spherical wavefront reconstruct a distribution of image points.

Fig. 9 shows an arrangement for recording the point hologram. A laser beam is splitted and focussed onto a pinhole acting as a reference source and a matrix of pinholes acting as signal sources. The spherical wavefronts emanating from these sources interfere. The interference pattern is photographically recorded.

This hologram can be considered as being the synthesized pupil of an imaging system whose point spread function may be described by a matrix of δ-functions. For this reason it has just to be put in the exit pupil of an ordinary imaging system in order to multiply the image information.

Fig. 10 shows the experimental arrangement. Fig. 11 shows an experimental result. Obviously, this principle even works in imaging systems employing spatially incoherent illumination, thus avoiding the typical disturbances of coherent imaging systems, as are speckle patterns, edge ringing and shifting. Fig. 12 is to demonstrate this difference.

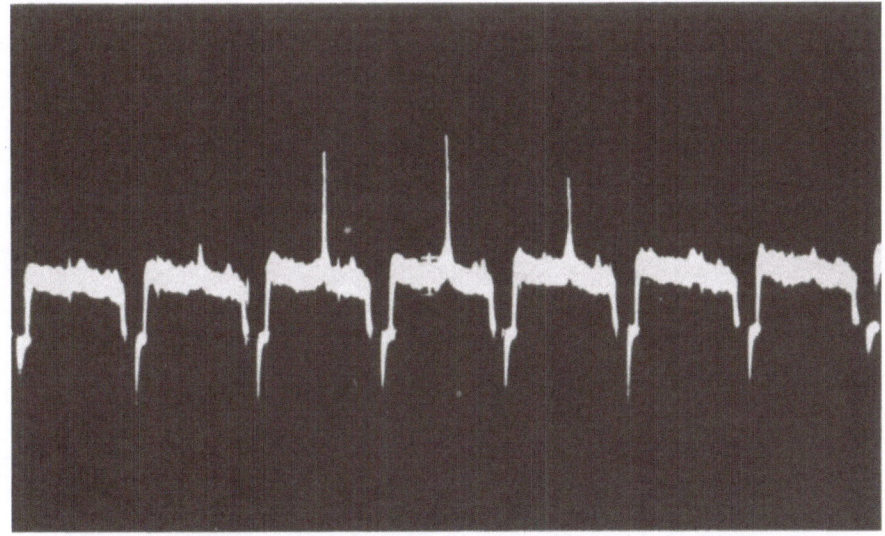

Fig. 7 Oszillogram of the videosignal of the autocorrelation peak

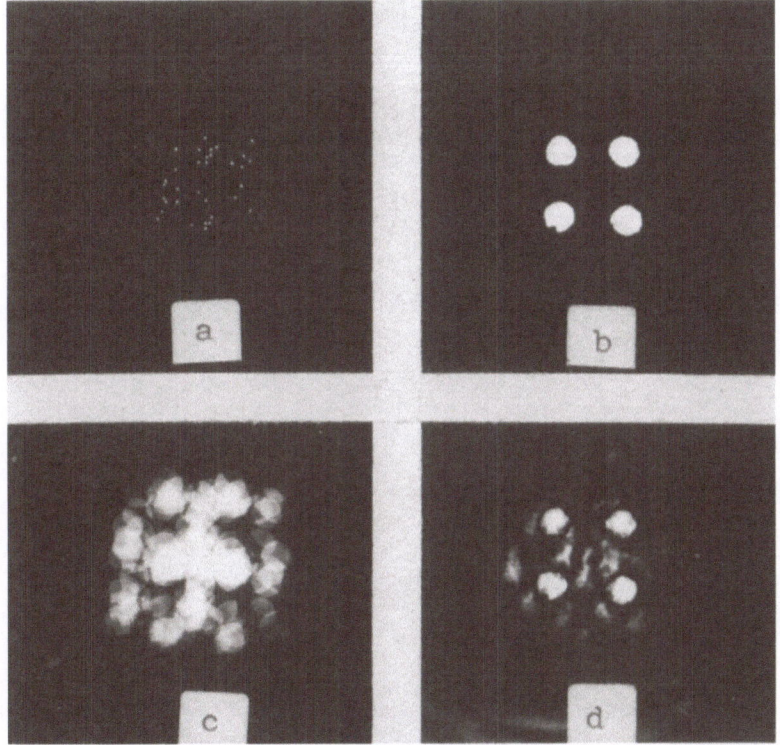

Fig. 8 Information retrieval from coded images
        a) point spread function of a generalized imaging system
        b) object, c) image formed by the system
        d) retrieved image

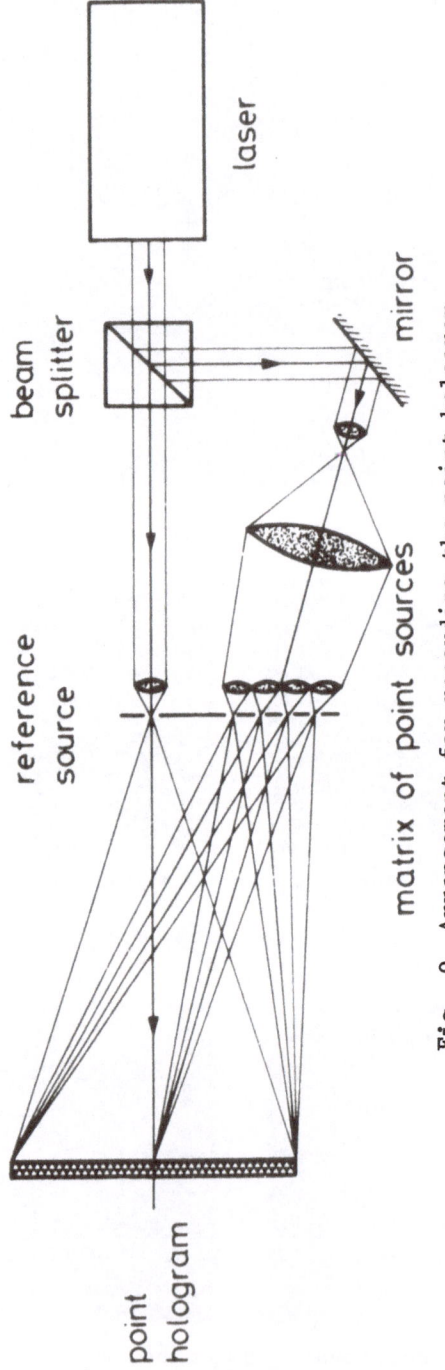

Fig. 9 Arrangement for recording the point hologram

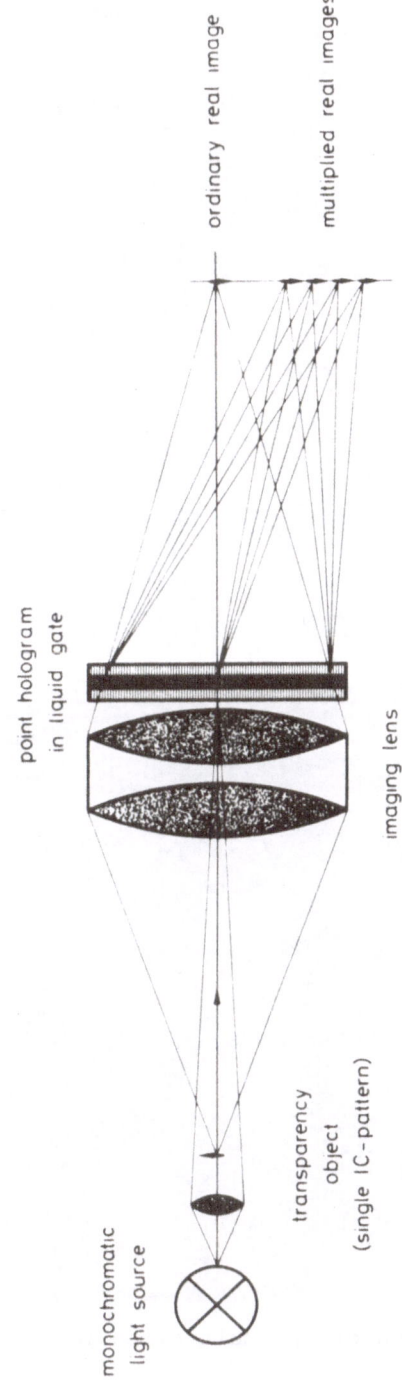

Fig. 10 Arrangement for multiple imaging by means of a point hologram

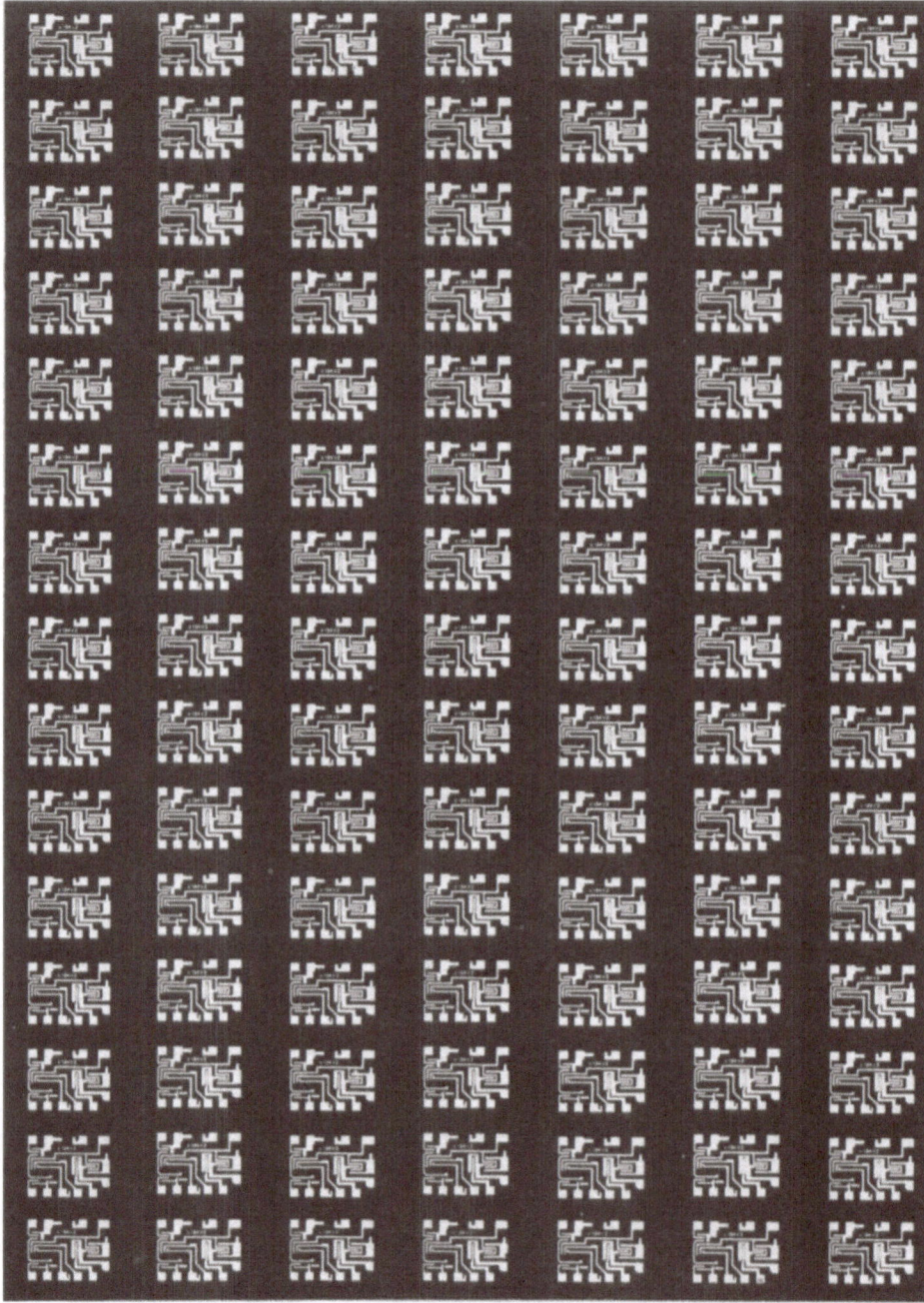

Fig. 11 Enlarged section of the multiplied image of an IC mask

Fig. 12   Enlarged multiplied image obtained with
         a) coherent
         b) incoherent illumination of the object

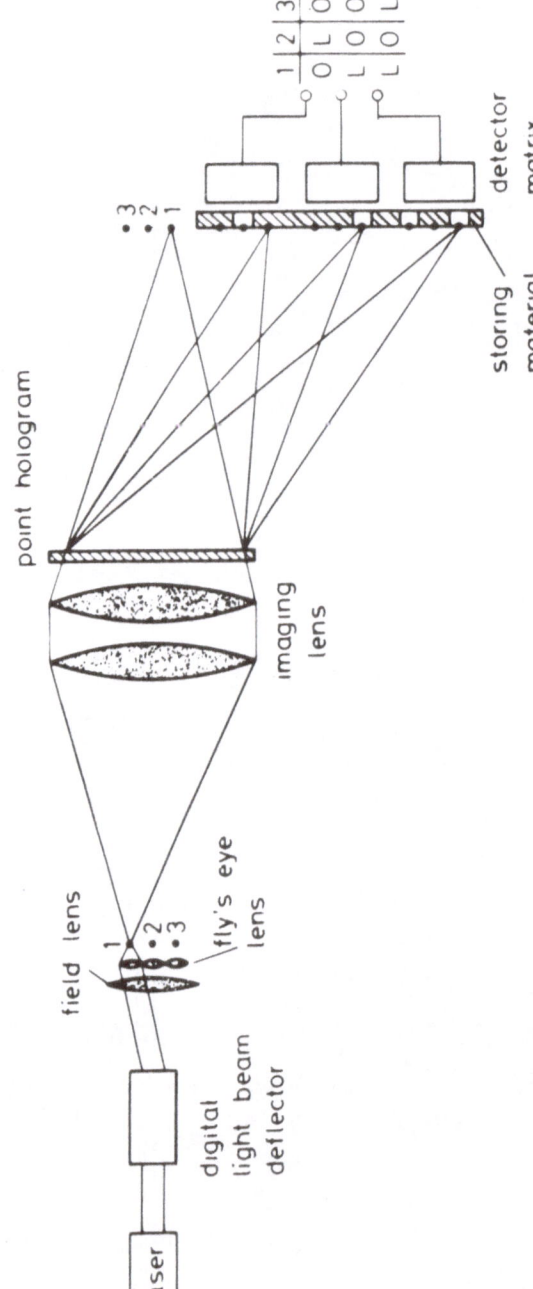

Fig. 13 Optical mass memory utilizing a point hologram

The same principle of multiple imaging can be utilized in a special arrangement (14) for an optical mass memory. A scheme of the system is shown in Fig. 13. The laser beam is deflected by a digital light beam deflector (15) and focussed by means of fly's eye lens. This field of focal spots is multiple imaged onto a storing material containing holes for the binary figure L. The light is detected by a detector matrix consisting of one element per multiplied field. Since a 20-stage light beam deflector with 1024 x 1024 deflection positions and a random access time of about 5 μsec is available and a multiplication factor of 1000 can be easily realized holographically, a read only optical mass memory of 10^9 bit total capacity has become feasible, in which addresses each containing 1000 bit can be addressed and read out in parallel within 5 μsec.

Another application of the point hologram in optical multiplexing is to be described in the next section.

MULTIPLEX ARRANGEMENT FOR PATTERN RECOGNITION
USING AN OPTICAL RELAY TUBE

Arrangements for pattern recognition using special holograms as complex spatial frequency filters are wellknown and have been extensively described in section 3 and reviewed in the literature (Ref. (16)). Though as a matter of principle these arrangements work rather well, there are two basic problems, which besides other difficulties have delayed direct practical applications of this method on a larger scale.

The first and most important one is due to the fact that the optical information is processed by using coherent light, i.e. the system works with the complex amplitude of the wavefront scattered at the object to be inspected rather than the intensity distribution one usually is interested in. For this reason it is almost impossible, e.g., to recognize a character or pattern printed on ordinary paper or even copied on photographic film. In most cases due to the lack of uniformity the information carrier, i.e. the paper or the film, itself modulates the phase and amplitude of the incident coherent light by far more significantly than the black and white structure to be investigated. Thus the signal-to-noise ratio of the output signal becomes terribly bad.

To avoid this difficulty in present day laboratory arrangements one usually employs photographic plates or films in liquid gates as object transparencies. Obviously, this procedure is not feasible for many applications where optical information has to be picked up and processed almost in real time.

For this reason there is an urgent need for an optically neutral, i.e. not scattering, information carrier on which the

images to be processed can be repeatedly written, erased and
replaced by other ones at rather high speed. The purpose of part
of this section is to describe an input device for optical data
processing systems which meets these requirements and which in
addition has some other advantageous properties regarding this
kind of application. The device is called TITUS (Tube Image á
Transparence Variable Spatiotemporelle) and was developed by
Marie (ref. (17) , (18)). It uses an electron-beam-addressed
crystal of deuterated KDP (KD_2PO_4).

Before, however, discussing this tube in more detail the other
basic difficulty mentioned above shall be briefly described. In
most applications one wants to look for a large number of characters or patterns rather than a single one. Or thinking, e.g., of
the most difficult problem of automatically reading handwritten
letters and figures one is facing the problem that a lot of patterns
of different shape, size and orientation have exactly the same
meaning. Since, however, ordinary holographic pattern recognition
arrangements can only look for one individual pattern at a time,
the holographic filters have to be quickly exchanged, which seems
to be not feasible taking into consideration the precision of the
necessary alignment.

To solve this problem an optical multiplex method utilizing
point holograms has been recently suggested (Ref. (19)), which
in addition offers the possibility of logically combining various
output signals thus reducing the wellknown ambiguity of the holographic process.

The integration of these methods and components in an arrangement for pattern recognition will be described. First of all,
however, the principles of operation of the individual components
have to be discussed.

The Fig. 14 illustrates the essential part of TITUS (Ref. (17) ,
(18)). It consists of a thin monocrystalline plate of deuterated KDP (KD_2PO_4) perpendicularly cut to its optical axis. The front
and back surfaces of the crystal are coated by an optically transparent, electrically conducting electrode and a dielectric mirror,
respectively. A collector grid is placed a short distance away
from this target. A scanning electron beam of sufficiently high
but constant energy passing the grid and impinging on the outmost
layer of the mirror produces secondary electrons which are collected by the grid and practically cause a short circuit between the
grid and the point of impact. When the video voltage of a TV
signal is applied between the grid and the conducting electrode
after one frame a spatially modulated electrical field will act
on the crystal. Since this one is operated near its Curie temperature the relaxation time is rather long (up to one hour) and
the electrical image remains scored until a new one is written in.

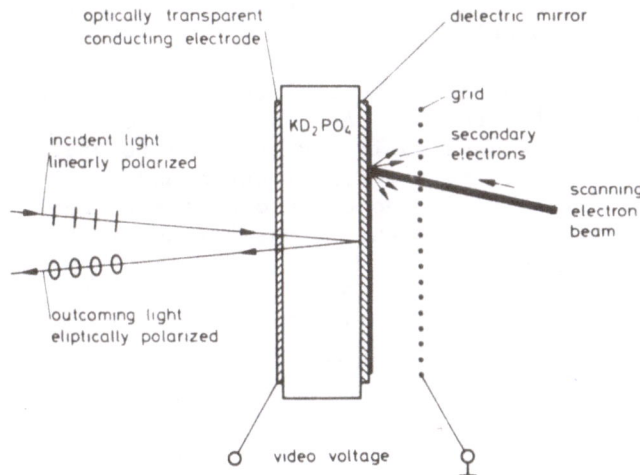

Fig. 14   Principle of operation of TITUS

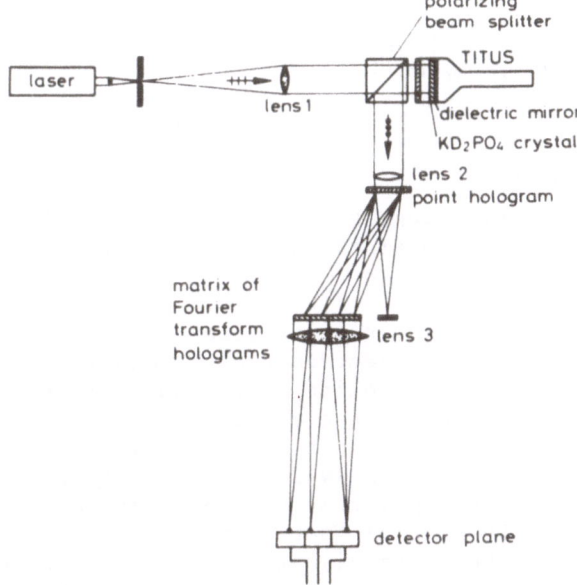

Fig. 15   Holographic multiplex system for pattern recognition using
          TITUS as an imput device

Due to the Pockels effect linearly polarized light illuminating
the crystal, being reflected at the mirror and passing the crystal
a second time will be spatially phase modulated or become more or
less elliptically polarized.

For coherent light only the amplitude transmittancy of the
KD_2PO_4 crystal is of interest. As has been discussed recently
(Ref. (20)), it is especially advantageous to operate the
crystal between crossed polarizers oriented under 45° with respect
to the induced main axis of the ellipsoid of the index of refraction. In this case the amplitude transmittancy τ becomes:

$$\tau\left(x,y\right) = \frac{1}{2}\left(e^{+ipU(x,y)} - e^{-ipU(x,y)}\right)$$
13.

or:

$$\tau(x,y) = i,\sin pU\left(x,y\right)$$
14.

U(x,y) being the voltage applied to the crystal in the point
(x,y); p is a constant containing the electrooptical coefficient
of KD_2PO_4.

In contradistinction to similar tubes (Ref. (21) and (22)),
where the intensity of the electron beam is modulated by the
videosignal, for TITUS it is quite easily possible to apply positive and negative voltages to the crystal. Thus positive and
negative values of the amplitude transmittancy can be realized
(equation 14). This is extremely favourable, since it offers the
possibility to suppress completely the undiffracted light, i.e. the
zero order term of the spatial Fourier spectrum, by simply blocking
the dc component of the video signal. Due to this fact, Fourier
transform holograms to be used as complex spatial frequency
filters can be recorded without running into the wellknown troubles
with the overexposed zero order term (Ref. (20)   ).

Another advantage is that using this mode of operation the

diffraction efficiency which obviously is directly related to light
economy of the total arrangement, is rather high. For simplicity
let us regard the sinusoidal grating:

$$U = U_m \sin kx$$ 15.

written in TITUS. Than the normalized intensity $|a_n|2$ of the
light diffracted into the n-th order is

$$a_n^{\ 2} = \frac{1}{4} \left| 1 - (-1)^n \right|^2 \cdot J_n^2 \left(pU_m \right)$$ 16.

as can be easily calculated from equations (13) and (15). J_n
denotes the n-th Bessel function. Since the whole information on
this grating is in the plus and minus first order light:

$$\left| a_1 \right|^2 + \left| a_{-1} \right|^2 = 2 J_n^2 (pU_m)$$ 17.

describes the modulation efficiency of TITUS when operated between
crossed polarizers. For the maximum modulation:

$$pU_m = \pi/2$$ 18.

this efficiency is 64%.

Even at this high modulation depth, however, the tube works almost exactly linear. This is due to the fact that according to equation (16) all even (including zero) order terms of the Fourier spectrum of the signal modulated at the sinusoidal characteristic of the tube compensate each other. The influence of the first non disappearing term is rather small:

$$\frac{J_3^2(\pi/2)}{J_1^2(\pi/2)} = 1.6 \cdot 10^{-2} \qquad\qquad 19.$$

A further advantage of TITUS is that the image contains no line structure as caused by the scanning electron beam. This is due to the special writing procedure (Ref. (23)) used here.

For producing a matrix of identical but spatially separated Fourier spectra a point hologram (19) is used in the exit pupil of the Fourier transform lens. (Lens 2 in Fig. 15.)

As has been shown earlier such a point hologram has the property of transforming one wavefront into a number of identical ones. This to a certain extent is even true for rather complicated wavefronts scattered at arbitrary objects. By this principle also the wavefront generating the Fourier spectrum in the back focal plane of a lens is multiplexed.

These identical Fourier spectra illuminate a corresponding matrix of filter holograms, each of which has been individually recorded from one of the patterns to be recognized. E.g. in the first column of the matrix there may be situated the Fourier holograms of all the different printing types of the letter "a", in the second column those of the letter "b" and so on. Regarding the applications it might be advantageous to give all the "a"-holograms a joint reference source and the "b"-holograms a likewise joint but spatially separated reference source.

Upon illumination with the matrix of identical Fourier spectra of the unknown object transparency each of the holograms reconstructs a wavefront corresponding to the virtual image of the cross correlation function of the object and the pattern stored in the hologram. The second lens projects the real images of the hologram responses onto a detector array where the light intensities are discriminated.

If the object contains e.g. one printing type of the letter "b" the corresponding hologram responds displaying the auto-correlation function. As in conventional filter experiments the position of the correlation peak describes the location of the character in the object. If the positions of different characters "a", "b", "c" etc. are to be discriminated the distances between the various reference sources have to be a little bit larger than the dimensions of the total object field in order to spatially separate the responses.

Instead of different printing types one can obviously store different rotational positions of a letter in part of the hologram matrix thus introducing some rotational invariance. The funda-mental property of this system is that it responds to one or seve-ral of a class of patterns. With this organisation of the filter holograms it does not distinguish between members of one class but discriminates different classes.

By slightly changing the arrangement of the reference sources of the filter holograms one can overcome another difficulty of pattern recognition as is the ambiguity of the correlation peak for patterns containing similar elementary structures. Obviously, the auto- and crosscorrelation signals of the letter "o" and "l" are rather distinct. Character identification, however, becomes difficult when e.g. the letters "b", "d", "p" or "q" are simul-taneously present. Utilizing the possibilities of the multiplex system it is advantageous to distribute the information e.g. on the letter "d" on two channels,one containing the elementary struc-ture "o" and the other the element "l". When recording the holo-grams two reference sources are used which are shifted with re-spect to each other by a distance which corresponds exactly to that of the correlation peaks of the elements "o" and "l" in the letter "d". Then for this letter and only for this the correlation signals of both channels are added resulting in a higher output signal.

More general, the multiplex system offers the possibility to form a linear combination of the output signals of various filter channels. By suitably arranging the reference sources those sig-nals only overlap when certain conditions regarding the geometrical arrangement of the elementary structures are fulfilled. Since, as a rule, the output signals are coherent on principle there is the possibility to introduce even complex weighting factors in the process. This, however, involves difficult alignment procedures.

The complete arrangement is shown in Fig. 15. Taking into consideration the previous sections it is self-explaining. The image to be written in TITUS is picked up by an ordinary TV-camera. In this part of the system also the more trivial adjustments are performed. I.e,, the scale factor is adapted by utilizing a zoom

lens and the image is rotated by using rotating deflection coils
for example. Moreover, the image information can be preprocessed
in the electronical part. In many cases it is, e.g., reasonable to
suppress the low frequencies containing only little information
already in this part. Though this, of course, can also be done in
the optical part it is often much easier to optimize the spatial
frequency filters for a preprocessed signal.

The feasibility of the methods used in this arrangement has
been demostrated in preliminary experiments. The tube used was
manufactured by Marie and his co-workers. It was capable of resolving more than 800 elements per line corresponding to 10 line
pairs per mm in a field of 30 x 40 mm^2. According to the frame
frequency of the TV-signal one image of this resolution can be
processed in 40 msec not including the rotation and the adjustment
of the scale factor. Due to the multiplexing point hologram up to
about 1000 filter operations can be performed in parallel. So that
it seems to be possible to look for about 1000 different patterns,
characters or their variants in this time.

A problem still are the rather low diffraction efficiencies
of the holograms used. For phase type point holograms the efficiency obtained was about 20%. Assuming 1000 filter channels and
taking into account $_6$50% efficiency for TITUS and 1% for each filter
hologram about 10 $^-$ of the incident light go into the individual
channel, i.e., when using a 1 W-gargon ion laser the light intensities to be detected are of the order of a fraction of one μW.

Fig. 16 shows one of the first experimental examples. In
this case six of the channels of the multiplex system has been
reserved for the letters a, s, l, k, o, e. The images on the
TV-monitor show the correlation peaks when the letters o or e are
fed into the system. The peak heights even of similarly shaped
letters like o and e can be clearly discriminated.

A completely different kind of optical information processing
will be discussed in the next section. The general problem is to
reconstruct three-dimensional images from a set of two-dimensional
photographs. The problem is solved by means of holography.

COMPOSED HOLOGRAMS FOR DISPLAYING
3-D IMAGES OF X-RAY OBJECTS

Due to the lack of imaging devices X-ray images are formed by
illuminating the object with X-rays from a point source and recording the shadow. Obviously, any information on the depth gets lost.
There exist several methods to retrieve the information on the
third dimension, the depth. Best known and most widely used is the
method of tomography (24). More recently holographic methods have
been proposed (25).... (29) .

In all cases the primary information is obtained by time sequentially illuminating the object from different directions thus
producing a set of X-ray images showing the object in different
perspectives (Fig. 17). This procedure might be interpreted as
generating a synthetic aperture whose width is determined by the
maximum distance of the positions of the X-ray source. Obviously,
the depth of focus in the retrieved 3-D image will be the smaller
the larger the diameter of this synthetic aperture is.

In the second step the individual X-ray photographs are
processed by storing them in spatially separated holograms whose
arrangement corresponds exactly to the distribution of information
in an ordinary hologram of a diffusely scattering object.

When recording the hologram the X-ray images of different
views are projected time sequentially onto a translucent diffusing
screen using coherent light (Fig. 18). The scattering characteristic of the diffusor has to be chosen broad enough so that each
point on the hologram receives light from every point on the screen.
The photographic emulsion is covered by an opaque diaphragm containing a sufficiently small aperture. For each exposure this aperture
is brought into a position with respect to the individual projected
image which apart from a constant magnification corresponds exactly
to the position of the X-ray source with respect to the shadow on
the fluorescent screen used when producing the X-ray images. That
is to say, if e.g., the X-ray source has been moved along a line
representing a section of a circle around a point in the fluorescent screen, the photographic emulsion recording the hologram has
to be situated on the corresponding cylindrical surface around a
point in the projection screen. Otherwise the object will be
reconstructed in a distorted space.

During the whole recording process the position of the photographic emulsion is fixed whereas the aperture is moved in steps
between exposures. Thus using a constant but arbitrarily shaped
reference wavefront, which itself has to illuminate the whole
number of holograms, a matrix of spatially separated or slightly
overlapping holograms of the projection screen containing different
silhouettes is produced.

On reconstruction with the reference wave this hologram matrix
exhibits properties very similar to those of an ordinary hologram.
The observer looking through the hologram matrix sees the 3-D
virtual image of the X-ray object in front of a brightly illuminated screen. He has the same impression as if he were observing the
object from the side of the X-ray source. This image is orthos-
copic. Of course, paralactic displacements are not continuous,
i.e. the image has the tendency to jump when the observer moves
his head. This phenomenon, however, can be made arbitrarily small
by increasing the number of matrix elements. We obtained good re-

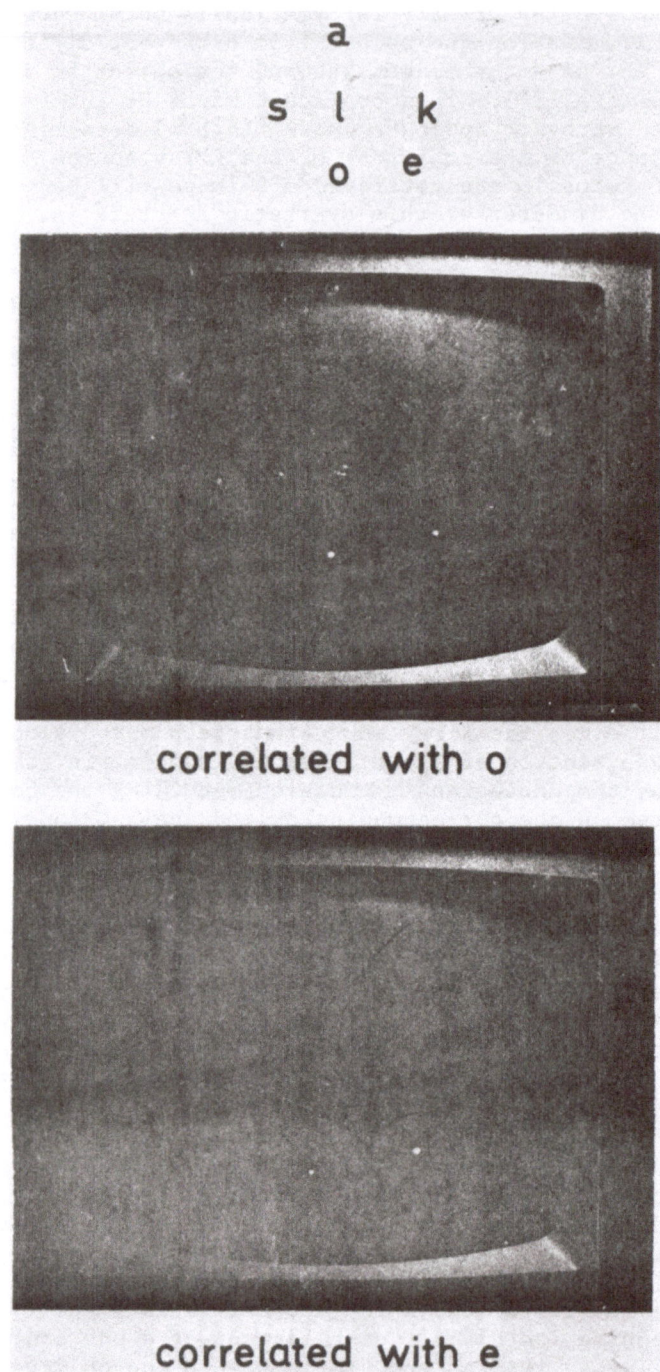

Fig. 16   Output signal of a multiplex correlator, Whose six
          channels are reserved for the letters a, s, 1, k, o, e

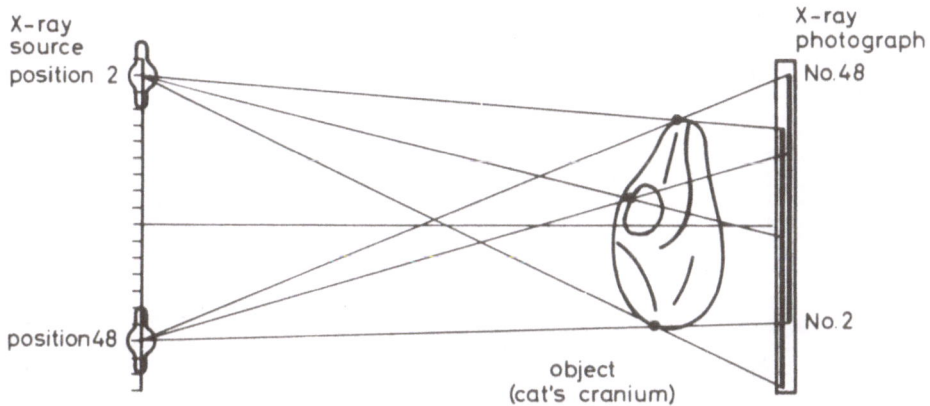

Fig. 17 Production of the X-ray photographs

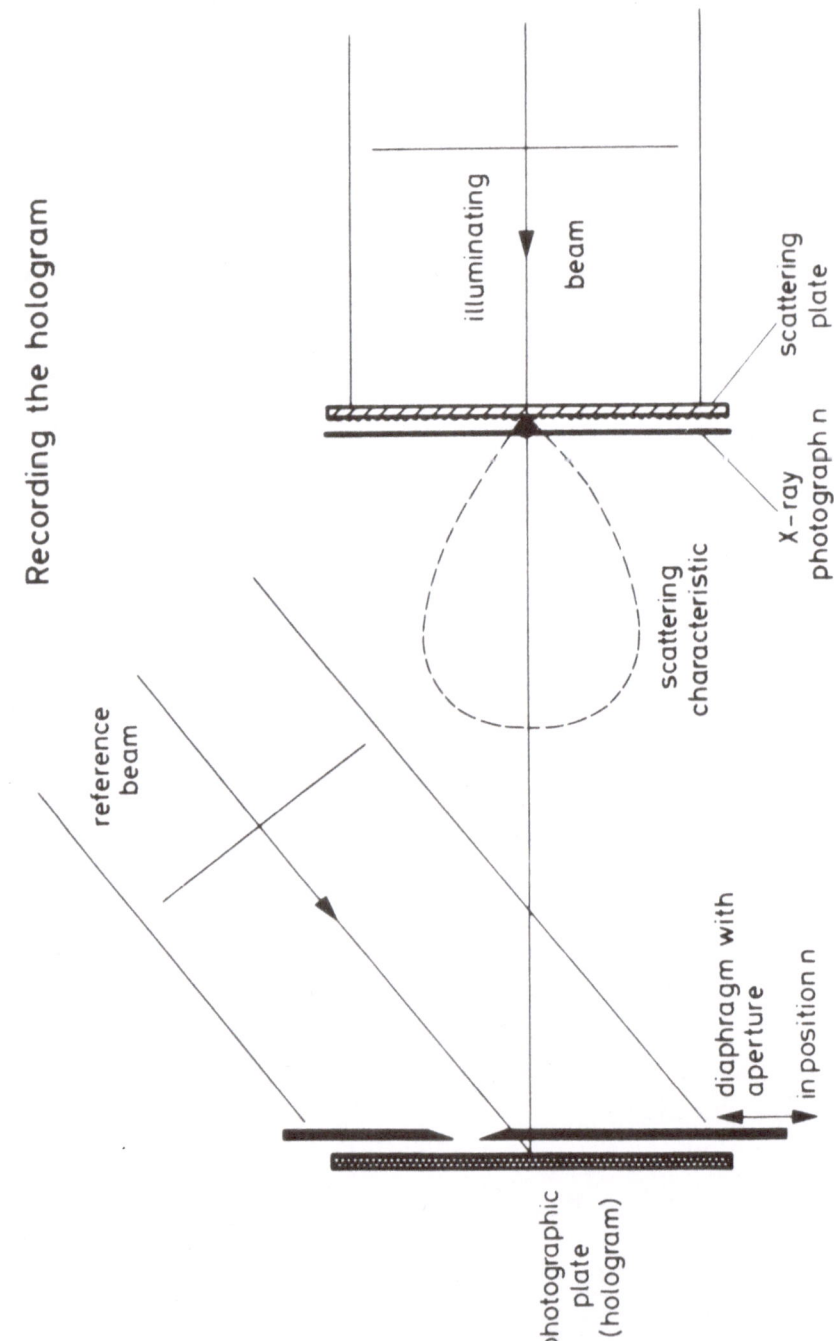

Recording the hologram

illuminating beam

scattering plate

X-ray photograph n

scattering characteristic

reference beam

diaphragm with aperture

in position n

photographic plate (hologram)

Fig. 18 Recording the composed hologram

Holographic tomography

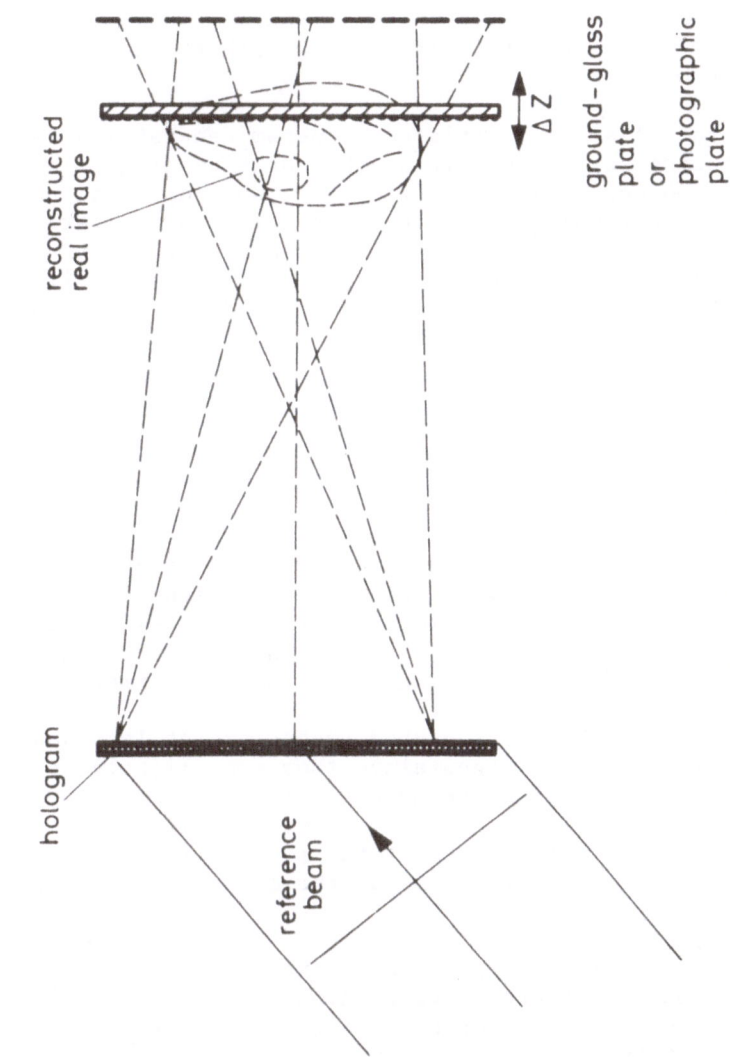

Fig. 19 Holographic tomography

sults with a number of 30 to 50 elements in one direction corresponding to an angle of about 0.5 degree between consecutive viewing
directions. The stereoscopic effect when looking with both eyes
through the hologram and moving the head is excellent. The depths
of details of the object can be estimated with about the same
accuracy as in ordinary holography, but can be determined still
more precisely by utilizing the real images, which are reconstructed
when illuminating the hologram matrix with the conjugate complex
reference wave (Fig. 19).

In this case each isolated hologram generates a real image of
the diffusing screen containing the corresponding silhouette of
the X-ray object. If the aperture of the individual hologram has
been chosen suitably small, the depth of focus is large enough
so that the silhouette is almost focused in front of the reconstructed diffusing screen. There, all these only slightly blurred
silhouettes are superimposed and can be observed for instance by
projecting them on to a translucent screen.

Due to the special kind of recording and due to integration
over all individual images only those parts of the silhouettes
appear focused which correspond to the cross-section through the
object in the plane of the screen, whereas all other parts of the
silhouettes appear smeared. In brief, those superimposed real
images as a whole exhibit similar properties as the real image of
a 3-D object reconstructed from an ordinary hologram. It can be
scanned in depth just by moving the projection screen. This technique of image integration leads to the same results as tomography.
The essential advantage, however, is that the observer can select
the depth at will after having taken the X-ray images.

Of course, these integrated images are of less quality than
the individual X-ray images since they are slightly defocused.
This blurr, however, can be made smaller by decreasing the size of
the aperture of the hologram. The optimum value is reached when
the diffraction limited size of the image point is of the same
order of magnitude as the size of the defocused geometrical image
point.

Fig. 20 shows an experimental example of holographic tomography. The photographs were obtained according to Fig. 19 by
directly recording the images in different depths. Fig. 19a shows
the plane of the brim of the human eye, Fig. 19b shows the plane
of the optic nerve hole.

A slightly modified arrangement for holographic tomography is
shown in Fig. 21. In this case the X-ray source was moved on a
circle when recording the primary X-ray photographs. Corresponding
ly the elementary holograms are circularly arranged. The advantage
is that the depth of focus is determined in two dimensions and can

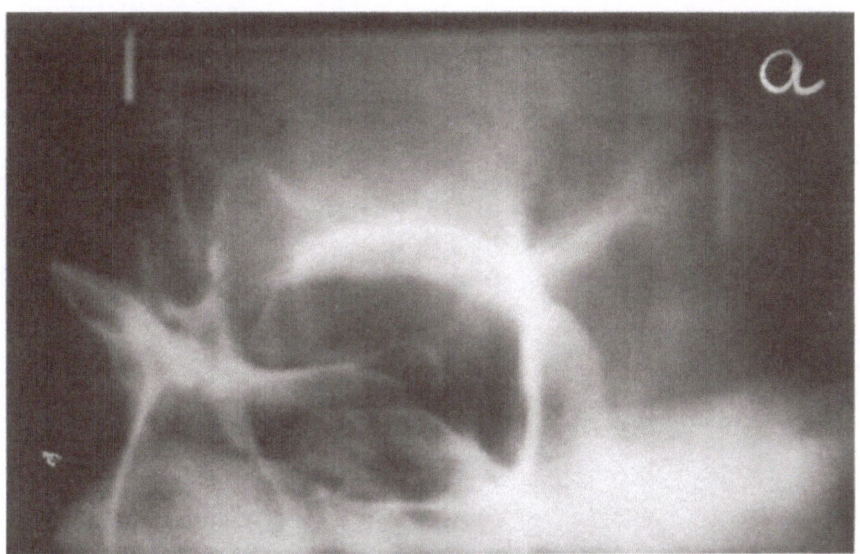

Fig. 20 Holographic tomography of part of human skull
   a) plane of the brim of a human eye
   b) plane of the optic nerve hole

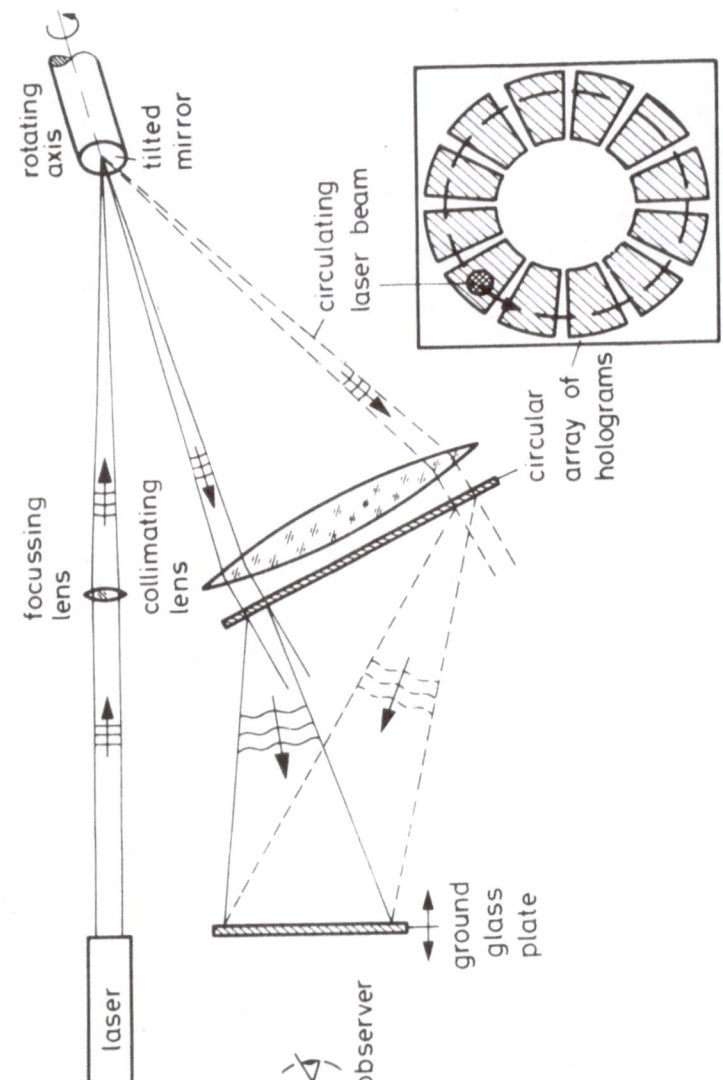

Fig. 21 Holographic tomography using a circular synthetic aperture

be measured more precisely in spite of using less primary X-ray photographs than for the linear arrangement of Fig. 19.

Moreover, the optical noise being a wellknown disturbance in coherent imaging systems can be considerably reduced by simply employing a rotating tilted mirror for generating the reference beam in the reconstruction process. Since the individual images are reconstructed time sequentially they are superimposed incoherently and part of the speckle patterns is averaged, when the revolution time is shorter than the exposure time of the photographic plate or the temporal resolution of the eye of the observer.

REFERENCES

(1) A. Papoulis: Systems and Transforms with Applications in Optics, Mc Graw Hill Book Comp., New York 1968.

(2) L.S. Watkins, Proc. IEEE 57 (1969) 1634.

(3) H. Dammann, M. Kock: Opt. Comm., to be published.

(4) A. Marechal, P. Croce, Compt. Rend. 237 (1953) 706.

(5) E.L. O'Neill, IRE Trans. - PIT 2 (1956) 56.

(6) A.B. van der Lugt, IEEE Trans. Inform. Theory IT-10:2 (1964).

(7) D. Gabor: Nature 208 (1965) 422.

(8) H.J. Caulfield, W.T. Maloney: Appl. Optics 8 (1969) 2354.

(9) G.W. Stroke: Optica Acta 16 (1969) 401.

(10) R.H. Dicke: Astrophys. J. 153 (1968) L 101.

(11) G. Groh, G.W. Stroke: Opt. Comm. 1 (1970) 339.

(12) M.J.E. Golay: JOSA 61 (1971) 272.

(13) Sun Lu: Proc. IEEE 56 (1968) 116. G. Groh: Holographie-Arbeitstagung im Battelle-Institut (1967). G. Groh: Appl. Opt. 7 (1968) 1643. G. Groh: Appl. Opt. 8 (1969) 967. S. Lowenthal, A. Wert, M. Rembault: C.R. Acad. Sc. Paris B267 (1968) 120.

(14) U. Schmidt: Private communication.

(15) U. Schmidt, W. Thust: Opto-Electronics 1 (1969) 21. U. Schmidt, et al.: Proceed. of Electrooptical Systems Design, Brighton 1971.

(16) A. van der Lugt: Optica Acta 15 (1968) 1-33. T.S. Huang: Opto-Electronics 1 (1969) 49-58. J.W. Goodman: Introduction to Fourier Optics, Mc Graw Hill, N.Y. (1968).

(17) G. Marie: Philips Res. Repts. 22 (1967) 110-132.

(18) G. Marie: Philips techn. Rev. 30 (1969) 292-298.

(19) G. Groh: Optics Comm. 1 (1970) 454-456.

(20) G. Groh; G. Marie: Optics Comm. in print.

(21) W.J. Poppelbaum, M. Faiman, Dr. Casasent, D.S. Sand: Proc. IEEE 56 (1968) 1744-1746.

(22) H. Wider, R.V. Pole, P.F. Heidrich: IBM J. Res. Dev. 13 (1969) 169-171.

(23) G. Marie: Private communication.

(24) F. Buchmann: Röntgenstrahlen (1967) H. 16, 20.

(25) J.D. Redman, W.P. Wolton, E. Shuttleworth: Nature 226 (1968) 58.

(26) T. Kasahara, J. Kimura, R. Hiohi, S. Tonaka: Japan. J. Appl.
     Phys. 8 (1969) 124.
(27) D.J. De Bitetto: Appl. Opt. 8 (1969) 1740.
(28) G. Groh, M. Kock: Appl. Opt. 9 (1970) 775.
(29) G. Groh, M. Kock: Röntgenblätter (1971) in print.

SYNTHETIC HOLOGRAMS AND KINOFORM

Jean J. Clair

Faculté des Sciences de Paris, Laboratoire d'Optique

Paris (France)

Some methods of synthesis of holograms will be described here.

At first, computed generated binary holograms which have been described by Lohman and Paris, in 1967 and, for example, by Lee in 1970. Then, a method in incoherent light and at least a new wavefront reconstruction device called Kinoform.

COMPUTER GENERATED BINARY HOLOGRAMS

An ordinary hologram is produced by recording the interference pattern caused by an object wave and a reference wave. As the photographic plate records only intensity variations, it may be possible to create such variations and therefore to synthetize holograms.

The object may not exist and the hologram allows us to see an imagined object: for example, spacial mathematical figures and manufacturing models. In other fields, it is possible to reconstruct well determined wavefronts as aspherical ones, whose applications appear in interferometric testing. Synthetic holograms offer at least an original solution for spatial frequency filtering, data storage and broadly speaking for data processing.

How do Computer Generated Holograms Work?

1) We have an object: we calculate the complex amplitude diffracted in a plane which is the plane of the holograms. Such amplitude is computed.

2) The computed amplitude is coded by the computer and transformed into a real and positive function. For example, the computer adds a complex amplitude to the one diffracted

111

by the object; it adds a coherent wavefront. The intensity is then positive and real.

3) A certain system, generated by computer, plots a diagram
   which reproduces this function. That system may be a cathodic tube, a printing machine or an optical system.

4) The drawing is reduced photographically and the negative
   is the synthetized hologram. The drawing must be very
   sharp and it is necessary to reduce it photographically.

How Is It Possible to Code the Information?

The complex amplitude diffracted by the object may be coded by
several manners:

a) Some plotters have half tone printing but grey levels are
   limited: 256 for cathodic tubes. Then, intensity of interference pattern may be simulated.

b) By binary holograms, which contain only black and white
   areas. The grey scale is simulated by the sizes of many
   fine dots and phase by their position. It is then possible to plot complex amplitude diffracted without introduc
   tion of a computed reference wavefront.

c) The phase of the wavefront is multiplied by a function to
   have a so called phase matching, that is a phase variation
   between 0 and 2π . This device operates only on the phase
   of the incident wave and form a single image by wavefront
   reconstruction without the unwanted diffraction orders.
   That new computer generated hologram is called kinoform.

In all these cases we must not forget that computers have limited possibilities and the previous computations are only possible
for a determined number of points. The object is then sampled and
diffraction calculated in a finite number of points of the hologram
plane.

Processing of Binary Holograms

For more information we will consider in details the processing of binary holograms.

I repeat that, in ordinary hologram, phase is recorded with a
coherent wavefront. There are interferences between diffracted
wavefront by the object and a reference wavefront. Phase shifts
are transformed in intensity variations recorded on photographic
plates. A mecanic reproduction of continuous variation of intensi
ty is generally impossible because of a finite number of grey level
of the plotters.

Binary holograms consist of many transparent dots on an opaque background. The grey scale is simulated by the sizes of many fine dots. The phase shifts are given by the position of that fine dots according to the phenomenon called "deteur phase effect" (see Fig.1).

To explain the detour phase effect we have to repeat the standard explanation of grating diffraction.

The M th diffraction order is characterised by a path difference of $M\lambda$ between two rays passing through adjacent grating slits.

If a_0 is incidence angle we have:

$$L_M = \left(\sin \theta \; - \; \sin a_0 \right) \varepsilon_0 = M \lambda$$

ε_0 being the distance between two slits.

If the lower part of the grating is displaced by an amount ε the path difference at the dislocation is:

$$L'_M = \left(\sin \theta - \sin a_0 \right) \left(\varepsilon_0 + \varepsilon \right)$$

Hence the plane wave emerging from the lower hand position of the grating is retarded by:

$$L'_M - M \lambda = \Delta_M$$

$$\Delta_M = \varepsilon \left(\sin \theta - \sin a_0 \right) = \frac{\varepsilon M \lambda}{\varepsilon_0}$$

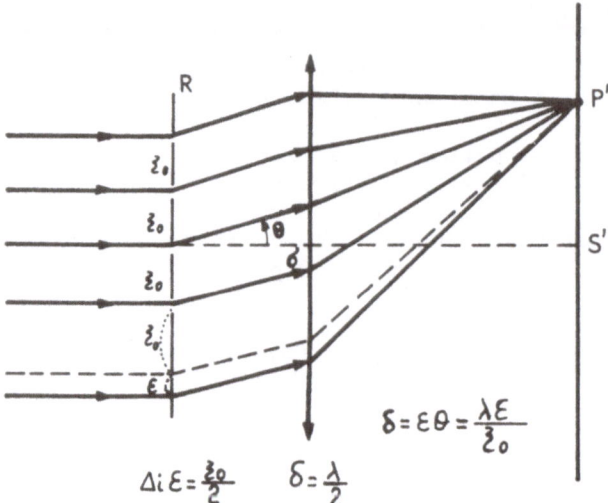

$$\delta = \varepsilon \theta = \frac{\lambda \varepsilon}{z_0}$$

$$\Delta i \, \varepsilon = \frac{z_0}{2} \qquad \delta = \frac{\lambda}{2}$$

Fig. 1   Detour phase effect. ε-dislocation; δ-phase difference
         introduced by ε; z_0-path of the grating R; θ-angle of
         diffraction.

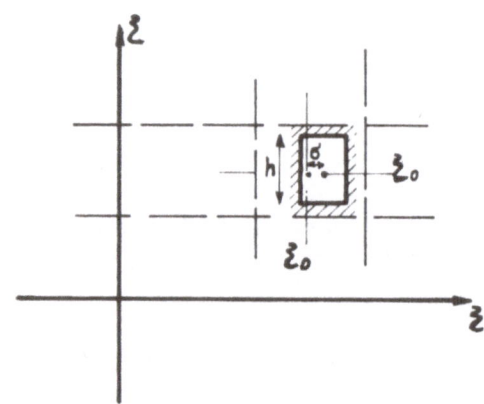

Fig. 2   Type of coding cell: (Lohmann, Paris). h-is proportional
         to the amplitude of diffracted wave; δ-is proportional
         to the phase; ξ_0-is the path of the cells pattern.

so:

$$\phi_M = \frac{2\pi\epsilon M\lambda}{\epsilon_0}$$

which does not depend on a_0.

How does it work for our purpose?

Consider the letter A. A is sampled and computer gives a Fourier transform which may be calculated with Cooley-Tuckey algorithm. The Fourier function in the plane M is sampled in a number of points. The plane is divided in cells (see Fig. 2) corresponding to those points. In each cell there is a dot; with each dot height, proportional to the amplitude and the position of the dot (rectangular opening), is displaced from the center of the cell by an amount proportional to ϕ_M according to "deteur phase effect". ϵ_0 is here the difference between two cells.

All this cells are drawn on a large scale by a computer guided plotter and then photographically reduced in size.

By illuminating this hologram we have a spot in F but in the first order; if has been chosen $\pi = 1$, we reconstruct the object and as the above formulation is the same for M = - 1 we have another image; and because of the periodical structure other noisy images in other orders (see Fig.3).

It is important to note that the binary plotting excludes difficulties from non linearity emulsions.

Lohman and others have realized filters, derivations effect and phase contrast filters. Other coding, system of dots, have been proposed for example by Lee.

To achieve our purpose on binary holograms, we give an easy example: it concerns the hologram of two points. The Fourier transform is a sinusoide and it is easy to sample it (see Fig. 4, 5,6).

If the path of cells is ϵ_0, to have negative amplitude we can consider that, at that point, the phase is π. With our formulation:

$$\phi = \pi \quad \text{if} \quad M = 1$$

$$\text{so} \quad \epsilon = \frac{\epsilon_0}{2}$$

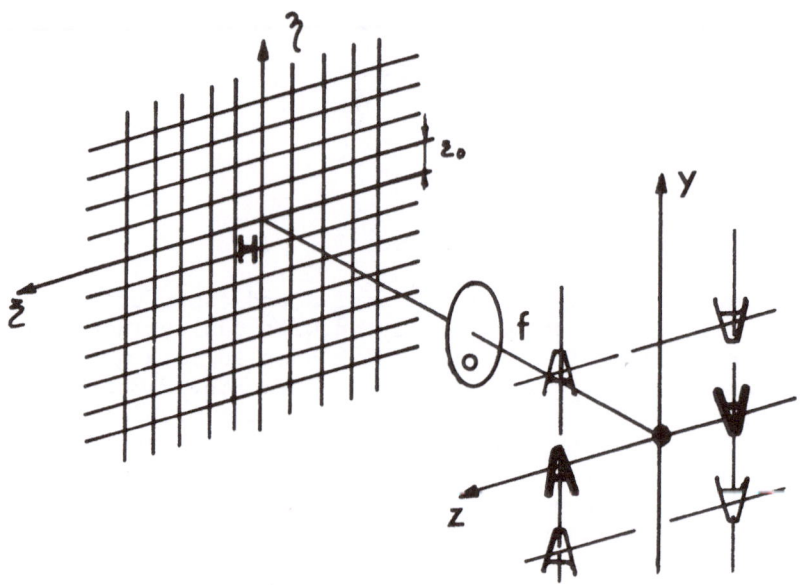

Fig. 3   Reconstruction of the letter A, given by a binary computer
         generated Hologram H.

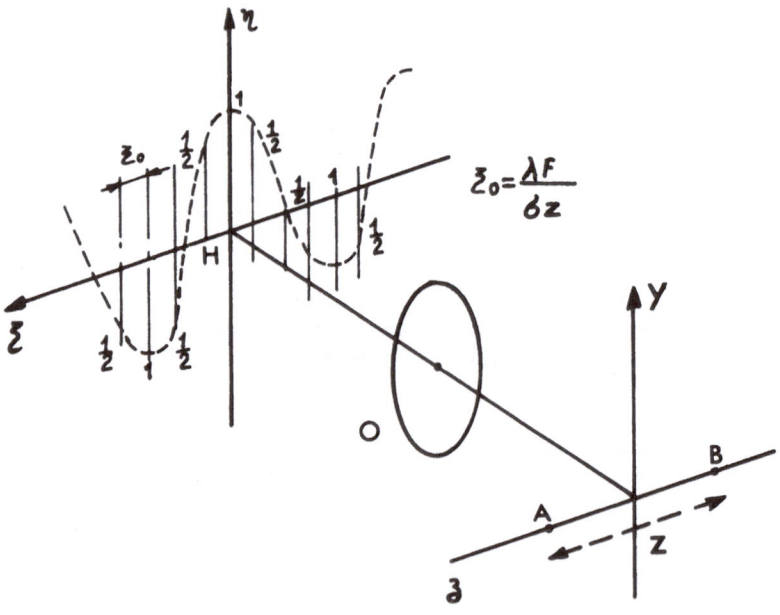

Fig. 4. Fourier transform of two points A,B sampling of sinusoidal
         function σ-samples.

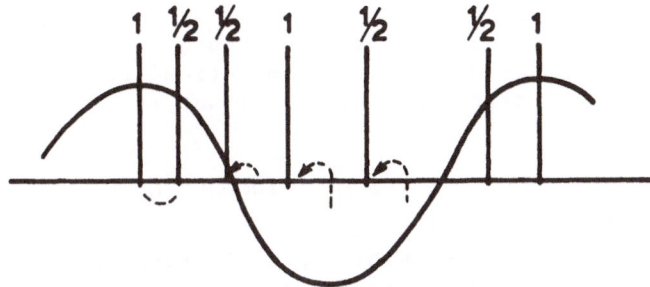

Fig. 5   Sampling and detour phase ($\swarrow^{\cdot}$) of $\mathcal{E}=\frac{\mathcal{E}_0}{2}$ to have negative
amplitudes.

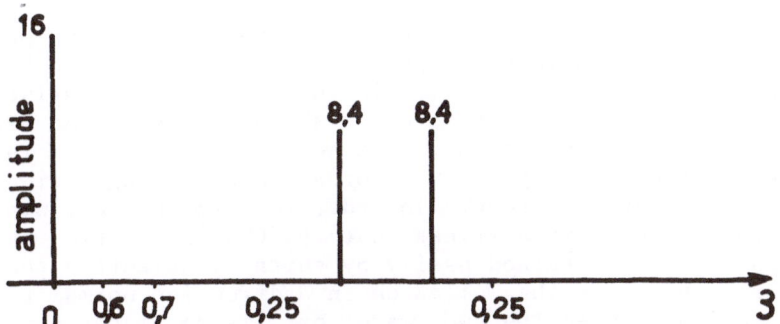

Fig. 6   Relative amplitudes given by the reconstruction of two
points A and B from the binary hologram.

Drawing is computer generated and the amplitudes of each sample is given here by the thickness of the line; the phase shifts π by a dislocation of $\frac{\varepsilon_0}{2}$

We reconstruct in the first orders two points which can be considered as two coherent sources and form fringe pattern.

Here we have chosen 6 samplings because the amplitudes are interesting values and we had found an optical trick to avoid a drawing of lines of different thickness; but it is only a particular case.

USE OF OPTICAL TRICK IN INCOHERENT LIGHT

After having briefly exposed binary holograms, we arrive at the second method, using suitable optical trick in incoherent light. We repeat the principle of holography with spatially incoherent light. The light from each object is split into two parts and each pair of waves of common origin interfere and form a fringe pattern. Each object is there encoded into a suitable system of fringes and image can be obtained.

Stroke and Restick (1), Stroke et al (2), Spitz and Werts (3) and Groh (4) have studied diverse methods for synthesizing holograms by multiple exposures. Besides Lohman (5) has shown how the birefringent elements such as Biot's lens can be employed to obtain holograms in three dimensions.

We describe a very simple arrangement for synthesizing Fourier holograms using a single point source and a Wollaston prism.

Fig. 7 illustrates the principle of the method. The point source S (laser) illuminates a photographic plate H through a Wollaston prism W. This element is placed between two polarizers which are fixed to it. These are not shown in Fig.7. An exposure is made under these conditions. The source S being split into two by the Wollaston W, the plate H records Young's fringes since the two images of S are coherent. In order to work in the linear region of the emulsion it suffices that the two intensities be different and this is obtained easily by suitably orienting the two polarizers. Displace the Wollaston in a direction normal to H. The separation between the two images of S is no longer the same and the fringe frequency on H is modified. The plate is developed and is illuminated by a point source. One observes, through the negative, a central luminous point 0 (zeroth order and two first order spectra formed of two points (Fig.8).

Consider the same experiment; two successive exposures are made; the Wollaston having been rotated about an axis perpendicular

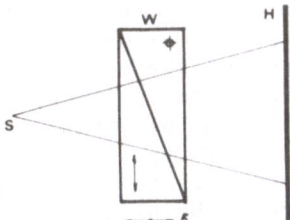

Fig. 7 Experimental set-up.

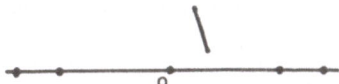

Fig. 8 Reconstruction of two luminous points when the Wollaston is displaced in a direction perpendicular to H between the two exposures.

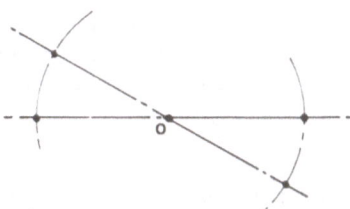

Fig. 9 Reconstruction of two luminous points when the Wollaston is rotated between the two exposures.

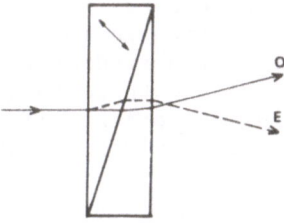

Fig. 10 Calcite-glass birefringent prism.

to H between the two exposures. After developing, a point source
is seen through the negative. One observes the central point O
(zeroth order) and the two spectra each composed of two luminous
points situated on a circumference with O as center (Fig. 9). It
is now evident that if one combines the longitudinal displacement
and rotation of Wollaston, any type of plane object can be reconstructed by successive exposures and by suitably adjusting the time of exposure. The source S being generally a laser, it is preferable to use a circular polarizer so that the intensities of the
two images of S remain constant irrespective of the Wollaston orientation.

For splitting the source S, the Wollaston prism may be replaced by a prism of the type calcite-glass (Fig.10) which gives large angles of splitting even when the thicknesses are small.

The prism we have used is 5 cm in diameter, 5 cm in thickness
and gives an angular splitting of 10 degrees.

Before speaking of kinoform, the third method, we may know
that with such a system, Wollaston, we have synthetized phase and
amplitude periodic filters or targets.

Let us consider a rectangular bar pattern having alternate
bright and dark portions. We form the image of this object on a
photographic plate H by means of an objective O and place a Wollaston prism W between the object and the objective (Fig. 11).
During the exposure, the Wollaston is rotated or displaced in a
direction perpendicular to H in a manner which is predetermined.
By the sweeping out process, one can obtain a desired intensity
profile.

For example, let us take an object in which the luminous band
is equal to the dark band. The distance of the Wollaston from the
object is such that the separation produced by it in the plane of
the object is equal to the width of a band. During the exposure,
let the Wollaston be displaced in a direction perpendicular to the
plate with a constant speed. The displacement of the Wollaston
must be such that a particular band is displaced through a distance
equal to its width. One records the autocorrelation function of
the object and the profile obtained is like that as shown in Fig.
12. This method may be utilized in making triangular wave targets,
whose use for transfer-function studies was first proposed by
Lohmann (6,7);Post (8) realized these targets by sharpening moiré
fringes. Without changing the object, if we rotate the Wollaston
uniformly, we shall get a transmission profile in $|\cos x|$ as in-
dicated in Fig. 13.

The use of a birefringent system like that of Wollaston is

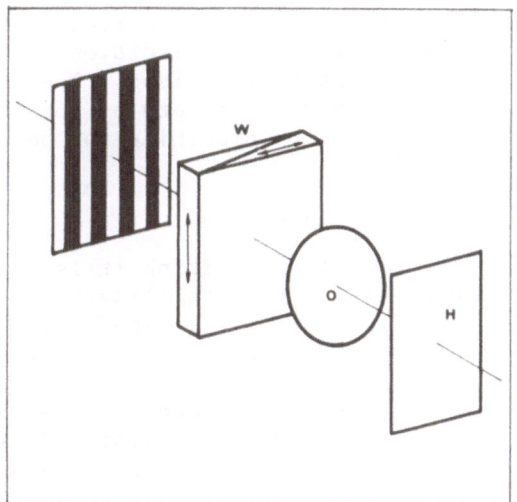

Fig. 11   Schematic diagram of the setup: W , Wollaston prism;
O, camera objective; H , photographic plate.

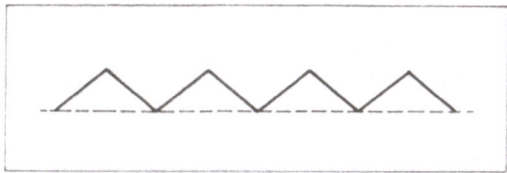

Fig. 12   Intensity transmission profile of a triangular wave
target.

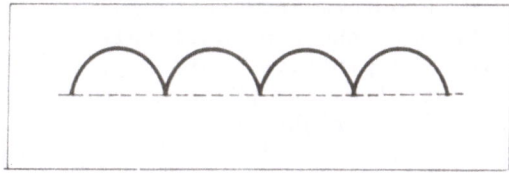

Fig. 13   Intensity transmission profile in $|\cos z|$.

particularly advantageous because one can realize a periodic profile
of high spatial frequency with great precision. Very small displace
ments may be obtained with large rotations. The diffent types of
birefringent systems offer a wide variety of possibilities. The
phase filters can be prepared by bleaching the negatives obtained
by this process.

KINOFORM

We spoke about bleaching and I think it is not necessary to
recall the bleaching effect, but the amelioration of bleaching has
introduced a new type of hologram: the kinoform, which will be the
last point developed in our lecture.

The kinoform is a new computer generated wavefront reconstruction. An illuminated kinoform yields a single diffraction order
and ideally all the incident light is used to reconstruct this i-
mage.

Computationally kinoform is fast because it is unnecessary to
introduce reference beams and image separation calculations.

The kinoform operates only on the phase of an incident wave,
being based on the assumption that only the phase information in
the scattered wavefront is required for the reconstruction. The
amplitude is assumed constant. Kinoform can be thought of as a
complex lens which transforms the known wavefront incident on it
into the wavefront, needed to form the image. Kinoform is a bleached hologram. The thickness after bleaching is to convert the
phase of the incident wavefront into the phase required (see Fig.
14).

In general, the amplitude of the scattered wavefront is not
constant, but a proper choice of the phase variation of the original object (illumination through diffuser) may give an amplitude
essentially uniform.

The object is sampled - Kirkhoff diffraction is calculated
and finite Fourier transform is calculated using Cooley-Tuckey
algorithm. The sampling in the plane of hologram must satisfy the
Nyquist sampling criterion. The calculated phase is plotted, but
reduced modulus: 2π . The phase of an off axis reference wave
can be added, but as kinoform involves only the phase, there is an
economy in time. We have the same difficulties with the finite
number of levels for the plotters. After plotting, the bleaching
must be carefully controlled for phase matching. With imperfect
phase matching there is a bright spot. About 75% of the beam is
used in constructing image-signal to noise ratio of 16.

Kinoform is used for filtering and in incoherent light filter

ing is thought of in terms of convolutional operation, but limited
to non negative impulse functions.

An optimized filter with all coefficients positive can be obtained by finding the filter that gives the best least square appro
ximation, under the contraint of non negative coefficient, to the
frequency characteristics of the perfect filter that has positive
and negative areas on its impulse response function.

Except filtering computer generated holograms and especially
kinoform are used in interferometric testing of aspherical optical
elements and more generally a kinoform may produce a spherical or
aspherical wavefront.

Let us consider the drawing intensity (see Fig. 15) proportio
nal to x^2. This drawing is photographically reduced and etched
and we get a phase variation surface - modulus 2π structure. We
have realized a real Fresnel lens. This lens will reconstruct a
point (see Fig. 16). We have realized the kinoform of a point or
a so called kinoform lens.

Generally, the plotting was computer generated one and may be
very long and difficult because of the finite number of grey levels.
We have proposed a method in order to have the plotting in a time,
by considering the pattern given by a Fabry-Perot. The repartition
of the rings is the same as the Fresnel zone. We photograph that
pattern. During the exposure we create a sort of scanning; the
thickness of the F- P has a variation of $\frac{\lambda}{2}$ so that each ring takes
the place of the previous or following one. If the scanning is
well calculated we may have a determined profile (see Fig. 17).

For a spherical lens:

$$\Delta e = \left(A \; _Y \right)^{\frac{1}{3}}$$

This scanning may be a mechanical of piezoelectrical one, but
it is necessary to control the parallelism of the F - P plates. We
have chosen the pression-scanning. The F - P is in an empty pan
and after vacuum we open an air admission by a band-controlled
gate valve.

The variation of optical thickness is given by the variation

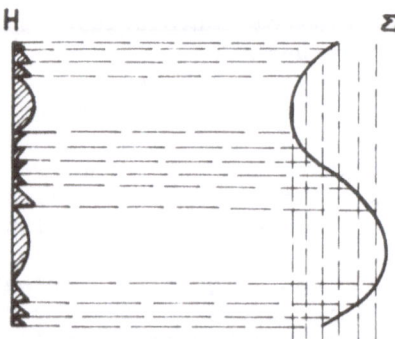

Fig. 14   Σ is the reconstructed wave from the phase plate H;   the
          thickness of H is proportional to the phase shift wanted;
          the phase shift is reduced modulus 2 π .   H is a kinoform.

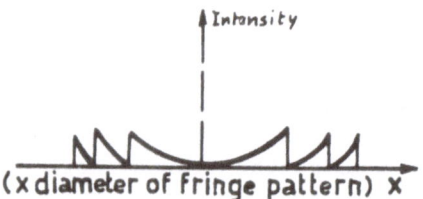

Fig. 15.   Intensity profile of the rings pattern.   The radii of
           the rings are proportional to $\sqrt{k}$, k being an integer.

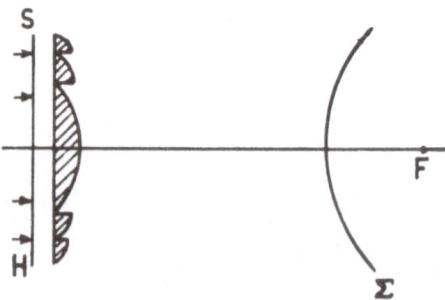

Fig. 16   The intensity profile is bleached: we obtain a kinoform
          lens and the wave front Σ is spherical (S plane incident
          wave front).

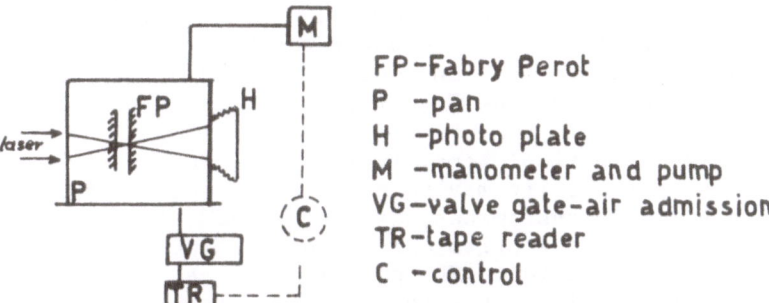

FP-Fabry Perot
P -pan
H -photo plate
M -manometer and pump
VG-valve gate-air admission
TR-tape reader
C -control

Fig. 17   Scheme of apparatus to obtain kinoform lens

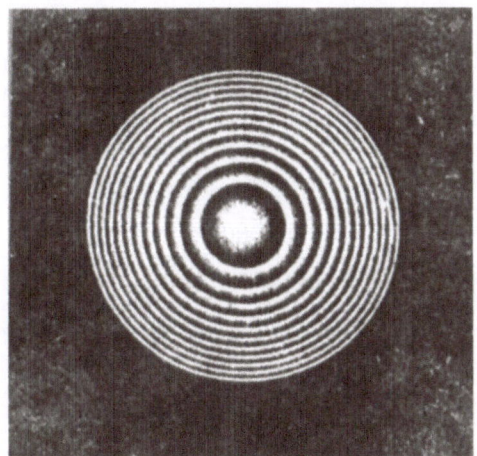

Fig. 18   Repartition of intensities

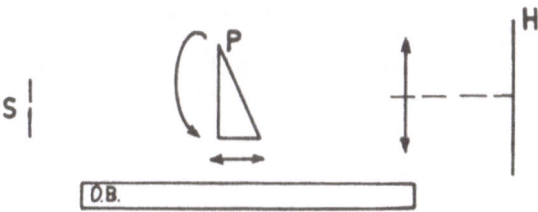

Fig. 19   Scheme to obtain aspherical lens. S-point source;
P-rotating and translating prism; H-photo plate;
O.B.-optical bench with motor driven system
controlled by band reader.

of index and we operate between 100mn and 700mn. The operation
time is about few minutes (by computer the plotting lasts one or
more hours). The photo is then reduced, etched or may be directly
recorded on photo-resin (see Fig. 18).

To realize aspherical lens, it is not necessary to have many
rings, because phase shift may be considered reported from a plane
wave. A very simple system has been developed. It consists on a
rotating and translating prism which from a point source gives a
ring image by the rotation. The position of the prism gives the
radius of the ring. The exposure time is variable with the position of the prism and we may have an infinite number of grey levels.
The rotation is a continuous one, but translation and exposure time
are computed and the whole is motor driven with tape readers (see
Fig. 19).

CONCLUSION

We have described some of the principal methods for synthetize
holograms and expose our modest contributions in the three principal one computer generated, kinoform and optical tricks. Our list
of references is not exhaustive, synthetised holograms and kinoform
have a promising future.

REFERENCES

(1)   G.W. Stroke and R.C. Restrick, Appl. Phys. Letters 7 (1965)
       299.
(2)   G.W. Stroke, F.H. Westervelt and R.G. Zech, Proc. IEEE 55
       (1967) 109.
(3)   E. Spitz and A. Wertz, Congrès sur les progrès récents en
       optique physique, Paris (1966).
(4)   G. Groh, I.E.R.E. Conf. Proc. G.B. 14 (1969) 264.
(5)   A.W. Lohmann, J. Opt. Soc. Am. 55 (1965) 1555.
(6)   A.W. Lohmann, Optik 14, 510 (1957).
(7)   A.W. Lohmann, Appl. Opt.5, 669 (1966).
(8)   D. Post, Exper. Mech. 7, 154 (1967).
(9)   J.J. Clair, M. Françon and P. K. Mondal, Fourier Hologram
       Synthesing Birefringent Elements, Optics Communications,
       Vol.2, n.4, Sept. 1970.
(10)   J.J. Clair, M. Françon, J. Kvapil and P.K. Mondal, Generation
       of Periodic Amplitude and Phase Filters, Appl. Opt., 9, 2585
       (1970).

CORRECTION OF OPTICAL INSTRUMENTS BY HOLOGRAPHY

V. Russo Checcacci

Istituto di Ricerca sulle Onde Elettromagnetiche del

C.N.R., Firenze (Italy)

INTRODUCTION

Many of the things I will talk about, concerning the correction of optical instruments, can be usefully applied also to correction of images coming out not from purely optical instruments. For this reason even if I will mention very often the term 'lens' I suggest that you think of it as a more general instrument which gives at the output a light distribution.

Sometimes I will use also a language (or a formalism) that is more familiar to electronics or to those working in communication theory, rather than to conventional optics. In fact, with the coming of the coherent optics many are the problems that can be solved with the same technique of the electric signal. Hence some of the more recent techniques that I will describe, are merely the transposition in coherent optics field of the techniques used in communication theory.

Let us define an optical instrument. In general it is a device which has the aim of reconstructing the image of an object as faithfully as possible. To do this is necessary that:

1) it collect as much information as possible:

2) the information should be transferred correctly (it is sufficient that information not be degraded to zero).

It follows that the aperture (angular aperture) must be as large as possible. We will see the reason for this. On the other hand, increasing the dimensions of the optical instruments produces aberrations.

Therefore we face two possibilities:

a) to use an instrument of small aperture, well corrected and therefore have an image corrected with a few corrected details;

b) to use an instrument of large aperture with aberrations and therefore have an image deteriorated, containing more details which may be retrieved.

If the image is made in monochromatic light, strumental aberrations, or aberrations coming from intermediate media can be compensated by corrector plates used in combination with the optical instrument (real time correction). If the image is taken in incoherent light, it can be corrected a posteriori by means of a filtering of spatial frequencies.

In both cases holography can be usefully applied.

HOLOGRAM AS A CORRECTOR PLATE (REAL TIME CORRECTION)

Let us recall that the complex amplitude of an image, a_i (monochromatic illumination) can be expressed by the following superposition integral

$$a_i(x_i) = \int_{-\infty}^{+\infty} a_o(x_o)\, h(x_i, x_o) \qquad 1.$$

where $a_o(x_o)$ is the complex amplitude of the object, and $h(x_i, x_o)$ is the amplitude at coordinates x_i in response to a point source object at x_o (impulse response or spread function).

The properties of the imaging system will be completely described if the impulse response, h, can be specified.

If the optical instrument (a convergent lens, for instance) has to produce high quality images, then a_i must be as similar as possible to a_o. Equivalently the impulse response should closely approximate

$$h = K\delta \qquad 2.$$

we can also say that a spherical wave should exit from the instrument and converge in an ideal point image, (Fig. 1a). We therefore shall specify, as the image plane, that plane where eq.2 is most closely approximated. There is a particular distance q from the lens, where this occurs. It is related to the distance object p and to the focal length f

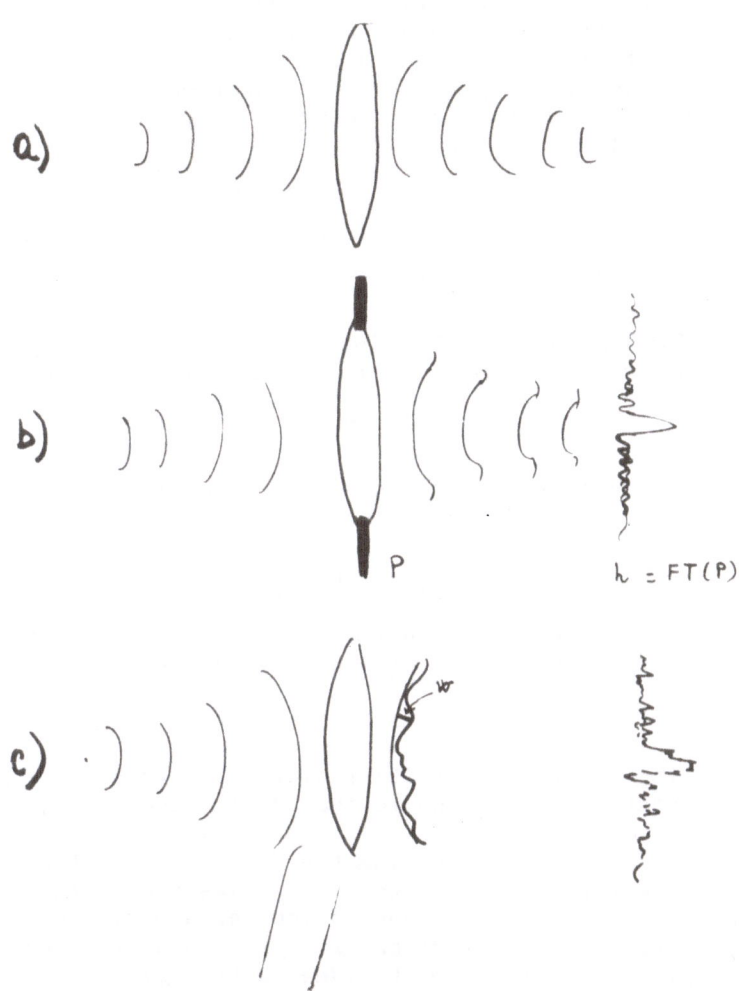

Fig. 1a. Spherical wave exit from the optical instrument.
Fig. 1b. Association of the lens with a pupil function.
Fig. 1c. An aberrated wavefront.

$$\frac{1}{p} + \frac{1}{q} = \frac{1}{f}$$

the well known lens law.

Lt us suppose the lens free from aberrations and examine the distribution on the image plane. One finds that the impulse response is given by the Fraunhofer diffraction pattern of the lens aperture that limits the extent of the spherical wave. The finite extent of the lens aperture can accounted for, by associating the lens with a pupil function P:

$$P \begin{cases} 1 & \text{inside the aperture} \\ 0 & \text{otherwise} \end{cases}$$

Recalling that the diffraction pattern of an aperture is the FT of the amplitude distribution on the aperture one has:

$$h = FT(P)$$

In analogy with the electric signal formalism, it may be useful to write eq.1 in the FT domain:

$$A_i = H \, A_o$$

defining H as the coherent transfer function of the optical instrument. Taking into account the preceding equation one has:

$$H \propto P$$

Larger is P, narrower is its radiation pattern ($a \simeq \lambda / D$) and we have a better approximation to the ideal image point, Fig.1b.

On the other hand, increasing the dimensions of the optical instrument, aberrations arise and are increased with the enlarging of the aperture. Suppose to have an aberrated wavefront like that one shown in Fig. 1c. Aberrations are defined as the departure of the exit pupil wavefront from the ideal spherical form.

Aberrations can arise from inherent properties of perfect spherical·lenses such as spherical aberration and from aberrating media interposed between the object and the lens.

A convenient artifice allow aberrations to be directly included in our previous results. When wavefront errors exist, we can

imagine that the exit pupil is illuminated by an ideal spherical
wave but that a phase-shifting plate exists within the aperture
thus deforming the wavefront that leaves the aperture.

If the phase error at the point (x) in the exit pupil is represented by kW(x), where $k = 2\pi/\lambda$ and W is an effective path
length error, then the complex transmittance $\bar{P}$ of the imaginary
phase shifting plate is given by

$$\bar{P} = P \exp \left[jk \ W \right]$$

The complex function $\bar{P}$ may be referred to as the generalized
pupil function. The impulse response of an aberrated coherent system is simply the Fraunhofer diffraction pattern of an aperture
with transmittance $\bar{P}$.

If one examine the effects of the aberrations on the transfer
function H one finds again:

$$\bar{H} = \bar{P} = P \exp \left[jkW \right]$$

The ideal compensation procedure might be conceived of, as
placing in the exit pupil a phase plate having the complex amplitude $\exp(-jkW)$.

Such an approach has been studied by Tsujuchi Marechal,
O'Neill (1)(2)(3). Unfortunately the required phase plate was expensive and difficult to construct. They used vacuum evaporation.

An alternative solution is to record a hologram of the system
aberrations. Since, its transmittance contains also the complex
conjugate of the system aberrations, the hologram can be used as a
compensating plate (4)(5)(6).

I will refer to the experiment made by Upatnieks et alia who
corrected the spherical aberration of a lens. The system used for
producing the hologram is shown in Fig. 2. A point source of monochromatic light is collimated by the aberrated lens. The hologram
of the aberrated wavefront is recorded using a plane reference
wave. The hologram transmittance which contributes to the real i-
mage is proportional to $\exp(-jKW)$ and can compensate the aberration of the lens:

$$a^x \exp(-jkW) \ \exp(jkW) = a^x$$

To accomplish this, the lens and the hologram are used as a
unit, Fig. 3, and the image formed by the appropriate diffraction
order will be corrected for spherical aberration.

Fig. 4a shows the on-axis point spread function of the lens. The irregular appeearence is caused by an irregular aperture and by three rodlike elements supporting the lens. The spread function when the hologram is used in combination with the lens is shown in Fig. 4b. In making the two photographs, the light levels and expo‾sure times were adjusted so that the peak of the spread function produced the same exposure on the plate in each case. The uncorrected spread function has 2 mm diam. and the corrector spread function has 5 μ . The corrected lens thus produces almost a diffraction limited on-axis response.

Figs. 5a and 5b show the effect of imaging a transparency with the lens alone, and with the lens and the corrector plate together. (The effect of spherical aberration is to produce an ambient background which reduces the image contrast).

Similar techniques have been used for compensating aberrations introduced by interposed media constant in time (7)(8), such as the ground glass shown in Fig.6 (7). The image corrected is shown in Fig. 7b compared with the object a) and the blurred image c).

When it is practically possible, another method can be applied. It consists of passing both the reference wave and the object wave through the same optical instrument (or aberrating medium).

Let us indicate with a exp(jkW) and b exp(jkW) the aberrated reference and object wave after passing through the lens. Interference of the two aberrated waves leaves an interference pattern which is unaffected by aberrations

$$t \propto I \simeq \left| a \exp(jkW) \; + \; b \exp(jkW) \right|^2 = \ldots ab^x + a^x b$$

This technique can be used only in working with a restricted field of view, since aberrations are supposed to influence in the same way as the reference and the object wave.

Some examples (9) are given in Fig.8, 9 where a radial grating affected by astigmatism Fig.8a and by spherical aberration Fig.8b are corrected Fig.9a, 9b. A similar technique is also used for compensating turbolent media (10).

The imaging system in which the hologram is used as corrector plate must be illuminated with monochromatic light. This requirement arises mostly from the diffraction grating nature of the corrector plate which dispers the incident light into its component colours. (Such a dispersion can be partially corrected by the use of a prism or another grating).

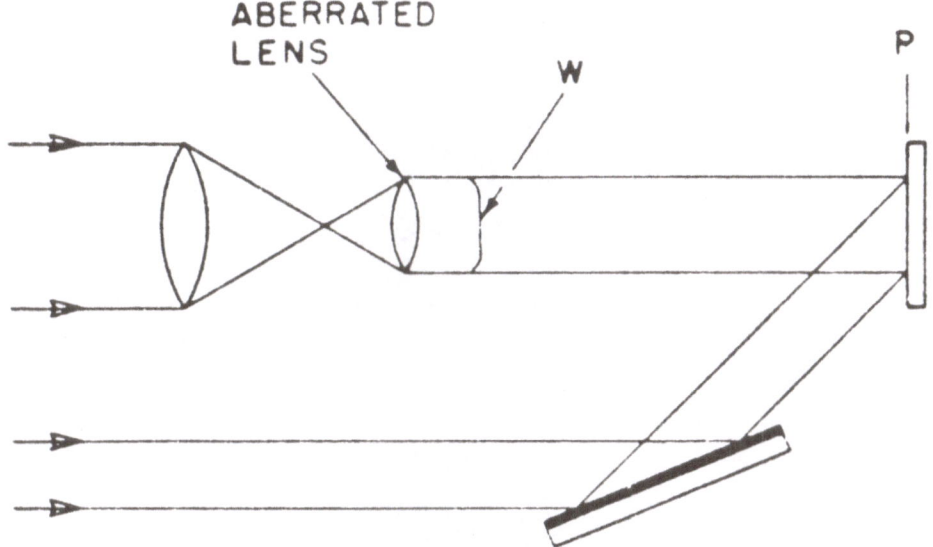

Fig. 2. Recording method of a hologram of the system aberrations, after Upatnieks et alia (1966).

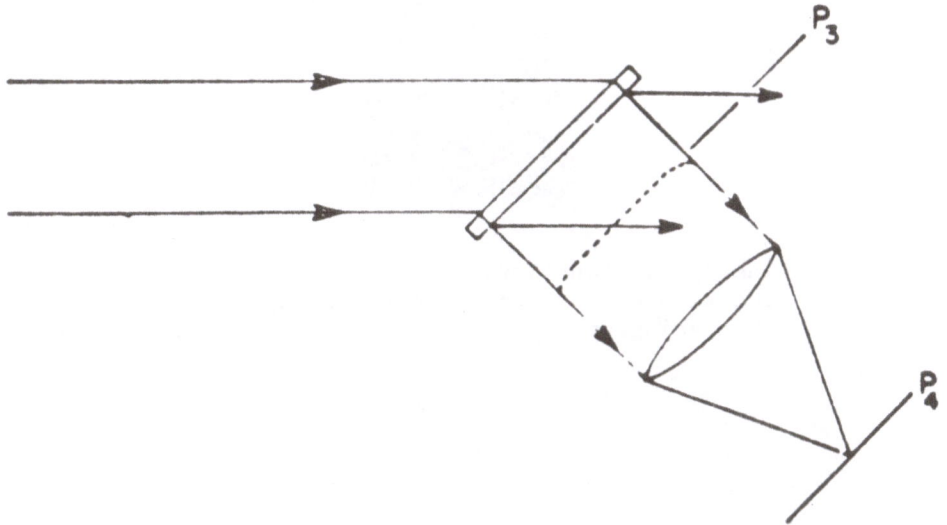

Fig. 3. Image correction for spherical aberration.

Fig. 4. The spread function. a) The on-axis point spread
         function of the lens; and b) the spread function
         when hologram is used in combination with the lens.

Fig. 5. Effect of imaging a transparency. a) With the lens alone.
b) With the lens and the corrector plate together.

Fig. 6. Method of compensating aberrations introduced by the
ground glass.

Fig. 7. Image correction from aberrations introduced by the ground glass. a) The object; b) the corrected image; and c) the blurred image.

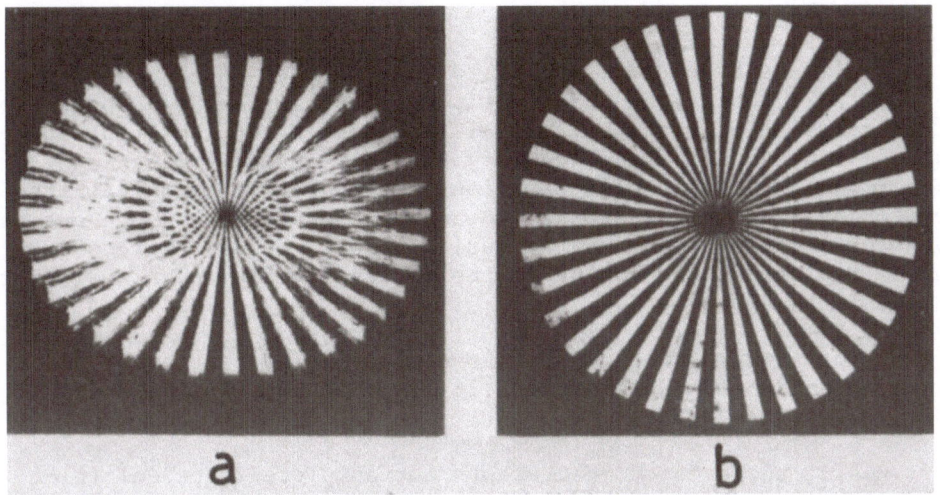

Fig. 8. A radial grating affected by: a) astigmatism, and
         b) spherical aberration.

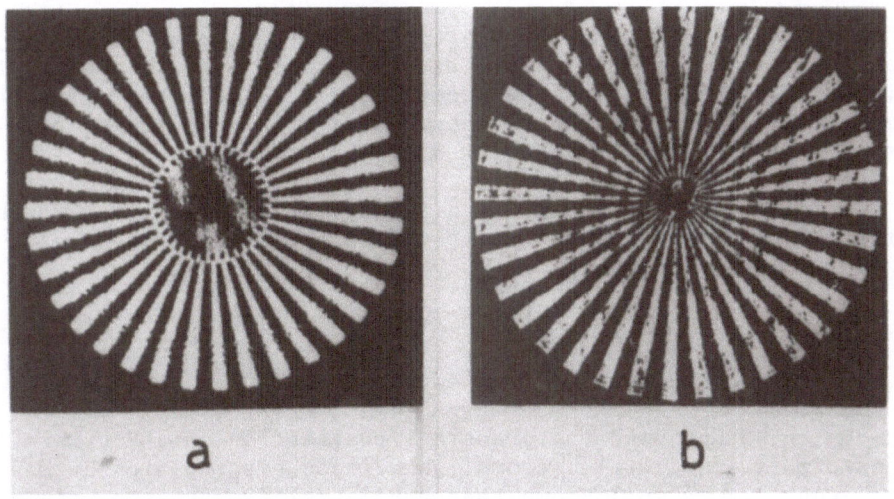

Fig. 9. Radial grating image corrected from: a) astigmatism,
         and b) spherical aberration.

A POSTERIORI RESTORATION BY SPATIAL FREQUENCIES FILTERING

A different and more general procedure with which one can cor
rect both incoherent and coherent photograph is that of a posterio
ri restoration by means of spatial frequencies retrieval (deconvolution). With this method one may also correct deteriorations occurring either because of aberrations inherent to the optical instrument or for other reasons. For example inadvertently an out of
focus or a camera motion may occur. In these cases the recorded
image is blurred, each point in the object space becomes a disc or
a line in the image space.

In other cases the pupil entrance may be of particular shape
(non conventional) and therefore its impulse response may be very
different from a point.

Beside there exist instruments not completely optical (electron microscope or X ray device) that give an output as a light
intensity distribution on a screen from which a photograph can be
made. The electron microscope is affected by spherical aberration,
so that the impulse response is a disc. The X ray device, by using
an extended source has an impulse response which is again a disc.
For this reason, physicists say that radiographs present penumbra
effects.

In all these cases correction can be made with the deconvolution method. Although a posteriori deconvolution systems exist in
space domain, I will speak only on that effected in the spatial
frequency domain.

Before illustrating the restoration method, I will remind you
that an image can be dealt with as an electric signal. We may fil
ter an image by modifying its spectrum. It is well known that a
complex amplitude a(x) can be decomposed in a infinite set of plane
waves (real and not real) having direction α (Fig.10).

It has been found that the distribution at infinity (Fraunhofer
diffraction pattern) may be written as the Fourier transform of
a(x):

$$A(u) \simeq A(\alpha)$$

where $u = \alpha/\lambda$ is the spatial frequency.

On the other hand the practical realization of a Fourier
Transform is immediately accomplished in coherent optics by means
of a lens. It is well known that a F.T. relation exists between
the focal planes of a lens. This can be explained by recalling
that the lens has the property of bringing the distribution at
infinity to its focal plane (Fig.11).

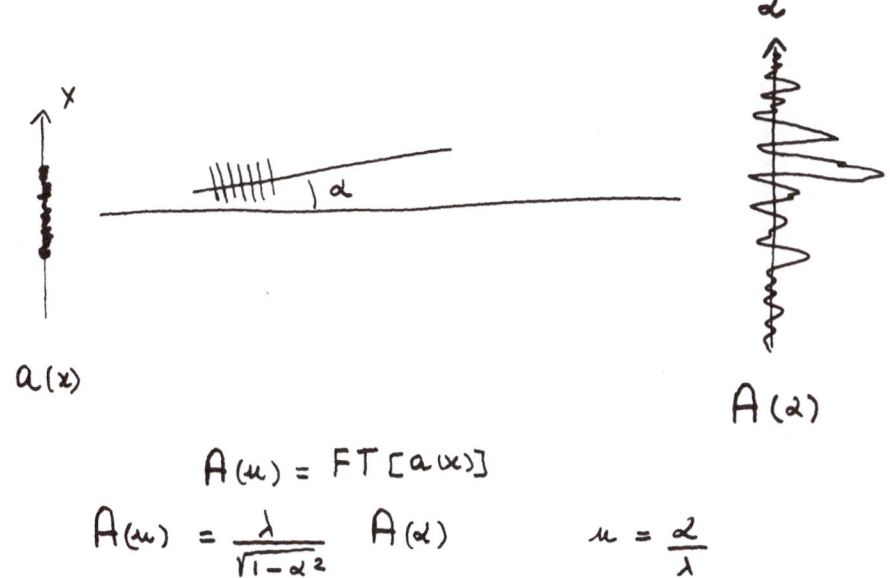

$$A(u) = FT[a(x)]$$

$$A(u) = \frac{\lambda}{\sqrt{1-\alpha^2}} \, A(\alpha) \qquad u = \frac{\alpha}{\lambda}$$

Fig. 10.   Decomposition of a complex amplitude a(x) in an
           infinite set of plane waves (real and not real).

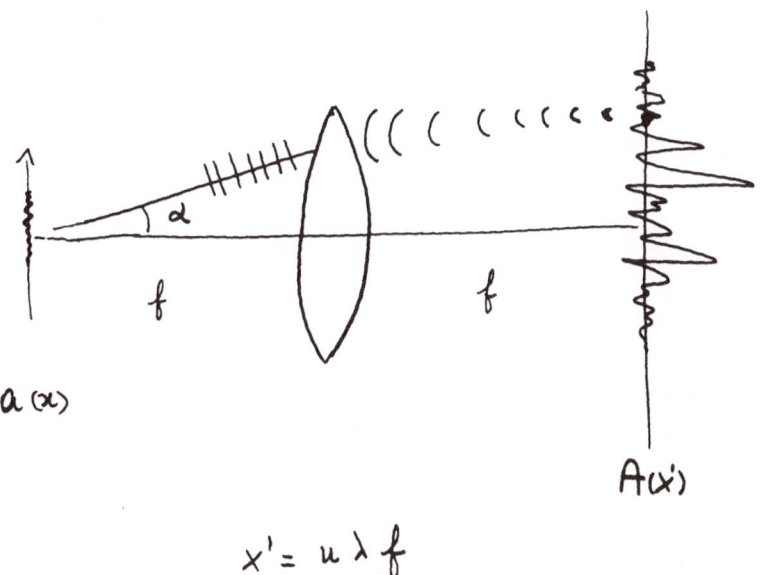

$$x' = u \lambda f$$

Fig. 11.   Lens distribution at infinity to its focal plane.

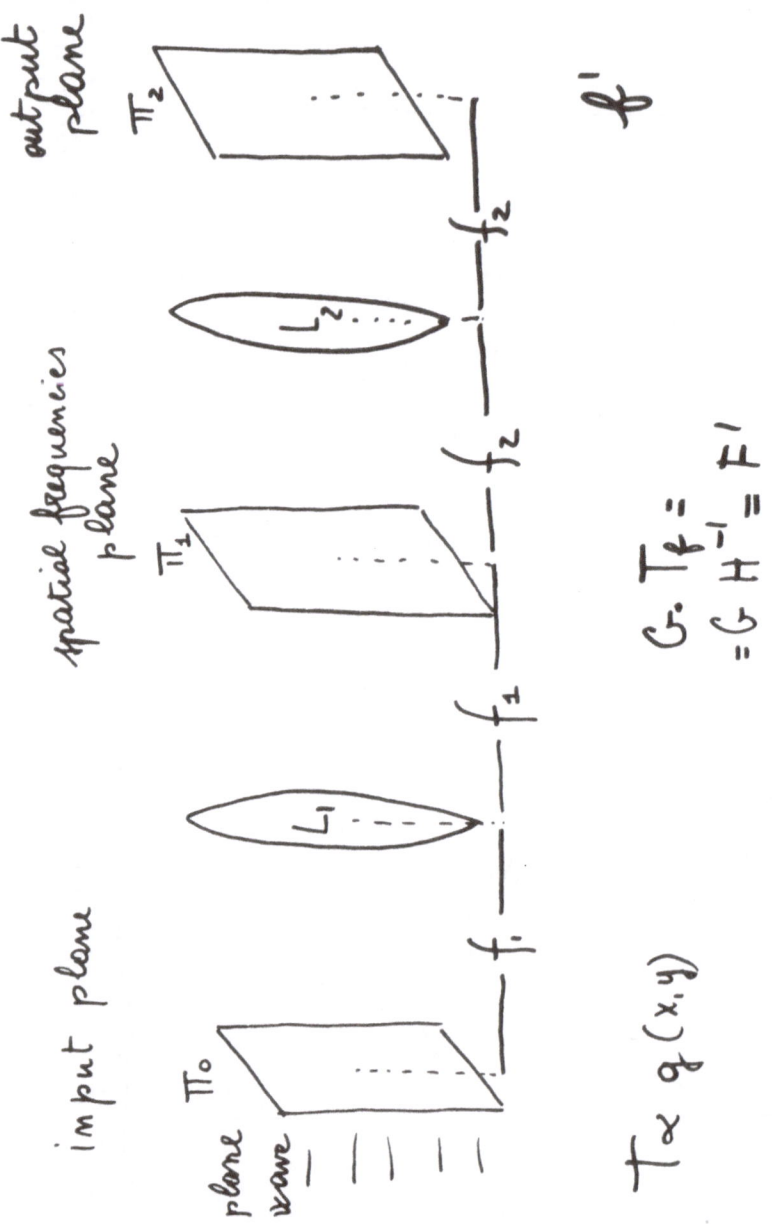

Fig. 12.   Double diffraction system.

If one places a transparency a(x) in the front focal plane of
a lens and illuminates with coherent light, in the rear focal plane
one obtains the F.T. of a, which in general is a complex function.

Proceding from this let us describe the principle of the a
posteriori restoration method by spatial frequency filtering. As
is generally done, the intensity distribution of the blurred image
in incoherent light is the convolution of the blurred point-image
intensity distribution h(x,y) and the difraction limited image intensity distribution f(x,y):

$$g(x,y) \quad = \quad f(x,y) \quad h(x,y) \qquad\qquad\qquad 3.$$

namely, f(x,y) represents the intensity distribution of the image
which would be formed by the given instrument in absence of aberrations or other blurring imperfections.

In order to retrieve the function f(x,y), one must deconvolve
g(x,y) from h(x,y). The deconvolution operation is easily accomplished in the Fourier transforms domain where the preceding equation may be written:

$$G(u,v) \quad = \quad F(u,v) \quad H(u,v) \qquad\qquad\qquad 4.$$

G,F,H denoting the Fourier transforms of g,f,h respectively. The
function F(u,v) is obtained from eq.(2) by division:

$$F'(u,v) \quad = \quad \frac{G(u,v)}{H(u,v)} \quad = \quad G(u,v)H^{-1}(u,v) \qquad\qquad 5.$$

Here, $F' \neq F$ indicates that F(u,v) cannot be recovered complete
ly.

As is evident, it is not possible to restore those spatial
frequencies which have been degraded to zero in the original blurred photograph.

The desired restored function f'(x,y) is obtained by inverse
Fourier transform from F'(u,v).

The Fourier transform division which appears in (5) is a spatial filtering operation which is usually performed in coherent op
tical systems such as the double diffraction system shown in Fig.
12. Such an arrangement operates in the following manner. A
transparency having amplitude transmittance T proportional to the
intensity distribution g(x,y) is placed in the input plane π_0
(front focal plane of the lens L_1) and illuminated by a plane monochromatic wave. In the rear focal plane of the lens L_1 one ob-

tains the Fourier transform of g,G. The spatial coordinates x_1, y_1
in π_1 are related to the spatial frequencies u,v by the following
equations:

$$x_1 = u \lambda f_1 \quad , \quad y_1 = v \lambda f_1$$

where λ is the wavelength and f the focal length of lens L.

A filter with amplitude transmittance T_f proportional to
$H^{-1}(u,v)$ is inserted in the plane π_1. The operation indicated by
(3) is performed in this plane and F' is retrieved. The restored
image is got in the output plane π_2.

The main problem is the construction of the inverse filter
which is generally a complex filter. Until now it had been obtain
ed by a combination of two filters, one absorption plus a phase
one (11).

Holography creates new possibilities of making easily any spa
tial filter of arbitrary amplitude and phase distribution. The
preparation of such filters was studied by G.W. Stroke and A.
Lohman among others (12)(13).

Stroke proposed that this filter can be realized using the
relation

$$\frac{1}{H} = \frac{H^*}{|H|^2}$$

The numerator of the above equation is obtained as the conjugate wavefront reconstructed from the F.T. hologram of the blurred
spread function h. The denominator is realized as the amplitude
filter whose amplitude transmission is proportional to the intensity distribution of H.

The F.T. hologram whose transmittance contains H^* is construct
ed as shown in Fig. 13.

The two plates constituting the filter H^{-1} are inserted in the
filtering apparatus in the spatial frequency plane π_1 Fig. 14. A
first result is shown in Fig. 15.

A difficulty in this technique is that in practice the intensity range of the Fraunhofer diffraction pattern H is much larger
than the dinamic range of the plate to be used of H holography. It
is very difficult to record the whole shape of H holographically.
This difficulty can be avoided by extending the dinamic range of
the available plate with a complicate process (14). Results obtained with extended range holograms are given in Fig. 16. The

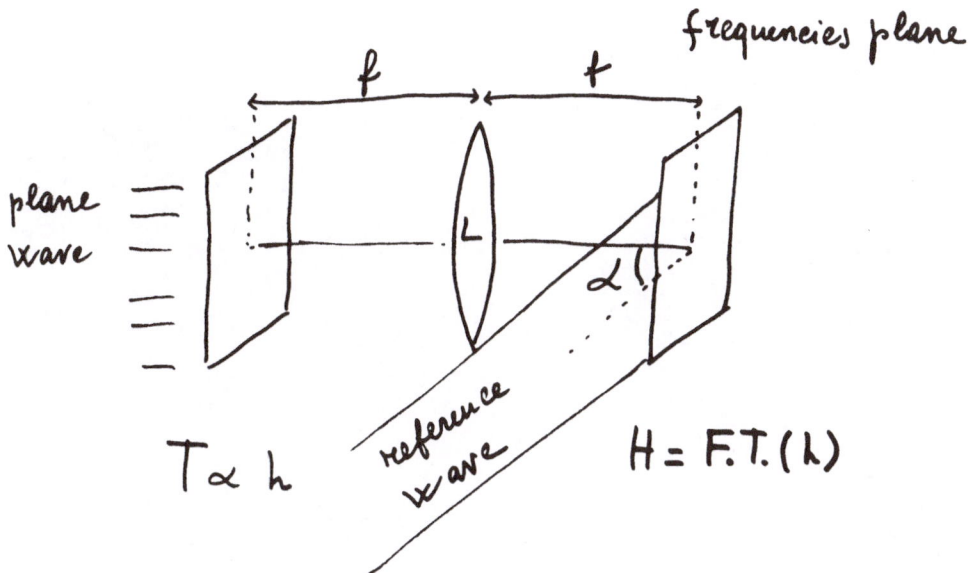

Fig. 13. Fourier transform hologram containing H*.

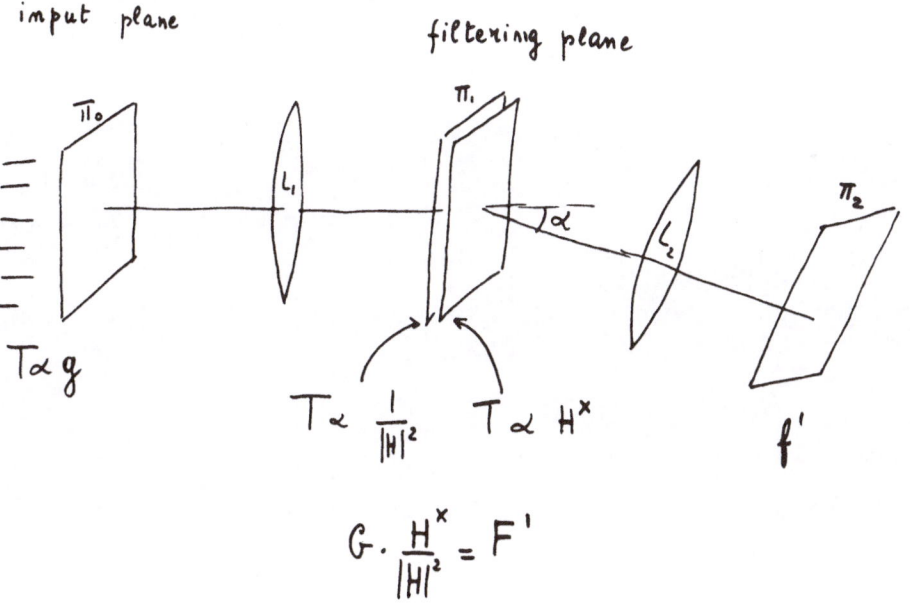

Fig. 14. Insertion of filter H^{-1} in the filtering apparatus
in the spatial frequency plane.

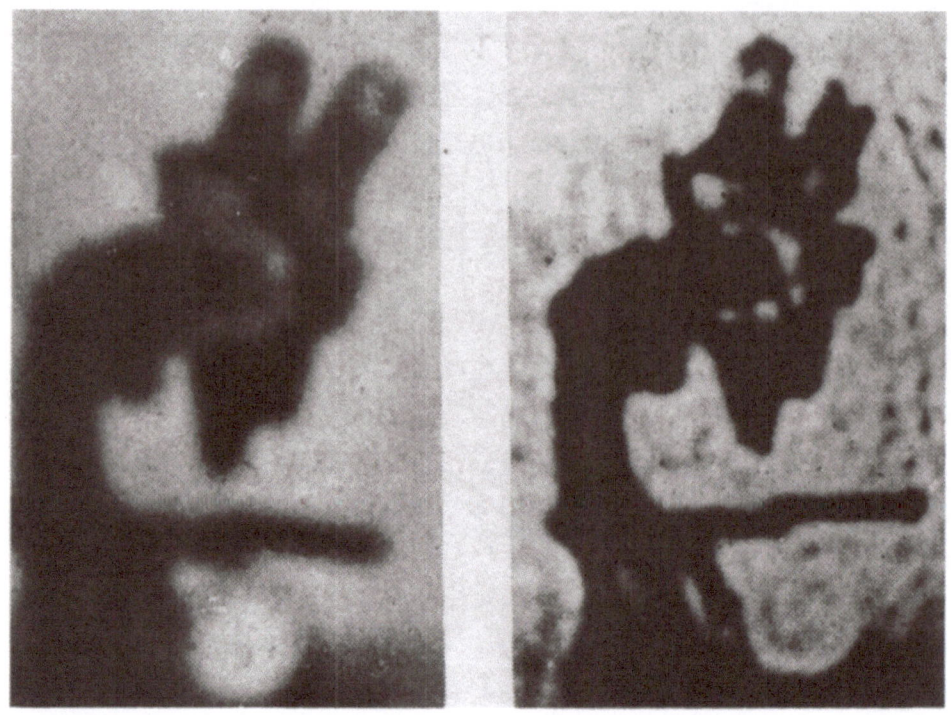

Fig. 15.   Result of insertion of filter H-1 in the filtering
apparatus in the spatial frequency plane.

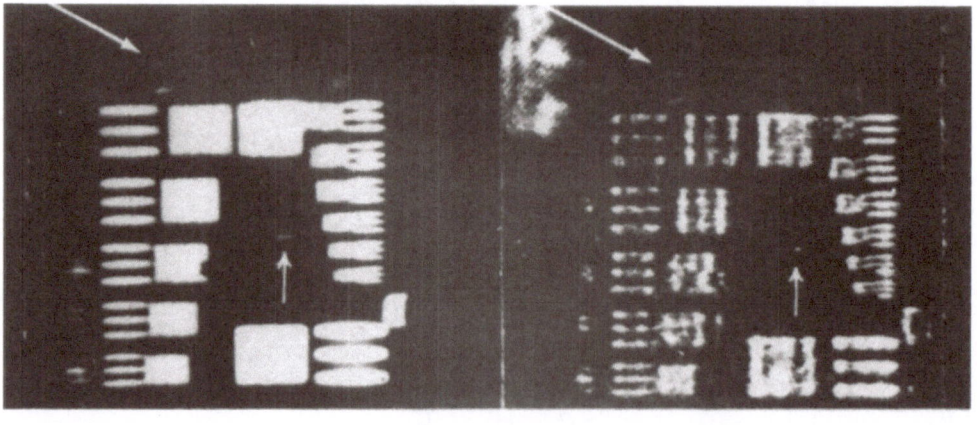

a   b

Fig. 16.   Results obtained with extended range holograms:
a) before blurring by motion, and b) after
blurring by motion.

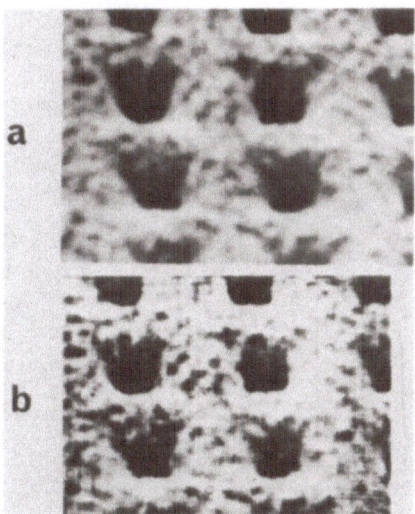

Fig. 17. Deblurring of a continuous tone microphotograph
          taken at a scanning electron microscope: a) before
          blurring, and b) after blurring.

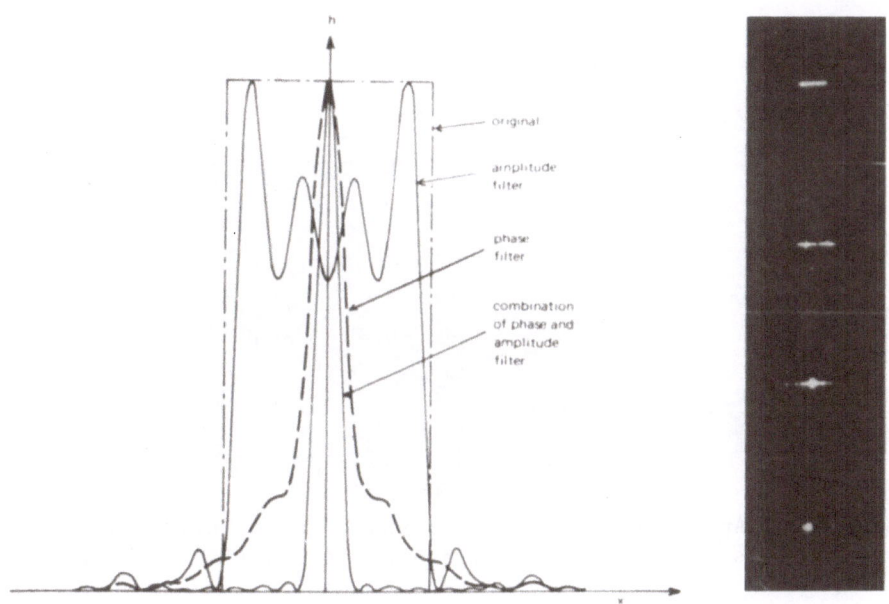

Fig. 18. Use of an artificial hologram filter for restoring
          point image blurred by motion.

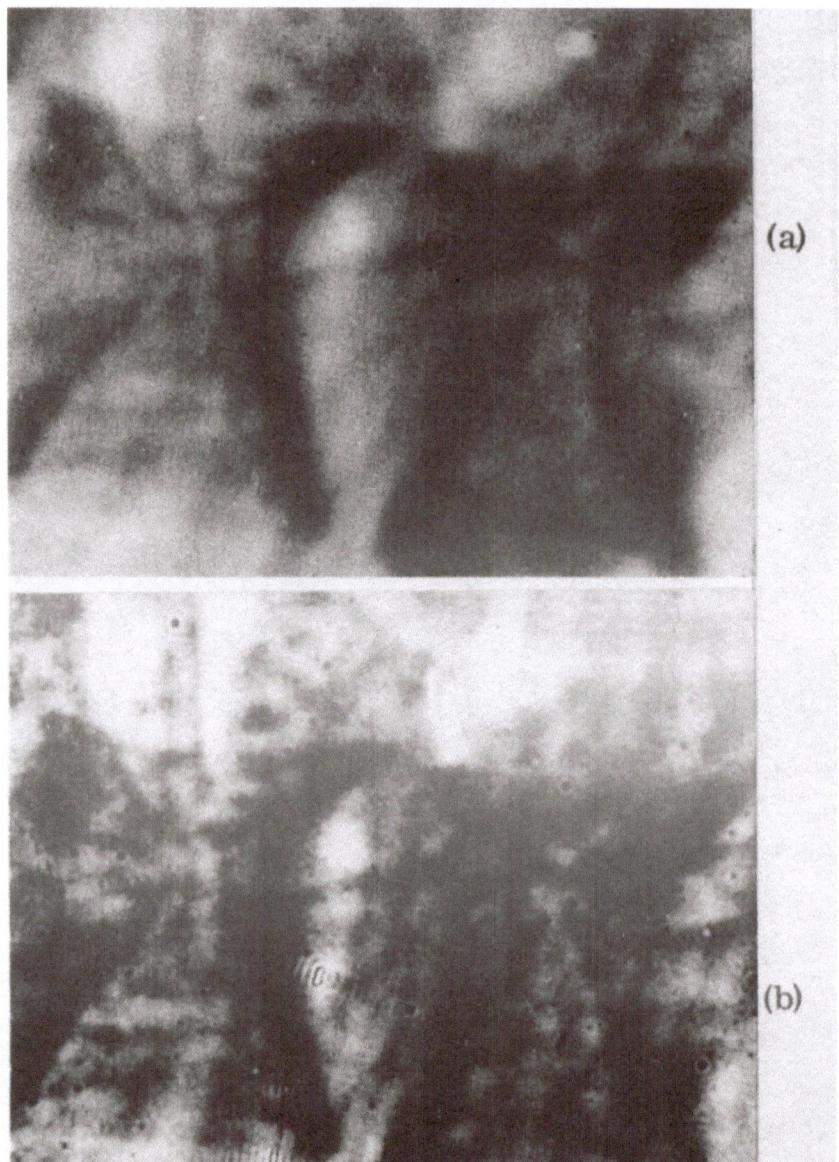

Fig. 19.   Application to radiographs of a posteriori filtering
system of spatial frequencies.   A fish skeleton.

test cart was deliberately blurred by motion. The small vertical
arrow indicates the blurred spread function.

The same method has been applied to deblurring a continuous
tone microphotograph taken at a scanning electron microscope (15)
Fig. 17.

An alternative technique for constructing a holographic filter
is constituted by an artificial hologram filter. Tsujuchi (16)
used a computer generated hologram as a phase filter in combination
with a photographic amplitude filter and then reduced it into small
size. Results are shown in Fig. 18 where this method is used for
restoring point image blurred by motion. This method has been applied in astronomy to separate the images of two adiacent stars.

The method of a posteriori filtering of spatial frequencies
has been applied also the radiographs (17). Some preliminary
tests are shown in Fig. 19, 20. The first radiographed object is
a fish skeleton (two vertebrae of the spinal coloumn). Contrast
and resolution appear improved. The second figure is a X ray of
a radiological test pattern. Converging lines that culminate in a
linear pattern whose resolution is 20 lines/mm.

The most recent method used by Stroke employes a different
technique (18). A hologram is made by using the blurred photograph
as object and the blurred impulse response as reference. The holo
gram is recorded through the 1/H transparency placed right in front
of it, Fig. 21. The wave impinging on photographic plate is:

$$(G+H) \cdot \frac{1}{|H|}$$

The transparency transmittance is:

$$T \ \alpha \ \frac{|G+H|^2}{|H|^2} \ = \ \dots \ \frac{GH^*}{|H|^2} \ = \ \dots$$

$$\dots \ \frac{FHH^*}{|H|^2} \ = \ F'$$

The deblurred image is obtained from F' by Fourier transformation.

A result obtained with this new technique is compared in
Fig. 22 with the preceding procedure b). Fig. 23 shows a X ray
photo of the sun deblurred with this method.

Fig. 20. Application to radiographs of a posteriori filtering
          system of spatial frequencies. An X-ray of a
          radiological test pattern.

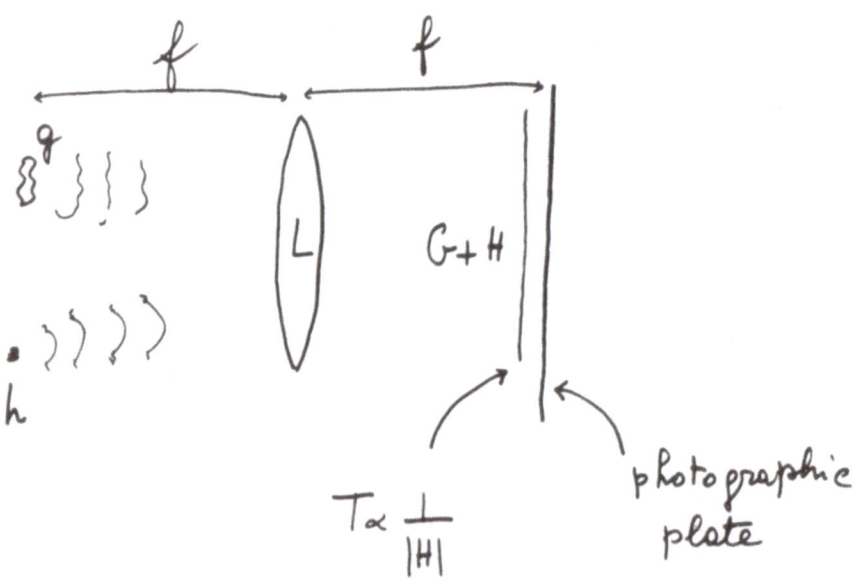

Fig. 21. Hologram made with method used by Stroke.

Fig. 22. Application of method used by Stroke. a) The test
image; b) the image obtained with method of a posteriori
filtering system of spatial frequencies; and c) the
image obtained with method used by Stroke.

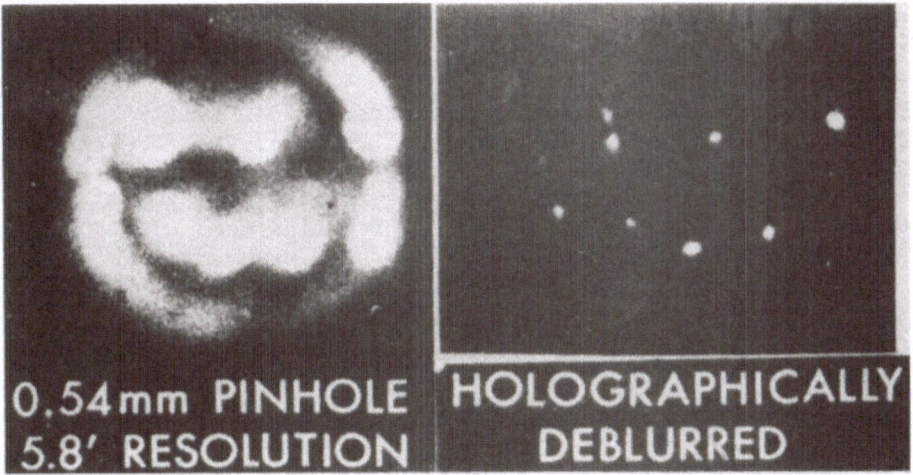

Fig. 23. X-ray photo of the sun deblurred with Stroke method.

CONCLUSION

As a conclusion we can say that holography is of use in optical data processing and in improving optical instruments.

The a posteriori restoration is a very effective method which can be utilized in different field of applications.

Of noticeable importance are the studies conducted on synthetic apertures whose images must be corrected with a posteriori method. In this respect I would like to recall that the famous radiotelescopes (Mill Cross, Culgora ring) are synthetic aperture antennas. Data coming from this type of antennas are processed a posteriori (by computer).

The same thing can be done with optical synthetic apertures and processing the image a posteriori.

REFERENCES

(1) A. Marechal, P. Croce: C.R. Acad.Sci., Paris 237, 607 (1953).
(2) E.L. O'Neill: Inst. Radio Engrs. Trans. 2, 56 (1956).
(3) J. Tsujiuchi: Progr. Opt. 2, 133 (1963).
(4) J. Upatnieks, A. Vander Lugt, E. Leith: Ap.Op.589, 5 (1966).
(5) J.E. Ward, D.C. Auth, F.P. Carlson: Ap. OP., 10, 896 (1971).
(6) T. Tsuruta, Y. Itch: Ap.Op. 70, 2138 (1968).
(7) E.N. Leith, J. Upatnieks: J. Op. Soc.Am., 56, 523 (1966).
(8) H. Kogelnik: Bell System Tech.J., 44, 2451 (1965).
(9) W. Mirande', J.Wzngartner, E. Menzel: Optics Com., 1, 315.
      (1970).
(10) J.W. Goodman: AppL. Phys. Lett., 8, 311 (1966).
(11) J. Tsujiuchi: Progress in Optics, 133 North Holland Publishing
      Co., Amsterdam.
(12) S.W. Stroke, R.G. Zech: Phys. Letters, 25A, 89 (1967).
(13) A.W. Lohmann, H.W. Werliel: Phys. Letters, 25A, 570 (1967).
(14) G.W. Stroke, F.Furrer, D. Lamberty: Optics Com., 1, 141 (1969).
(15) D.J. Evins et alia: Phys. Letters, 33A, 377 (1970).
(16) J. Tsujiuchi, T. Honda, T. Fukaya: Optics Com. 1, 379 (1970).
(17) G.A. Krusos, S.K. Hilal, W.B.Seamon, G.H. Myers: Appl.Phys.
      Letters, 16, 37 (1970).
(18) G.W. Stroke, M. Haliova: Phys. Letters, 33A, 3 (1970).

THE WAVE LENGTH DEPENDENCE OF THE TRANSFER PROPERTIES OF

PHOTOGRAPHIC MATERIALS FOR HOLOGRAPHY

H. T. Buschmann

AGFA-GEVAERT AG, Leverkusen - Bayerwerk (West Germany)

INTRODUCTION

Since the invention of the laser and the evolution of holography a lot of materials has been proposed for the recording of holograms. In spite of many disadvantages, photographic materials on
silver halide basis have the outstanding feature of high sensitivity and therefore are mostly used for holographic purposes.

For holographic work up to now preferably the wavelengths of
He-Ne-and ruby lasers (633 and 694 nm resp.) have been used. Agfa-
Gevaert Scientia materials (8E75 and 10E75) can be used with
success in this spectral region because they are specially sensitiz
ed for red light.

In the meantime, however, powerful lasers with emissions in
the short wavelength region are available (e.q. Argon and Krypton
laser), and there is a need for materials suitable for holographic
work in the blue and green range of the visible spectrum.

The transfer properties of the materials mentioned before are,
however, insufficient at shorter wavelengths, as will be shown
later on. For this spectral region Agfa-Gevaert offers, however,
emulsions specially sensitized for the short wavelength range,
namely Scientia 8E56 and 10E56.

All materials were developed in Agfa-Gevaert G3p for five
minutes at 20°C.

AMPLITUDE TRANSMITTANCE CURVE

The relation between density D and exposure E of a photogra-

phic layer is usually given by the characteristic curve $D = D(E)$.
In holography the developed layer is used as a diffraction pattern
for the incident wave front. Instead of spatial changes of density
the primarily decisive property is in this case the spatial change
of the amplitude transmittance (1) . The amplitude transmittance
T_a is defined as the ratio of the amplitudes of a monochromatic
plane wave after and before transition through the emulsion layer.
T_a is in general a complex quantity, since in passing through the
layer not only the amplitude, but also the phase of the incident
wave is changed. The change of the phase may well depend on the
amount of exposure (phase hologram). However, merely the intensity
transmission $T_i = T_a$. T_a^* of developed layers can easily be
measured. From this the value of the amplitude transmittance
$|T_a|$ is obtained.

Amplitude transmittance curves for the materials in question
are shown in Fig. 1. The plates were exposed by the light of a
Krypton laser with wavelengths of 647, 521 and 476 nm.

Generally, the materials 10E56/10E75 are faster than the
materials 8E56/8E75. This is due to the different mean grain diameter of the silver halide grains in the emulsion, which is about
80 nm for the 10E- and 50 nm for the 8E-materials respectively.

The sensitivity of 10E56 at 476 and 521 nm is nearly equal to
that of 10E75 at 647 nm. The sensitivity of 8E56 at 476 and
521 nm, however, is less than that of 8E75 at 647 nm, a fact which
is (probably) due to the use of an additional screening dye in the
emulsion.

In Fig. 2 the amplitude transmittance curves of an experimental material named Pan 300 are shown for three wavelengths. The
mean grain diameter of this emulsion is about 30 nm, and the material is sensitized for whole of the visible spectrum. This material is available on special order and was developed for use in
color holography. The sensitivity is about ten times smaller than
that of the above mentioned materials. The very fine grain diameter leads to excellent transfer properties, which is due to very
small scattering cross sections. It is well known (2) , that the
optimal brightness of amplitude holograms is got at exposures
around the bias point $|T_a| = 0.5$. In Fig. 1 the values
$a_{max} = -d |T_a| /d \lg E$ (for $|T_a| = 0.5$) are included, for the
brightness is proportional to a^2 , if the input modulation is
sufficiently low to assure linear recording.

DIFFRACTION EFFICIENCIES

For a quantitative determination of the transfer properties of
holographic materials, it is advisable to apply interferometric
methods, (3) . For this purpose, an apparatus has been constructed

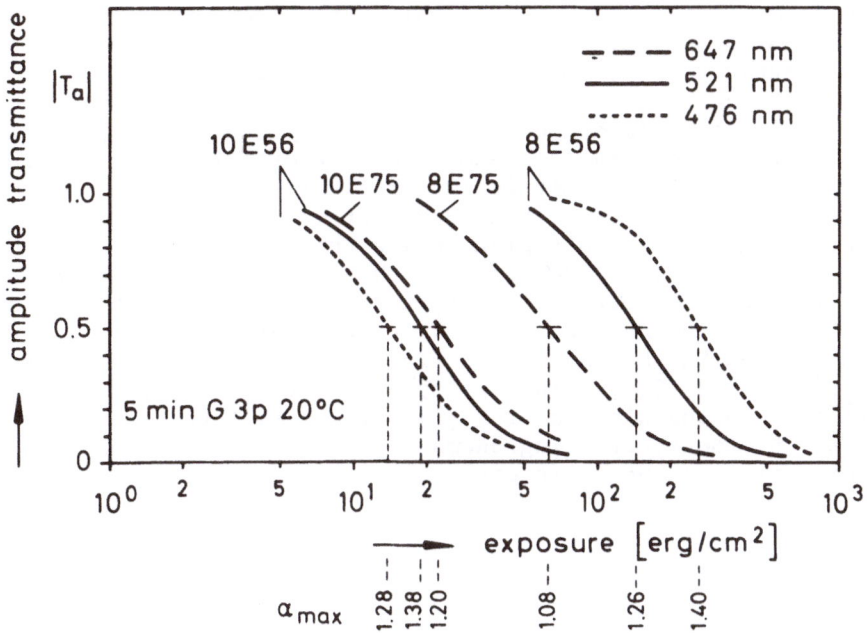

Fig. 1. Amplitude transmittance curves for different wavelength.

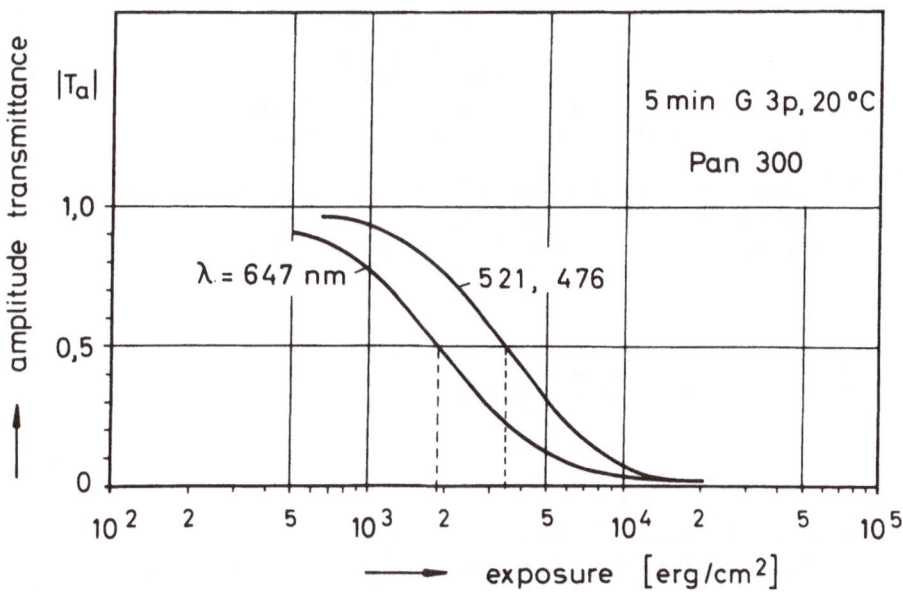

Fig. 2. Amplitude transmittance curves for different wavelength.

which permits the exposure of double beam interference patterns
and a subsequent evalution of the resultant diffraction gratings.
The arrangement is schematically shown in Fig. 3.

The original laser beam is expanded by means of a pinhole and
a collimator to a parallel beam of approximately 4 cm diam. This
beam is slit in two geometrically nearly identical beams by means
of the mirror prism S_0. These beams are reflected by mirrors S_1
and S_2 towards the sample plane where they interfere in an area
of 1 x 1.5 cm^2. The E-vector of both beams is normal to the plane
of incidence. The mirrors S_0, S_1 and S_2 are mounted on a sled
which can be moved perpendicularly to the sample plane. Mirrors
S_1 and S_2 can be shifted on the sled between the limiting positions
S_1, S_1', and S_2, S_2', respectively, and pivoted about their axis
vertically to the plane of the diagram, so that the angle θ
between the two interfering beams can be varied between 3 and 160°.
The spatial frequency v is given by

$$v = \frac{2}{\lambda} \sin \left(\frac{\theta}{2}\right)$$
3.1.

λ is the wavelength in air.

The intensitities I_1 and I_2 can be varied independently within
the two beams by inserting appropriate absorbing filters. A change
of the line frequency by an eventual shrinkage of the photographic
material is avoided due to the symmetric incidence of the two
beams relative to the normal to the photographic layer.

The principal arrangement of exposure and reconstruction is
illustrated in Fig. 4. After adjustment of a desired intensity
ratio I_2/I_1 and thus of the degree of modulation

$$m = 2 \sqrt{I_1 \cdot I_2} / (I_1 + I_2)$$

the photographic plate is brought into the sample plane for exposure. With an adjustable plate holder up to 14 exposures, e.g.,
with different exposure times, may be made in rapid sequence on a
9 x 12 cm^2 plate. Immediately after exposure the plate is developed and processed as usual.

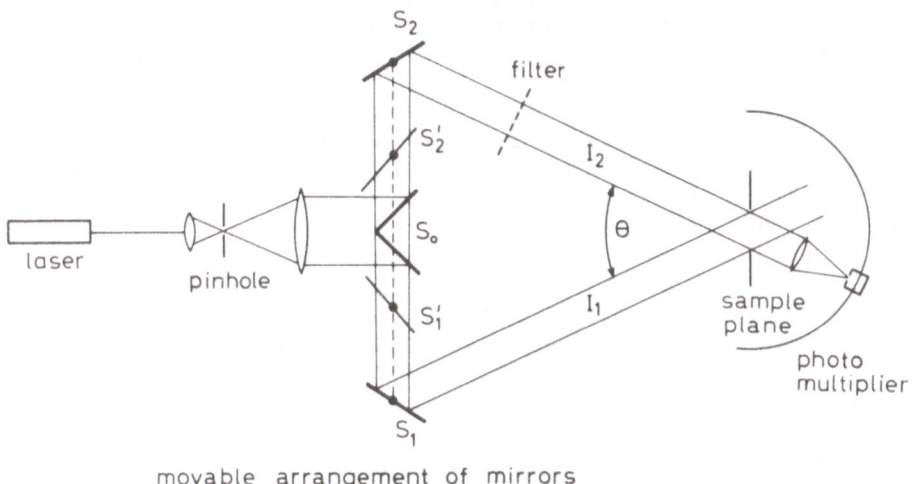

Fig. 3.   Apparatus for exposure and evaluation of two-beam
         interference patterns (schematic).

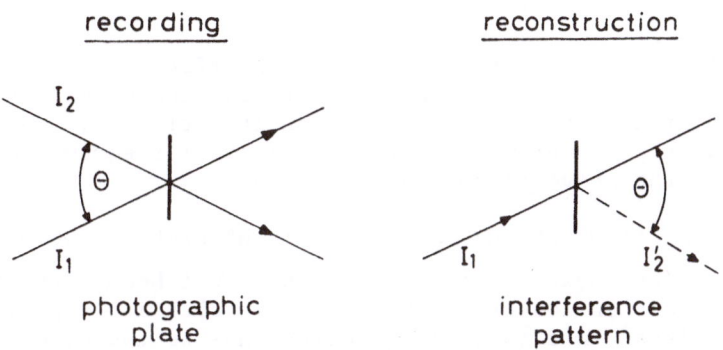

Fig. 4.   Recording and reconstruction (schematic).

After processing, the plate is again placed in the sample plane and is illuminated with beam 1 only. Thus beam 2' is reconstructed. The intensity I_2' is measured behind the sample plane by a photo multiplier which can be moved along a calibrated angular scale. When the sample layer is removed from the plate holder the intensity I_1 of the incident beam can be measured in the same manner.

The ratio I_2'/I_1 of the reconstructed and incident beam defines the diffraction efficiency η. By varying the exposure at constant spatial frequency the maximum value η_{max} of the diffraction efficiency was determined, which is reached for amplitude transmittance near 0.5. The modulation was chosen m = 1.

Fig. 5 shows the maximum diffraction efficiency η_{max} vs. spatial frequency ν for three wavelengths of the Krypton laser, namely λ = 647, 521, 476 nm and the material 8E75. Obviously the performance of this material is much better for red light than for green und blue light, due to the special sensitization for the red light. This applies to the material 10E75, too. These materials cannot be recommended for holographic work in the short wave length range. Fig. 6 presents the maximum diffraction efficiency η_{max} vs. spatial frequency for λ = 521 and 476 nm for 10E56 and 8E56. For comparison the curves for 10E75 and 8E75 for λ = 647 nm are included.

Obviously the transfer properties of the materials of the 56 series are in the short wavelength range equal if not better than the respective values of the materials of the 75 series for red light.

Finally Fig. 7 shows the diffraction efficiencies for Pan 300. The values are nearly independent from the wavelength as can be expected by the panchromatic sensitization. In addition the curves drop slower with increasing spatial frequency as in the case of the above discussed materials.

DETERMINATION OF THE MODULATION TRANSFER-FUNCTION MTF

The curves shown in Fig. 6 and 7 have not been corrected for Fresnel reflection losses at the boundary faces air - photographic layer and glass - air (4). This corresponds to the case usually met in practice. The true transfer properties of the photographic layers, however, are better. The situation in question is represented in Fig. 8.

In the first place, the reconstructing beam of intensity I_1

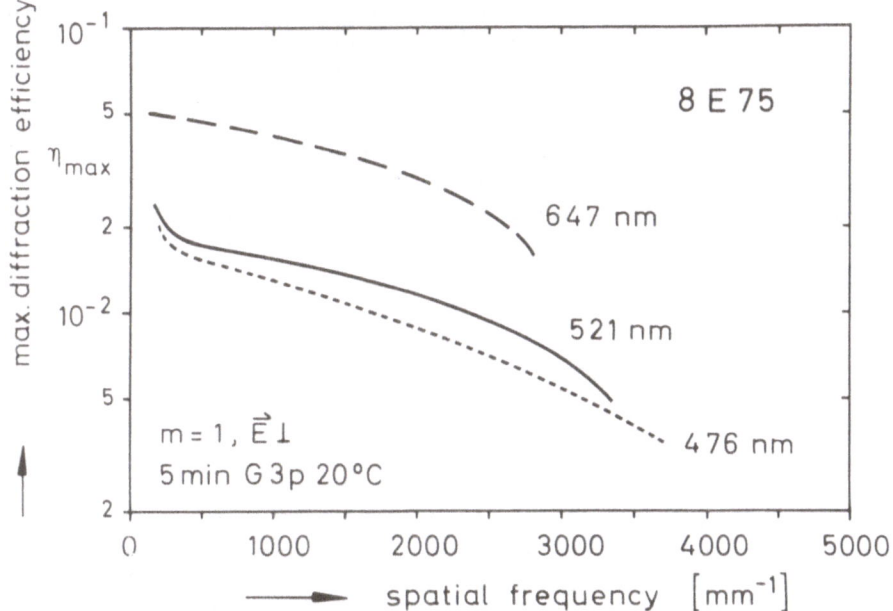

Fig. 5. Optimal diffraction efficiency of Scientia 8 E 75 for different wavelength.

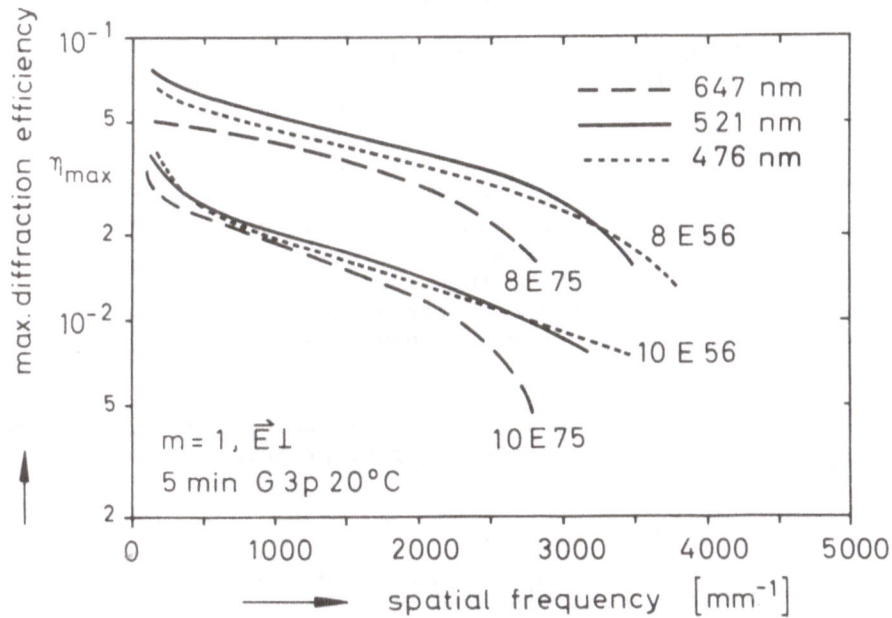

Fig. 6. Optimal diffraction efficiences of Scientia 8 E 75, 8 E 56, 10 E 75, and 10 E 56, for different wavelength.

is attenuated by a factor, which depends from angle $\theta/2$. In the second place, the true diffracted intensity $I_2^{i\prime}$ is attenuated by reflection at the boundary 3-4, so that the measured intensity I_2' is too small.

If multiple reflections and the reflection at the boundary photographic layer - glass are neglected, we obtain for the corrected diffraction efficiency of the photographic layer:

$$\frac{I_2^{i\prime}}{I_1^{i}} = \frac{I_2'}{I_1} \; \frac{1}{(1-R_{12})\,(1-R_{34})} \qquad\qquad 4.1$$

where R_{12} and R_{34} are the reflectivities at the boundaries 1-2 (air - layer) and 3-4 (glass - air). The reflectivity R depends upon polarization, upon the angle $\theta/2$ and the refractive indices of the layers. As refrective indices we applied n = 1.535 for the developed photographic layer and n = 1.514 for the glass plate.

For the case that the $\bar{E}$-vector of the light is perpendicular to the plane of incidence the deviations at large angles are considerable. For $\theta = 140°$ they constitute approximately a factor 2.

In Fig. 9 this correction was done for three materials and red laser light. The modulation in this case was m = 0.45 in order to avoid nonlinear recording. The result is: the diffraction efficiency of the finest grained material Pan 300 is completely independent from spatial frequency, the corresponding curve of Scientia 8E75 is nearly independent from spatial frequency whereas the curve of 10E75 decreases somewhat faster with increasing spatial frequency. This behaviour can be well understood by regarding the diffusion of the light by the finite diameter of the silver-halide grains as shall be done lateron. Now we state, that such weakly scattering emulsions should have nearly constant MTFs over a wide range of spatial frequencies.

Frieser (5) has given an analytical expression for the line spread function of weakly scattering photographic materials (Fig.10):

$$\phi(x) = \varrho\,\delta(x) + (1-\varrho)\,\frac{2.3}{k}\,e^{-\frac{4.6|x|}{k}} \qquad\qquad 4.2 .$$

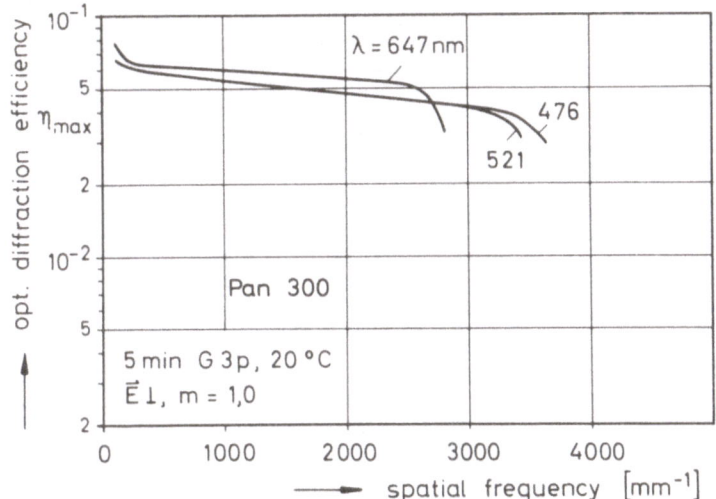

Fig. 7.   Optimal diffraction efficiences of Scientia Pan 300
          for different wavelength.

reconstruction

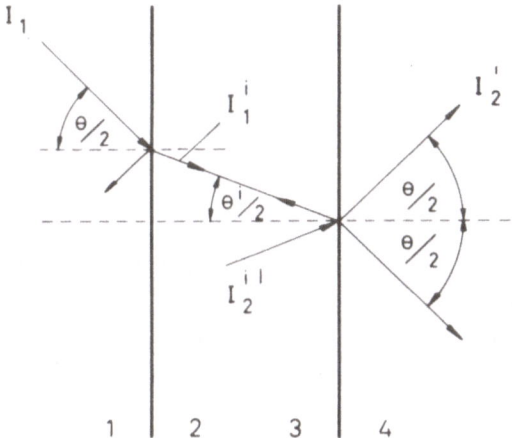

interference pattern

Fig. 8.   Influence of the Fresnel-reflection losses on the
          diffraction efficiency at the boundaries.

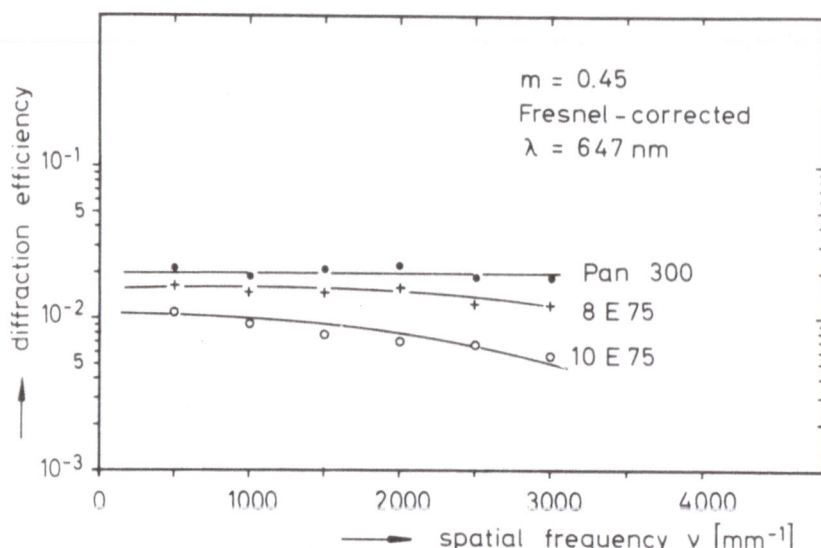

Fig. 9. Diffraction efficiency corrected due to Fresnel-
reflection losses for λ = 647 nm and different
materials.

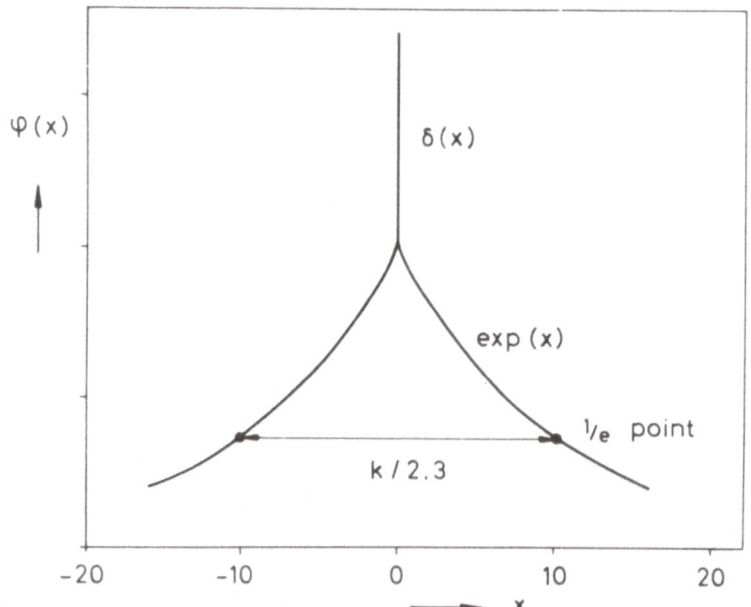

Fig. 10. The line spread function for a weakly scattering
material.

where $\delta(x) = 0$ for $x \neq 0$ and $\int_{-\infty}^{+\infty} \delta(x) \, dx = 1$.

The first term describes the unscattered absorbed light and
the second term the light diffusion caused by scattering. This
light diffusion is well approximated by an exponential function
and k is Friesers k-value. The value ϱ can be defined as the
ratio of specular that means unscattered absorption to total absorp
tion $\varrho = A''/A_t$.

The MTF is the Fourier transform of the line spread function:

$$M(\nu) = \int_{-\infty}^{+\infty} \phi(x) e^{-i\omega x} \, dx \qquad\qquad \omega = 2\pi\nu \qquad\qquad 4.3.$$

For this special line spread function one gets:

$$M(\nu) = \varrho + (1-\varrho) \frac{1}{1 + \left(\dfrac{\pi \, k \, \nu}{2.3}\right)^2} \qquad\qquad 4.4.$$

This relation is plotted in Fig. 11 for special values of ϱ
and k.

Since the k-values for weakly scattering materials are in the
order of 50 - 100 μm, the function drops fast with increasing
spatial frequency. This means the MTF reaches its constant value
ϱ at frequencies of 100 1/mm or less: $M(\nu) \approx \varrho(\lambda)$.
Possibly the most interesting characteristic of a holographic material is the function $\varrho(\lambda)$. One can try to determine $\varrho(\lambda)$ by
measuring the diffraction efficiency as a function of the wave-
length λ, since the relation holds for $m \leq 0.4$:

$$\eta(\lambda) = C \, (m\alpha \, \varrho(\lambda))^2 \qquad\qquad C = \text{const.} \qquad\qquad 4.5.$$

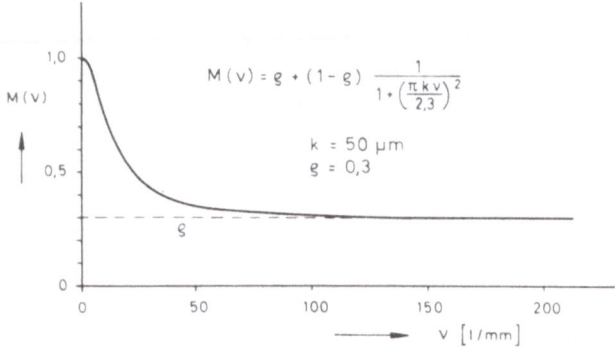

Fig. 11. The modulation transfer function for a weakly scattering
material.

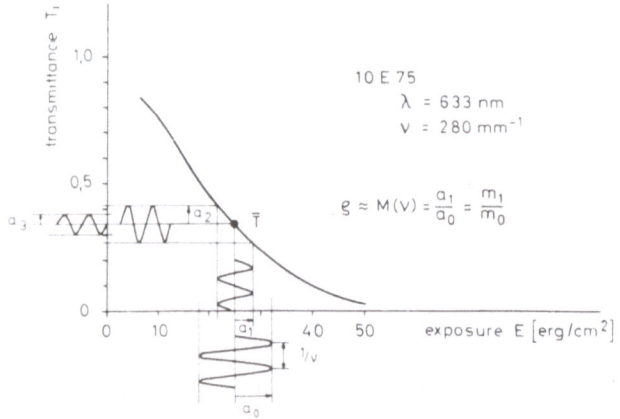

Fig. 12. Principle of the determination of the modulation transfer
transfer function.

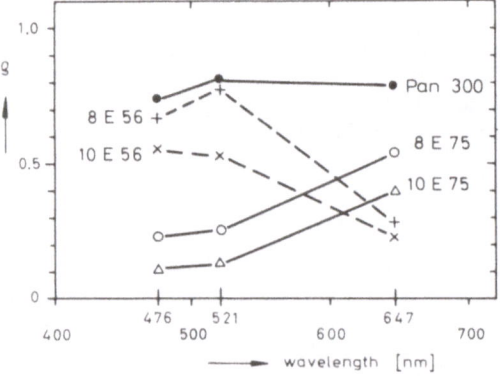

Fig. 13. ρ-values as a function of the wavelength.

where m is the input modulation and $a = d \left| T_a \right| / d \lg E$. It is well known, however, that holograms taken on silver halide layers usually are mixtures of amplitude and phase-holograms. The phase parts are not considered in this equation, therefore the values obtained for ϱ are usually too large, if one tries to determine them from measurements of a, m and η according to this equation.

We have tried a more conventional method for evaluation of the ϱ- values for these materials (6). This method includes laser exposure of the sample with two beam interferences of known modulation m, and measurement of the actual modulation on the processed plate with a microdensitometer (MDM).

Fig. 12 describes this kind of measurement in principle. The characteristic curve is given here in the form of intensity transmittance vs. exposure in a linear scale. a_3 is the amplitude of the processed layer as measured in the MDM. This amplitude rises to a_2 by application of the correction due to the MTF of the MDM. The amplitude a_2 is further transformed into the amplitude a_1 of the effective exposure using the characteristic curve T (E). Finally the transfer factor ϱ is given by the ration a_1 / a_0; a_0 being the amplitude of the input signal, which is to be known in each case.

The input modulation was less than 0.4 in our measurements to assure linear transfer. The spatial frequencies used ranged from 100 to 300 1/mm. In this range the MTF has already reached a constant value ϱ and the MDM corrections can be done without difficulty (7).

Fig. 13 shows the ϱ- values obtained in this manner for five materials at the wavelengths 476, 521, 647 nm of the Krypton laser. As one expects, the ϱ- values of the materials are well correlated to the respective diffraction efficiencies shown before.

Generally the ϱ- values for the materials of the 8E-class are higher than those for the 10E-class. This can be understood by remembering that the mean grain diameter of the 8E-materials is about 50 nm, whereas the diameter of the 10E-materials is 80 nm. The larger diameter causes higher scattering power and decreased MTF values. Expecially the 8E56 material reaches transfer factors as high as 0.8 at 521 nm; that means that this material can hardly be improved from this point of view. The ϱ- values for Pan 300 are nearly independent from the wavelength and very high, this is achieved by the very small grain diameter of 30 nm and the corresponding small scattering power.

We therefore conclude, that an essential condition, to obtain a high diffraction efficiency is to produce materials with high values of ϱ in the spectral region considered. High values of ϱ

emulsion which is decreased by reducing the particle size of the
silver halide. This is the situation met with the material Pan
300. It is, however, necessarily coupled with a loss in sensitivity.

The other way of enlarging ϱ is to increase the absorption power of the emulsion, which can be achieved by optimal sensitization
and eventually further addition of screening dyes.

THE INFLUENCE OF FINITE GRAIN DIAMETER ON THE MTF

Up to here, we have neglected the influence of the finite
grain diameter on the MTF. We have theoretically described the MTF
by two terms: firstly we have considered the light diffusion caused
by the scattering of the photons. The light distribution was approximated by an exponential function, which is governed by the k-
value. Secondly the most interesting part of light, which is ab -
sorbed without preceding scattering acts, was taken into account by
ϱ - values and a ϱ - function. That means for sufficiently high
spatial frequencies the MTF is constant. That is of importance for
holography since the brightness of the reconstructed image becomes
independent from spatial frequency. But in praxis, we have seen
this is only an approximation. The actually measured diffraction
efficiencies decrease somewhat with increasing spatial frequencies.
This effect is more distinct for the emulsions with the larger
grain diameters as can be expected.

We shall now try, to consider the influence of the finite
grain diameter on the MTF. This can be done only in a simplified
manner.

At first we consider the silver halide crystals of the emulsion here in question. Fig. 14 shows an electron micrograph of the
emulsion 8E75. The figure represents a 1000 Å thick ultra thin
section in order to see whether the grains are conglomerated or not.
The grain distribution cannot be jugded from this micrograph,be-
cause some of the grains have been cut by the knife of the micro-
tome.

Nextly we study the influence of the developer. Fig. 15 represents an electron micrograph of a 1000 Å thick ultra thin section of a developed 8E75 layer. It was developed for 5 minutes in
developer $G3_p$ at 20°C. The density of the sample was 0.6, which is
equal to an amplitude transmittance of 0.5. Onc can say, from every grain, which has become developable by exposure, a mostly curv-
ed filament arises by development.

In Fig. 16 and 17 electron micrographs are shown of the same
material, but now the material was prepared by removing enzymatic-
ally the gelatin and the grains and silver-filaments respectively

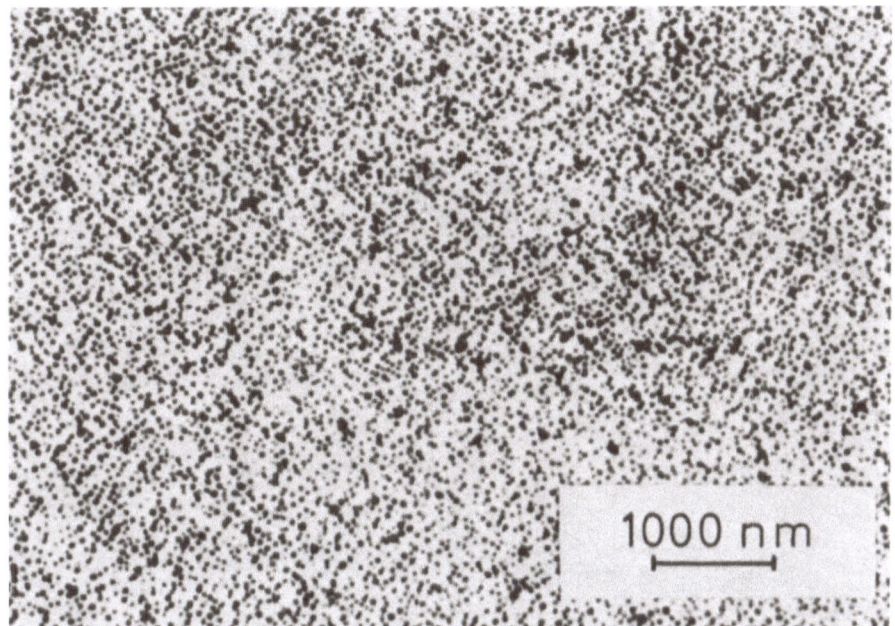

Fig. 14. Electron micrograph of undeveloped Scientia 8 E 75 emulsion, (ultra-thin section).

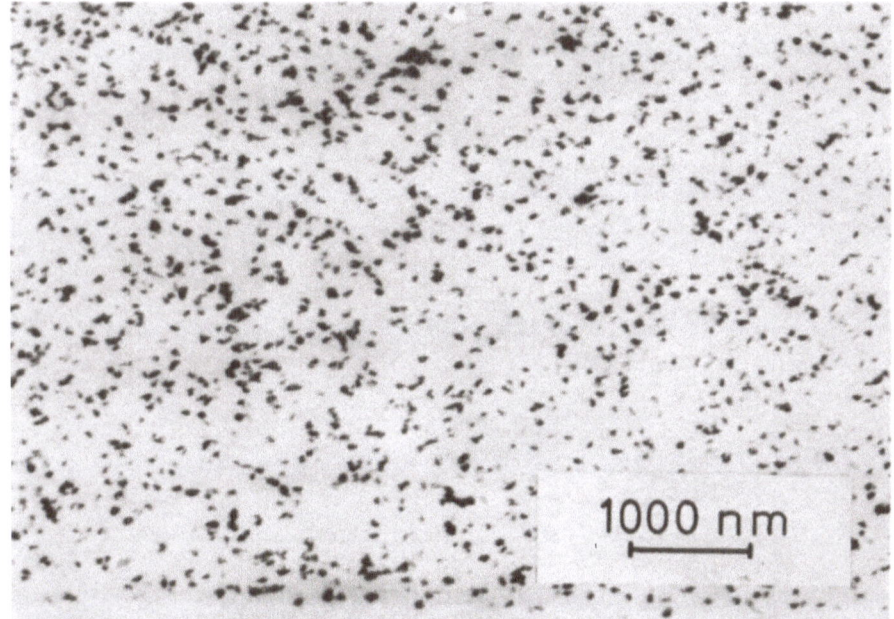

Fig. 15. Electron micrograph of developed Scientia 8 E 75, 5 min G3$_p$ at 20°C, the density was 0.6, (ultrathin section).

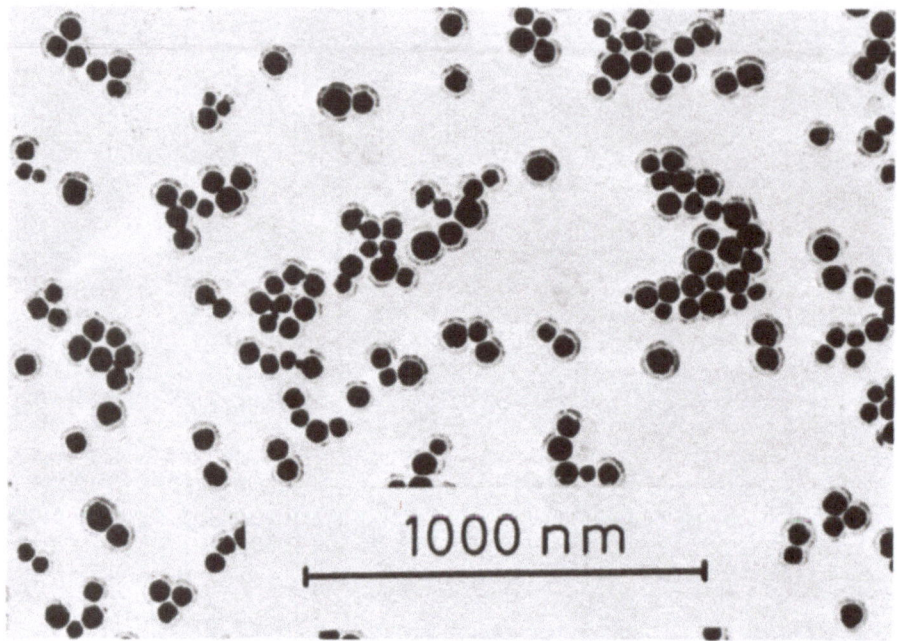

Fig. 16.   Electron micrograph of undeveloped Scientia 8 E 75,
the gelatin was removed by enzymatic digestion, the
grains are evaporated with carbon.

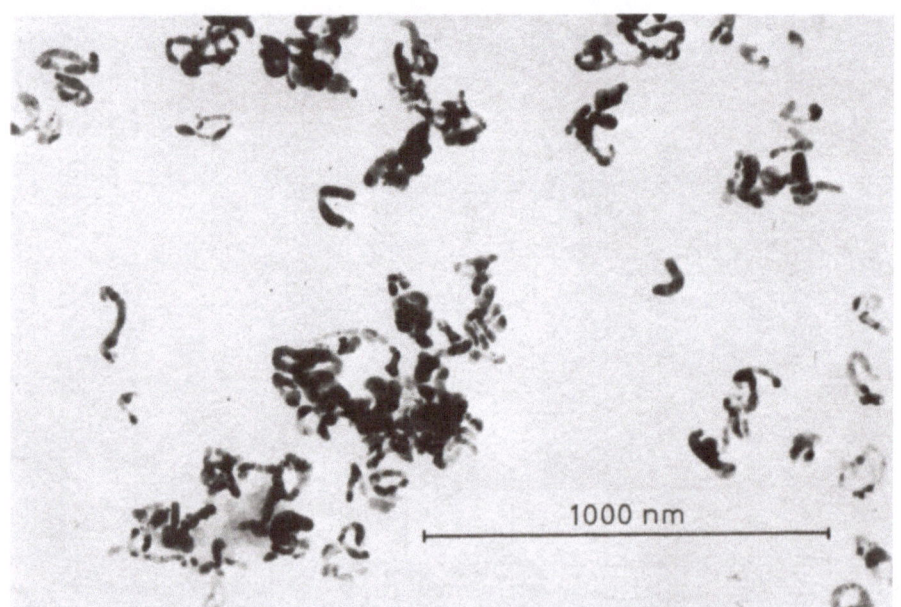

Fig. 17.   Electron micrograph of developed Scientia 8 E 75, 5 min
G3$_p$ at 20°C, the gelatin was removed by enzymatic
digestion.

were evaporated with carbon to avoid destruction of the samples by
the electrons. Now one can judge the habit of the crystals and
the grain distribution , which is very narrow. In good approximation one can consider the silver halide particles being spheres.
In Fig. 17 the filament structure of the developed silver is clearly visible.

For this consideration the influence of the development is neglected. The grains are assumed being spheres and having equal
diameters. An infinite small slit exposure represented by a δ -
function is spread by the finite diameter of the grains, Fig. 18.
The spheres are statistically distributed.

Then all particles being in an interval $<- 2r, 2r >$, r is
the radius of the spheres, are hit by photons. The simplest model
is accepted by the assumption: that all spheres hit by photons become developable, which means that all spheres with centers X_m in
the interval $<-r,r>$ contribute to the line spread of the light.
In Fig. 18 the grains which are exposed are hatched. The line
spread $\phi(x)$ should be proportional to the mass of the silver halide particles exposed by light.

The mass at the point x results in:

$$dm(x) = \pi N_o \left[r^2 - (x - x_m)^2 \right] dx_m \qquad\qquad 5.1.$$

N_o being the number of the grains for $x = 0$. $m(x)$ is received by
integration from $x-r$ to r of this equation:

$$m(x) = \frac{4}{3} \pi r^3 N_o \left[1 - \frac{3}{4} \left(\frac{x}{r} \right)^2 + \frac{1}{4} \left(\frac{|x|}{r} \right)^3 \right] \qquad\qquad 5.2.$$

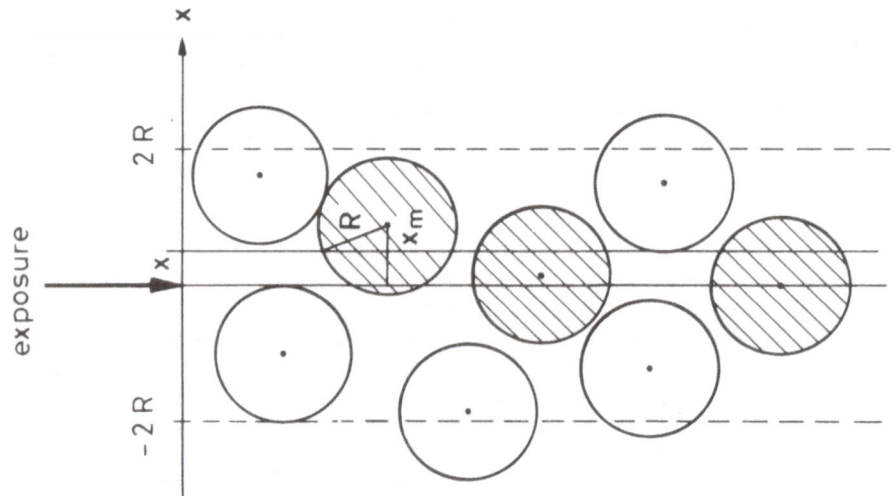

<u>Fig. 18</u>   Influence on the line spread by the finite grain
              diameter, the grains were approximated by spheres

$$m\,(x) \;=\; \frac{4}{3}\,\pi r\, N_0 \left[1 - \frac{3}{4}\left(\frac{x}{r}\right)^2 + \frac{1}{4}\left(\frac{|x|}{r}\right)^3 \right]$$ 5.2.

The normalized spread function is then given by:

$$\phi\,(x) = 1 - \frac{3}{4}\left(\frac{x}{r}\right)^2 + \frac{1}{4}\left(\frac{|x|}{r}\right)^3. \qquad |x| \leqslant 2r$$ 5.3.

      Finally, the MTF is defined as the Fourier transform of the
line spread function:

$$M(\omega) = \int_{-2r}^{+2r} \phi(x)\, e^{-i\omega x}\, dx \qquad\qquad 5.4.$$

K is a constant and can be determined from $M(0) = 1, \omega\, 2\pi\nu$,
ν spatial frequency. The evaluation of the integral leads to:

$$M(\omega) = \frac{3}{2}\ \frac{1}{(\omega r)^4}\left[2\sin^2(\omega r) - \omega r\ \sin(2\omega r)\right] \qquad\qquad 5.5.$$

A more realistic model, however, should be accepted by the
assumption, that the probability of the light absorption is proportional to the volume or the surface of the grains which is cross
ed by the light. The first case should be proved right for the
range of the intrinsic sensitivity of silver bromide, that means
wavelengths less than 430 nm. The second case should be true
for the range of the optical sensitization and this range is more
important, for most of the lasers used at the present time have
wavelengths above 430 nm.

In this case one can derive for the spread function:

$$\phi(x) = \int_{x-r}^{x} N(x)\left[r^2 - (x - x_m)^2\right]\, dx_m \qquad\qquad 5.6.$$

with:

$$N(x) = N_0\left[1 - \left(\frac{x_m}{r}\right)^2\right]^{\frac{1}{2}} \qquad\qquad 5.7.$$

The integration results in (normalized):

$$\phi(x) = \frac{8}{3\pi} \left[\left(\frac{3}{8} - \frac{1}{2} \left(\frac{x}{r} \right)^2 \right) \left(\frac{\pi}{2} - \arcsin \left(\frac{|x| - r}{r} \right) \right) - \right.$$

$$- \left(\frac{3}{8} - \frac{1}{2} \left(\frac{x}{r} \right)^2 \right) \frac{|x| - r}{r} \left(1 - \left(\frac{x - r}{r} \right)^2 \right)^{\frac{1}{2}} -$$

$$\left. - \left(\frac{1}{4} \frac{|x| - r}{r} - \frac{2|x|}{3\,r} \right) \left(1 - \left(\frac{|x| - r}{r} \right)^2 \right) \left(1 - \left(\frac{x - r}{r} \right)^2 \right)^{\frac{1}{2}} \right] \qquad 5.8.$$

The results for both cases are presented in Fig. 19. The spread function, which includes the surface term, decreases some-what faster with distance x. The Fourier transform of equation (5.8) was numerically calculated by using Simpson's integration rule.

Since the diffraction efficiency for linear recording can be assumed proportional to the square of the MTF, these calculations are plotted in Fig. 20 for both models and three different radii. As one can expect, the curves with the surface term drops slower with increasing spatial frequencies. The function is nearly con-stant for a radius of 15 nm up to 4000 lines:mm, whereas the de-pendence from spatial frequency becomes greater and greater for increasing grain diameter. Since we have neglected the influence of the development and the differences between both models are small, we should use the simpler case.

So we can now give a complete expression for the MTF of weakly scattering materiais by combining equation (4,3) and (5.5): the δ-function of equation (4.1) is substituted by the spread function represented in equation (5.3):

$$M(\omega) = \varrho \frac{3}{2} \frac{1}{(\omega r)^4} \left[2 \sin^2 (\omega r) - \omega r \sin (2 \omega r) \right] + (1 - \varrho) \frac{1}{1 + \left(\frac{k\,\omega}{4.6} \right)^2} \qquad 5.9.$$

with $\omega = 2\pi\nu$

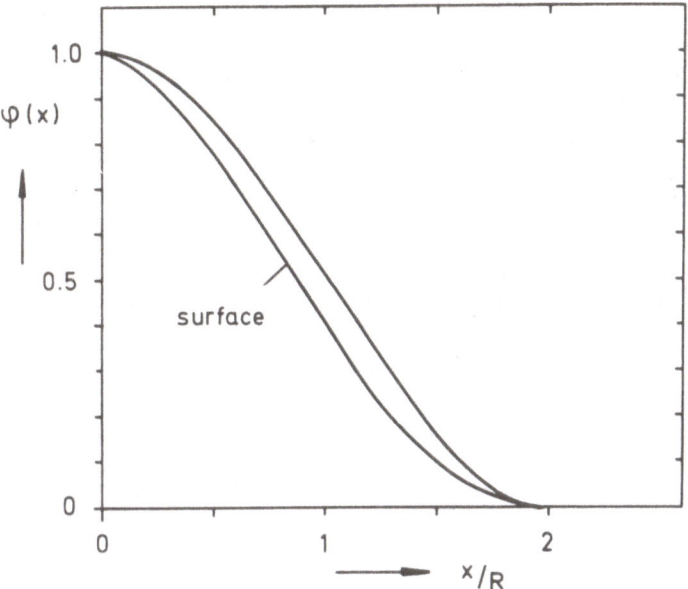

Fig. 19. The line spread function due to finite grain diameter.

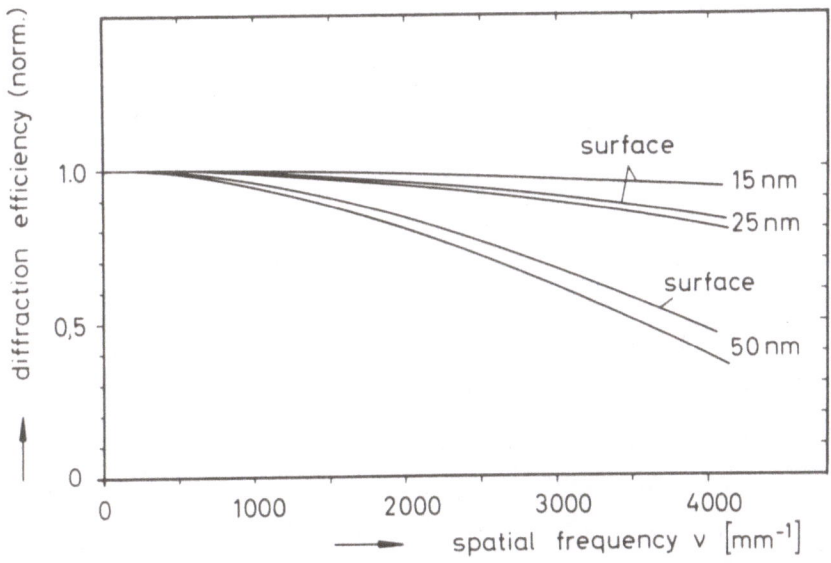

Fig. 20. Calculated dependance of the diffraction efficiencies
          (normalized) as a function of spatial frequency and
          grain diameter.

A comparison of the results of Fig. 20 with the measured diffraction efficiencies of Fig. 9 shows satisfying agreement concerning the dependence from spatial frequency and grain diameter. For k-values of 50 - 100 μn and spatial frequencies between 100 and 500 mm^{-1} M (ω) $\approx$ ϱ , so that our method of determining the values should be correct.

So the light scattering and absorption is regarded as the most important step of recording holographic images. The part of unscattered absorbed light can well be approximated by a wavelength dependent constant ϱ (λ), if the grain diameters of the emulsion are sufficiently small.

REFERENCES

(1)   D.G. Falconer, Phot.Sci.Eng. 10 (1966) 133.
(2)   K. Biedermann, Optik 28 (1968) 160.
(3)   A.A. Friesem, A. Kosma a. G.F. Adams, Appl. Optics 6 (1967)851.
(4)   H. Nassenstein, H.T. Buschmann a. J. Geldmacher, Optik 30 (1970) 527.
(5)   H. Frieser, Phot. Korr. 91 (1955) 69 - Phot. Korr. 92 (1956)51 Phot.Sci.Eng. 4 (1960) 324.
(6)   H.T. Buschmann a. H.J. Metz, Optics Communications 2 (1971)373.
(7)   H. Frieser a. H. Kramer, Phot. Korr. 102 (1966) 69

A REVERSIBLE OPTICAL RECORDING MEDIUM

G.K. Megla

Director, Electronic Research, Technical Staff Division

Corning Glass Works, Raleigh, North Carolina (U.S.A.)

INTRODUCTION

A variety of reversible recording media have been developed in the United States in the last few years such as photochromic, cathodochromic, thermochromic, ferroelectric materials and liquid crystals.

This paper deals mainly with photochromic materials, their characteristics, configurations and applications. This material was originally developed by Corning Glass Works [1] and has found an increasing interest as prescription lenses for sun protective eye glasses. These glasses darken when exposed to sunlight and return to transparency again when no ultraviolet irradiation is present. Because of this application, the optical density has to decrease quickly to regain full transparency; such fast fading glass was not suitable for optical recording purposes. Therefore, a number of slow-fading photochromic glasses were developed that retain the optical image over a long time or as long as no external bleaching energy is applied. The retentivity characteristics of these glasses permit a wide range of different applications.

Depending on the composition, some glasses retain the image over hours if not exposed to ultraviolet light; some do not fade if stored in a dark environment at room temperature; and others, if cooled to about −20°C, show no fading at all. Also, depending on the particular composition, the glasses darken when exposed to light in the 300-400 millimicron wavelength range and can be bleached (the information erased) by a red light of 600-700 millimicrons wavelength.

In the following Chapter, the relevant properties of photochromic glass are described in some detail. After that, various
configurations that improve the resolution of images will be shown,
followed by a description of a storage cathode ray tube using a
photochromic fiber optic plate as a target. After a discussion of
the optical systems that are necessary to optimize the image forma
tion in photochromic glass, some established and possible applications are given.

MATERIAL PROPERTIES

For photochromic recording storage media, the following eight
parameters have to be known to characterize the material for record
ing applications: spectral sensitivity, exposure time, spectral
transmittance, image contrast, resolution, retentivity, reversibility, and gamma value. Some of these parameters have been published for general applications in detail by employees of Corning
Glass Works [2] , [3] . I will, therefore, concentrate only on
those properties and concepts that are of interest for optical
recording applications.

Since photochromic glass is a reversible optical recording
medium, it is necessary to know the optimum light wavelengths to be
applied for the image forming and the erasing functions. Darkening
and bleaching are achieved by using light of different frequencies;
therefore, an optimum reading or probing frequency has to be found,
which does not affect the image contrast. The darkening of photochromic glass is caused by the generation of color centers that
absorb energy in the visible spectrum. If the glass is exposed to
red light, the reverse action takes place: the darkened parts of
the image become clear and regain their original transparency. The
large variety of photochromic glasses that have been developed by
Corning Glass Works show a peak response for darkening at wavelengths between 310 and 400 millimicrons and for bleaching between
530 and 630 millimicrons. Between these two wavebands for darkening and bleaching, there is a "neutral wavelength" that has little
or no effect on the image contrast. This optimum wavelength for
probing depends on the type of material and lies between 430 and
540 millimicrons. The typical permissible bandwidth for probing
is about 10 millimicrons.

The sensitivity of photochromic glass can best be described by
the light energy density at the optimum wavelength that is necessary
to obtain an optical density change of 1 dB of darkening or of
bleaching. The required energy density for darkening lies between
3 and 25 $mJ/cm^2/dB$ and for bleaching between 30 and 60 $mJ/cm^2/dB$.
These values indicate that photochromic glass is not sensitive
enough to be used as photographic material for camera applications.
It can be shown that a conventional photographic film with a speed
of photographic emulsions has, of course, been bought by the dis-

advantage of requiring a chemical development process after image formation.

The values for sensitivity, discussed above, can be employed to determine the exposure time for a given light source energy density. If an ultraviolet phosphor of a cathode ray tube is used as the darkening light source, the energy density can easily be determined if the spectral radiation efficiency of the phosphor is known. A practical example of such a device will be discussed later in Chapter: Photochromic and Cathodromic Storage Tubes.

Much higher energy densities can be attained by using lasers for the writing and bleaching function. In this case, the beam can be focused to about twice the diffraction limit which results in very large power densities. In that case, writing speeds of less than 1 microsecond per dot can be achieved (see Chapter: Random-Access Memory).

The spectral transmittance of a typical photochromic glass is shown in Figure 1. The difference between the optical density of the unexposed glass (bleached glass in Figure 1) and that of the darkened material (different degree of activation in Figure 1) is a measure for the attainable contrast. As can be seen the maximum contrast is obtained at a wavelength of about 475 millimicrons. This shows that the neutral wavelength, which lies between the darkening and bleaching wavelengths and has therefore the smallest effect on the image contrast, results also in the highest contrast ratio.

One of the most important requirements for optical recording media is the retention time of the information. R.J. Araujo [4] investigated the kinetics of bleaching and thermal fading of photochromic glass and described the process with the equation:

$$\frac{d\,(x_1 + x_2)}{d\,t} = n_1 x_1^2 + n_2\,x_2^2 + n_3 I\,(x_1 + x_2) \qquad\qquad 1.$$

where n_1, n_2, n_3 are rate constants; I is the intensity of the bleaching light and x_1, x_2 are concentrations of two different types of color centers at a certain instant of time. The last term of the right-hand side of equation (1) relates to the optical bleaching, whereas the first two terms are related to thermal bleaching. This equation can be explained if one assumes that the thermal bleaching is caused by thermal excitation of the trapped electrons (color centers), and that the optical bleaching is caused by excitation of holes.

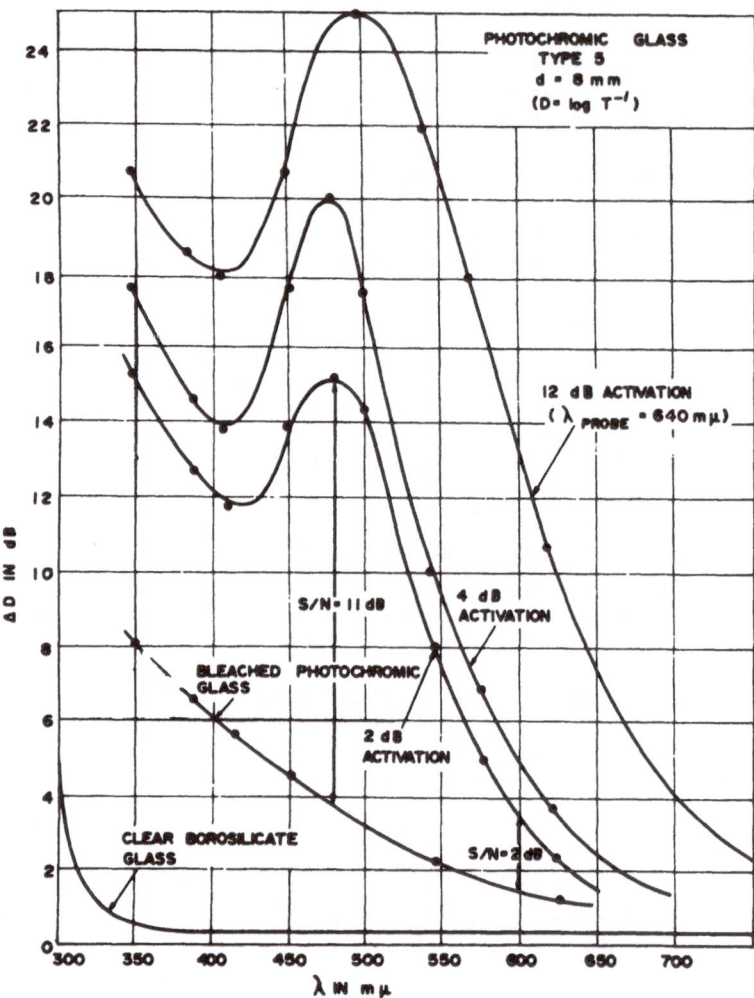

Fig. 1   Absorption Spectra of Darkened and Bleached Photochromic
Glass

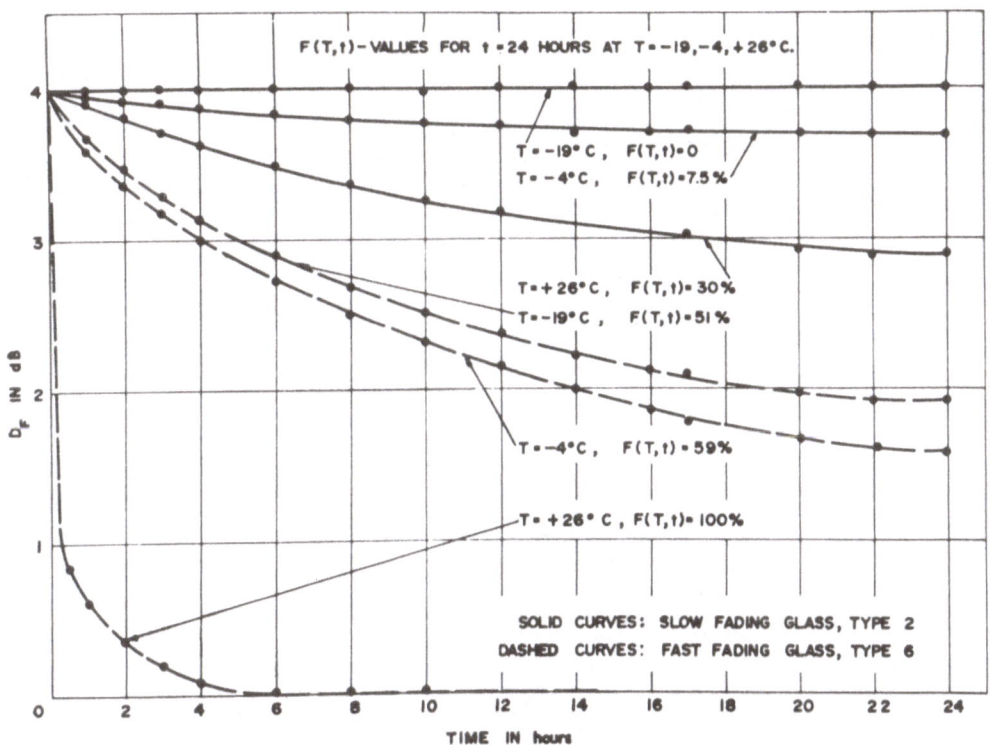

Fig. 2   Fading Rates of Slow- and Fast-Fading Photochromic Glass at
Low Temperatures

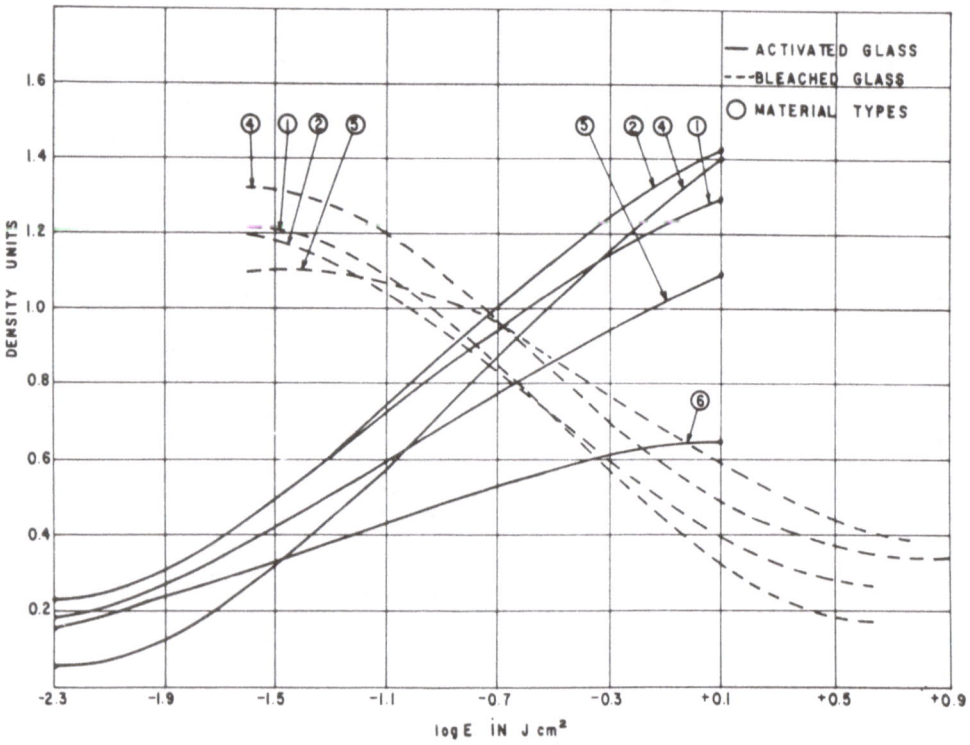

Fig. 3 D-log E Curves for Darkened and Bleached Photochromic Glass

Figure 2 shows the fading rate for two extreme types of photo-chromic glass. As can be seen, the fading rate decreases rapidly at lower temperature which results in a longer storage time of the information. The upper curve shows that no fading at all occurred if the material is cooled to -19°C. Other glasses have been de-veloped which retain the information at room temperature over years if they are stored in a dark environment. If, however, the glass is exposed to the probing light, the contrast gradually fades to its half value in 15 minutes to 3 hours, depending on the light energy of the probing source.

The linearity of optical density change versus the exposure energy is shown in Figure 3 for some typical photochromic glasses.

The gamma value, defined by

$$\gamma = (D_2 - D_1)/(\log E_2 - \log E_1) \qquad 2.$$

(where D_1 and D_2 are the optical densities in density units; and E_1 and E_2, the exposure energy densities) can, as for photographic materials, be used to describe the linearity of optical density change. However, one must account for these basic differences:

  a) photochromic glass has a constant gamma value since no development process after exposure is required and

  b) for photochromic materials, two gamma values exist describing the darkening and bleaching behavior.

The results of the corresponding gamma values are compiled in Table 1. The linear gamma range for photochromic glass is a direct measure of the attainable dynamic range and determines, as in nonreversible recording media, the gray scale quality.

Immunity to fatigue is an important advantage of photochromic glasses. An automatic test advice that allowed alternately darkening and bleaching showed that more than 3000,000 accumulated cycles still did not affect the photochromic properties [2]. The applied darkening energy was 60 mJ/cm^2 and the bleaching energy, 800 mJ/cm^2. Applying the same energies to organic photochromic films resulted in total loss of photochromism after 50 to 200 cycles. Although the test was terminated after 3000,000 cycles, no evidence of fatique has ever been detected in other experiments to date.

Type	ACT	BL
1	0.59	0.71
2	0.61	0.6
4	0.74	0.64
5	0.27	0.16
6	0.23	0.16

Table 1.   Gamma Numbers of Darkened (γ ACT) and
            Bleached (γ BL) Photochromic Glasses

MATERIAL CONFIGURATIONS

Bulk Photochromic Glass

The diagrams shown in Figures 1 and 3 were measured on photochromic glass plates of several millimeters thicknesses. This bulk
photochromic glass has a similar distribution and size of silver
halide crystals as the fast-fading photochromic material that is
being used for reversible sun protective eye glasses known under
the Corning Glass Works trade names of PHOTOGRAY^{TM} and PHOTOSUN^{TM}.
As will be discussed in Chapter: Random-Access Memory inherent
resolution of the glass itself is extremely good since the sizes
of the silver halide crystals that are embedded in the glassy
matrix are only around 50 to 100 Å and their center-to-center
distances are approximately 500-1000 Å . Hence, the crystal size
is considerably smaller than the crystals in photographic material
where emulsions of low resolution contain silver halide crystals
of 20,000 Å and materials of high resolution contain crystals of
around 1000 Å diameter.

The required thickness of photochromic glass prevents high
resolution if exposed to non-collimated light. Whereas the layer
thickness of organic photochromic films and photographic emulsions
are in the micron range, photochromic glass needs thicknesses in
the millimeter range to obtain comparable contrast. That means
that the lateral propagation of light, that is negligible in thin
films, becomes for photochromic glass too large for high resolution
image recording except if highly collimated light with large

f-number optics is used, as discussed in Chapter: Optical Systems
for Image Formation.

Thin Layer Photochromic Glass

One way to improve the definition of the image is, of course,
to try to decrease the thickness while retaining a high available
optical density without imparing the mechanical stability. One
method of achieving this is to introduce silver into a thin layer
of relatively thick glass by means of an ion exchange technique.
Subsequently, the silver halide in this thin layer was precipitated
by heat treatment. A sample of these thin-layer photochromic
glasses is shown in Figure 4b.

Since the photochromic layer thickness for a saturation density
of about 1.8 density units (18 db) had to be only 100 to 200
microns, a far better resolution was attained than with the bulk
glass of several millimeter thickness. The penetration depth of
bulk and ion exchanged thin layer photochromic glass can be seen
by the sideviews of Figure 4a and b. The sensitivity of these
thin- layer photochromic glasses is around 200 $mJ/cm^2/dB$ and their
reversibility is poor, but they appear to be quite useful for a
nonreversible recording medium since no additional chemical develop
ment process for contrast enhancement is required.

Another method tried was the fusing of a photochromic microsheet of about 150 micron thickness to a glass substrate. The
photochromic microsheet, that contained a much higher concentration
of silver halide crystals than in bulk glass, was prepared by employing a chilling process, developed by S.D. Stookey [5] . This
glass showed values for sensitivity and fading comparable to those
obtained with bulk photochromic glass. A sample is shown in
Figure 4c.

The effect of thin-layer material can also be achieved by
reflection from opaque photochromic glass ceramics. Some opaque
glass ceramics designed for cathodochromism also exhibited photochromism.

These materials look slightly reddish-brown with perceptible
mottling (about 1/3 millimeter grain size) when viewed with the
unaided eye.

In opaque photochromic glass ceramics the darkening causes
reduction of the reflection coefficient. For this material the
peak darkening wavelength is about 390 millimicrons with a half
sensitivity bandwidth of about 100 millimicrons. The absorption
band of the darkening extends from 400 nanometers into the infrared.
Most of the darkening fades rapidly but some remains for over an
hour.

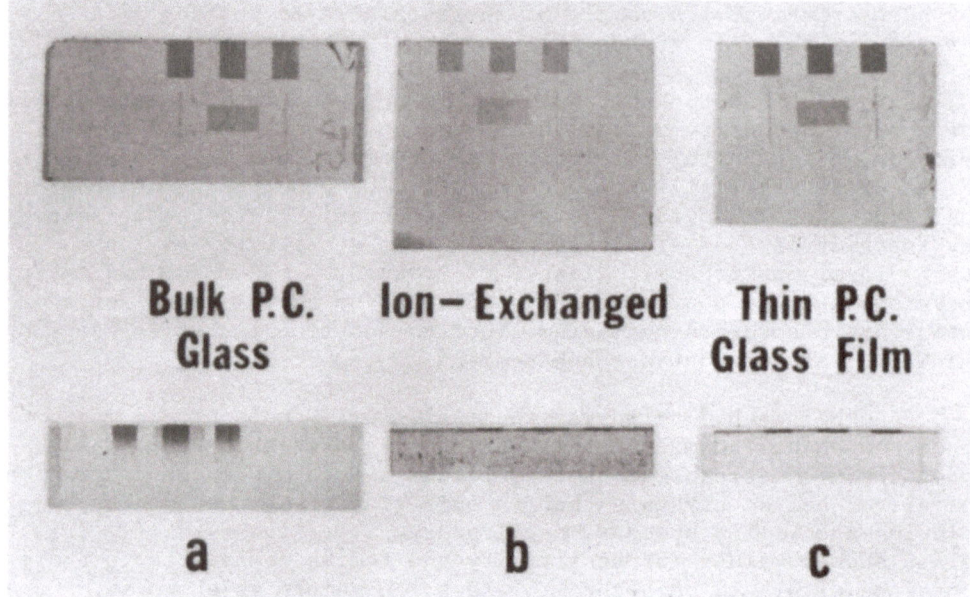

Fig. 4   Irradiation of Bulk-, Thin Layer- and Thin Film- Photochromic Glass

Photochromic Fiber Optic Plates

A concept resulting in high resolution for which sensitivity
and reversibility was the same as (or even better than) for bulk
photochromic glass was developed, employing fiber optic techniques.
Here, the recording medium is a fiber optic plate in which the
cores of each fiber are photochromic. In order to achieve high
packaging density, the cross sections of the cores and claddings
are not circular, but square shaped. The side length of the cores
is about 10 microns and the thickness of the cladding, 1 to 2
microns.

The values for the refractive index of a typical photochromic
fiber optic plate are n_{CO} =1.63 for the core material and
n_{CL} =1.53 for the cladding. For these values, the acceptance angle
given by:

$$N. A. = \sin \theta = (n_{co}^2 - n_{cL}^2)^{\frac{1}{2}} \qquad\qquad 3.$$

is a 34° semicone angle corresponding to a numerical aperture of
0.56.

The optical properties of the fiber optic plates must be evaluated for design into an optical system. Interesting phenomena
that provide much information about the optical properties of
photochromic fiber optics, are observed when the plate is illuminated by a laser beam. In Figure 5, the laser beam was applied normal
to the fiber plate and the light transmitted through the fiber
plate fell directly on a photographic plate. The central patch of
light, which is due to light entering and exiting the cores,
contains an array of light spots caused by the grating diffraction
pattern of the fiber array. The outer ring of light is the light
which entered claddings and exited cores at the fiber trapping
angle. The light in the cross is due to the diffraction pattern of
the square exit apertures of the cores.

In Figure 6, the fiber plate was tilted about ten degrees.
The central portion has become four separate light patches, each
with its own cross, and the exit angle of the ring has increased.
The central portion constitutes the useful light for system applications. Figure 7 shows the relation between the entrance beam
and the four exit beams which occurs with square shaped fibers.
The incident angle can be resolved into the components θ_H and θ_V
in the planes of the fiber walls. The exit directions are at the

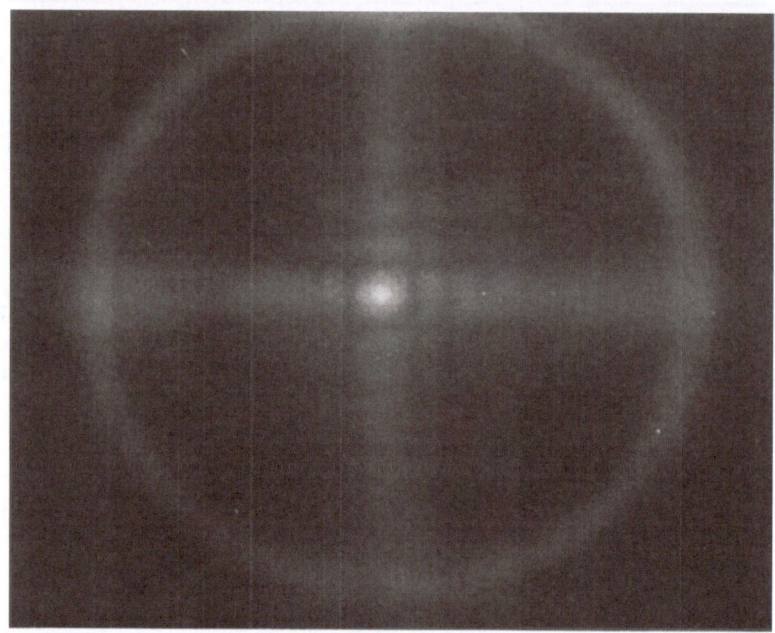

Fig. 5   Diffraction Pattern Caused by Normal Incidence of Laser
Beam on Photochromic Fiber Optic Plate

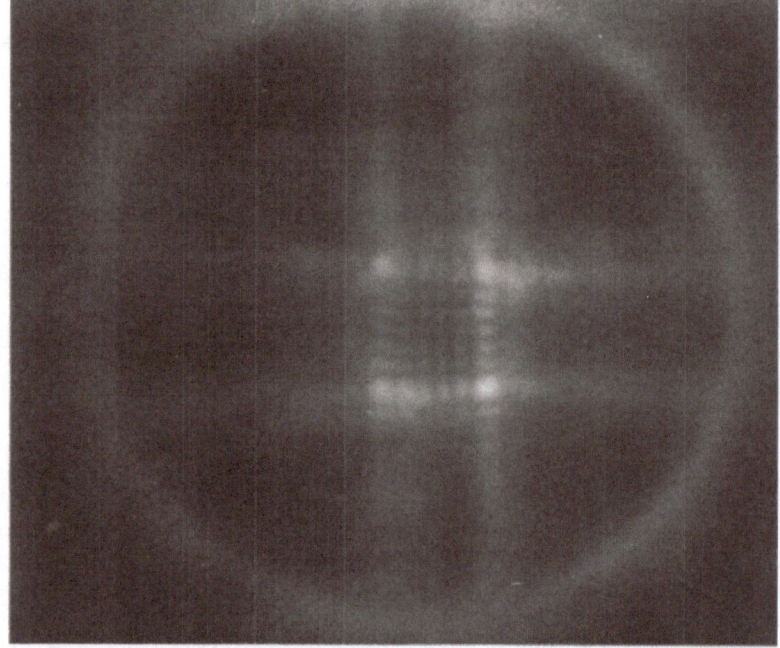

Fig. 6   Diffraction Pattern Caused by 10° Tilted Photochromic
Fiber Optic Plate

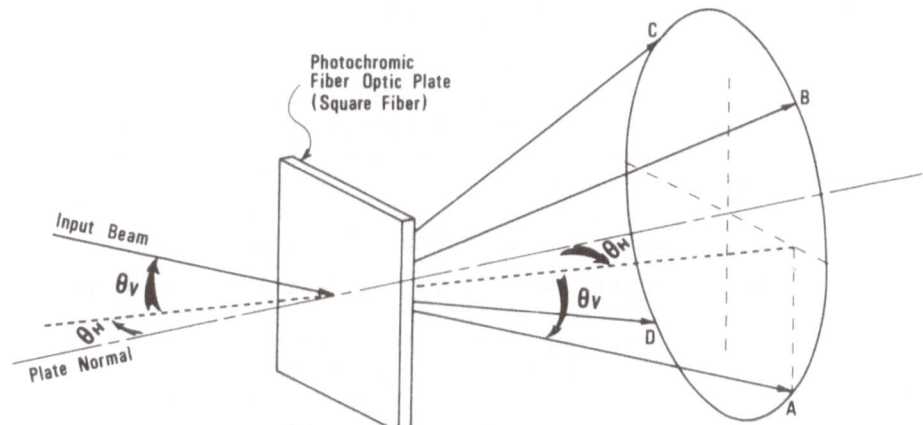

Fig. 7 Exit Direction of Core Light Incident on Fiber Optic Plate

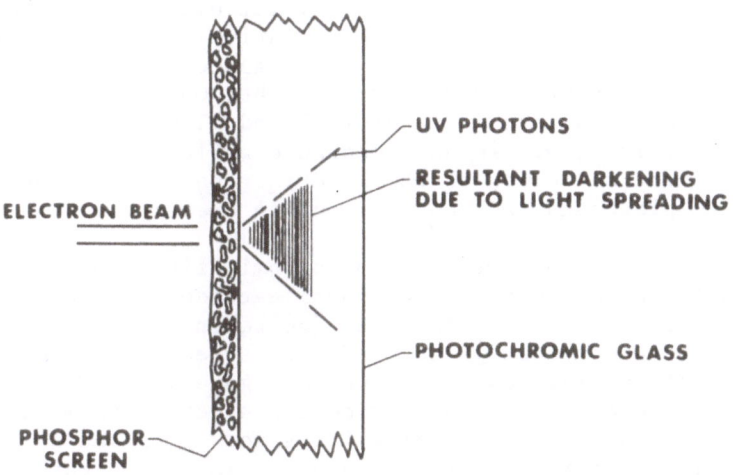

Fig. 8 Light Spreading in Bulk Photochromic Glass

four directions given by $\theta_H\theta_V$, $\theta_H-\theta_V$, $-\theta_H-\theta_V$, and $-\theta_H+\theta_V$, indicated by rays A, B, C, and D, respectively. It will be noted that
ray A, in effect, passes straight through the plate. For thick
plates or deformed fibers the four exit directions merge into a
ring passing through points A, B, C and D.

By making both the light source and projection lens pupil
telecentric with respect to the fiber optic plate, light spreading
and projection lens size are both minimized while light uniformity
in the projected image is optimal (see also Chapter: Optical
Systems for Image Formation).

Photochromic and Cathodochromic Storage Tubes

A photochromic fiber optic plate, as described in Chapter:
Photochromic Fiber Optic Plates can be used as a target in cathode
ray tubes. In that case, an ultraviolet energy emitting phosphor
deposited on the plate will, when irradiated by the electron beam,
darken the photochromic cores. In bulk photochromic glass, the
Lambertian propagation pattern generated by the phosphor would
cause ultraviolet spreading which results in an image of poor resolution as indicated in Figure 8. However, by using the described
photochromic fiber optic, the ultraviolet energy emitted from the
phosphor is trapped and confined to a few cores as shown in Figure
9. Consequently, the spreading is reduced and resolution significantly improved. Since the overall efficiency depends on the
numerical aperture of the fiber plate -- which means the larger
the acceptance angle, the more ultraviolet emitted from the phosphor that will enter the fibers -- a relatively large difference
between the refractive indecies of core and cladding is required.
As mentioned in Chapter: Photochromic Fiber Optic Plates, the
values of the refractive index of 1.63 and 1.53 for core and cladd
ing respectively result in an acceptance angle of 34° for which
Fresnel reflection losses at the interface between the photochromic
fiber optic plate and the phosphor screen are small.

A cathode ray tube with a photochromic fiber optic plate has
an obvious advantage over conventional cathode ray tubes: the
image has only to be generated once (no regenerating is necessary)
to obtain a flicker-free picture. A cross-section of such a
"storage" cathode ray tube is shown in Figure 10. This tube is
used as a storage and display device for a remote terminal that is
now being manufactured by Corning Glass Works [6] and is briefly
described in Chapter: Computer Driven Display. The cathode ray
tube shown in Figure 11 has as target a photochromic fiber optic
plate of 4" x 4" active area that is 1.5 millimeters thick on
which an especially developed ultraviolet phosphor, radiating with
maximum efficiency at a wavelength of 350 millimicrons, is deposit
ed.

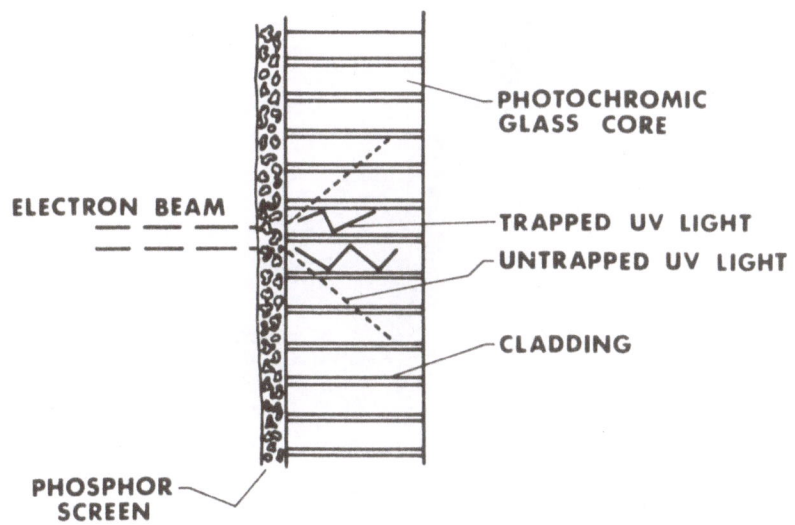

Fig. 9 Light Trapping in Fiber Optics with Photochromic Cores

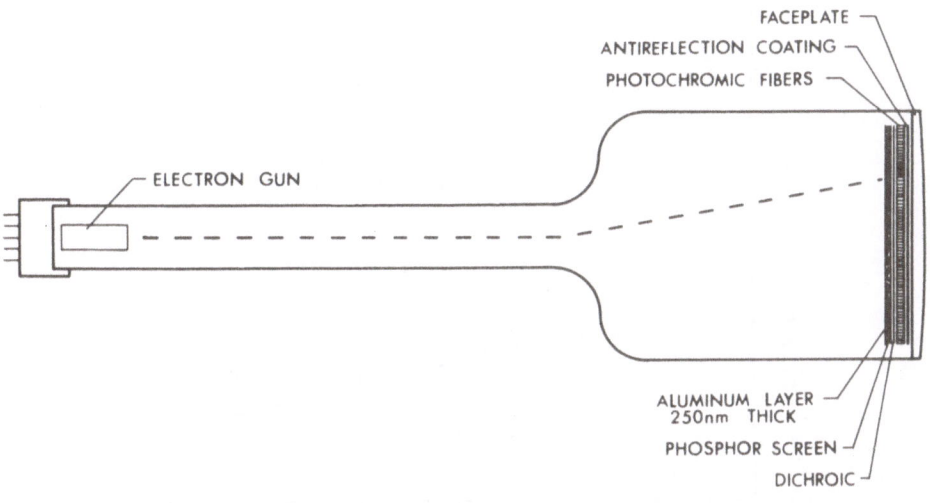

Photochromic CRT

Fig. 10 Cross Section of Storage-Display Cathode Ray Tube with
Photochromic Fiber Optic Target

Fig. 11   Storage-Display Cathode Ray Tube with 4" x 4" Photochromic
          Raget

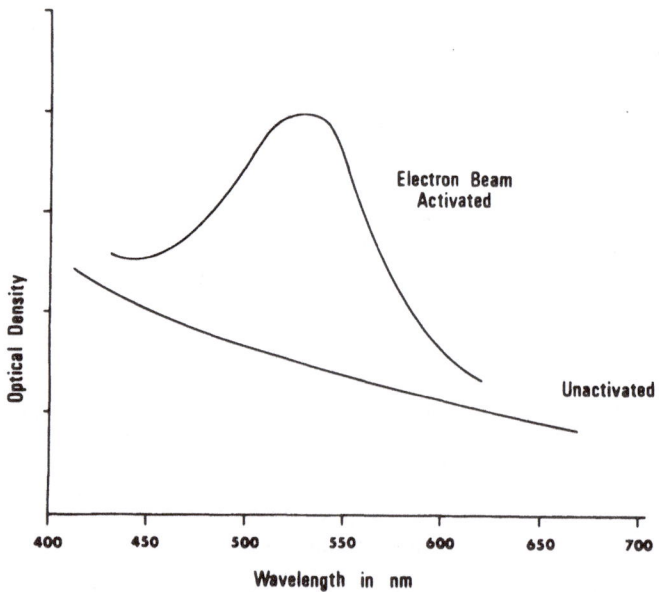

Fig. 12   Absorption Spectrum of Cathodochromic Material

Similar storage and display cathode ray tubes have been developed in the United States that use as the target, cathodochromic materials of the sodalite family such as $Na_4Al_3Si_3O_{12}Cl$ or titanite hackmanite (a sulfur containing sodalite)(7). It is believed that the change in color under electron bombardment is the result of transferring electrons to vacancies in the crystaline lattice of the material, such as a chlorine vacancy in the sodalite lattice. Increased optical absorption usually occurs in the visible part of the spectrum as shown by Figure 12.

The new materials allow longer viewing times under high illumination, and can be used in a projection made similar to that used with a photochromic glass cathode ray tube, the concept of which will be described in next Chapter,Figure 24. Although cathodochromic screens can be erased optically, heat is normally being applied for this purpose. In one technique, the screen is heated by radiation from a heating coil (inside the tube) surrounding the screen (Figure 13).

The cathodochromic material is either evaporated on the viewing substrate or it is deposited onto it by techniques similar to normal phosphor deposition. In both cases, the screens are opaque.

The major difference in utilizing cathodochromics and photochromics in cathode ray tubes is that photochromics require an intermediary phosphor screen, whereas cathodochromics are directly activated by the electron beam. Since typical ultraviolet-emitting phosphors are only about 4-5% efficient, less energy is available for writing in the photochromics than is available for the cathodochromic materials resulting in about equal writing speeds for both materials. Typical writing times and other key parameters are shown in the Table 2 for both the cathodochromic and photochromic cathode ray tubes. As can be seen, the performance differences between the two concepts are negligible.

Cathodochromic material can only be used in conjunction with cathode ray tubes, whereas photochromic glass can be used for any image formation if light sources of proper wavelengths are available. The optical erasure of images on photochromic glass as mentioned in Chapter: Material Properties, is, of course, preferable to the thermal erasure mode of cathodochromic material since thermal erasure shortens the life-time of the tube.

OPTICAL SYSTEMS FOR IMAGE FORMATION

Since a number of applications are emerging that use photochromic glass, not necessarily as targets in cathode ray tubes as described in Chapter: Photochromic and Cathodochromic Storage Tubes, some details about the optical systems necessary for the writing and reading fuctions for these applications will be discussed.

Table 2.   Comparison of Photochromic and Cathodochromic
Cathode Ray Tubes

STORAGE DEVICE	RESOLUTION	TYPICAL NO. OF CHARACTERS ON DISPLAY SCREEN	VIEWING TIME ——— RETENTION TIME	WRITING TIME PER DOT	BULK ERASE TIME
PHOTO-CHROMIC STORAGE CRT	13 LINES PER MM	>4000	15 MIN-1 HR ——— WEEKS	300 µsec.	4-6 sec.
CATHODO-CHROMIC STORAGE CRT	13 LINES PER MM	>4000	15 MIN-1 HR ——— WEEKS	300 µsec.	3-6 sec.

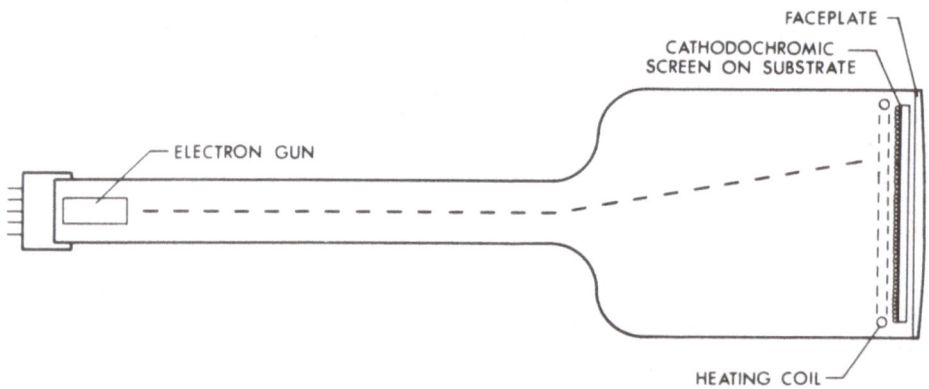

Cathodochromic CRT

Fig. 13   Cross Section of Storage Display Tube with Cathodochromic
Target

In general, good resolution of a stored image in bulk photo-chromic glass is possible provided two conditions are satisfied:

a) the image readout light should follow the same path through the material as that traversed by the recording light and

b) the f-number of the darkening ultraviolet should be approximately duplicated by the f-number of the projection light.

The actually attainable resolution depends on the thickness of material and the f-number of the optical system.

Figure 14 shows the basic scheme for image recording in bulk photochromic glass. The apparent distance from the photochromic glass plate to the darkening and readout lens pupils, as well as the apparent pupil sizes, should be identical. It should be noted that the apparent pupil position and sizes are as viewed from the photochromic material. Consideration of storing and readout of a pinhole image satisfies the two rules mentioned above. The higher ultraviolet refractive index requires the readout pupil to be slightly further away from the glass than the darkening pupil; i.e., the apparent pupil distances should be identical when viewed from inside the photochromic glass. Using opposing directions of light propagation for darkening and readout evades problems caused by same way propagation (Figure 14).

The same type of considerations, and nearly the same numbers, apply to determine the actual resolution attainable as for the spot size in bulk photochromic glass digital memories, such as discussed in Chapter: Random-Access Memory. To achieve high resolution in thick material, high f-number optics are necessary which reduce the speed of darkening by reducing the available ultraviolet inten-sity at the photochromic material as compared to low f-number (higher "speed") optics. In order to generate dots of 35 microns diameter in 2 millimeter thick photochromic glass, an f-number of 35 is required.

There are many possible variations of the basic scheme of Figure 14. If the object were a transparency, it could be contact printed in the photochromic glass. The transparency would be transferred from the indicated object position to the surface of the photochromic glass. The rest of the ultraviolet optics remains as shown with the projection lens acting as a spatial filter to control the f-number of the beam (for high f-number systems, a simple pupil could be used instead of the projection lens). Contact printing involves some loss of resolution as compared to imaging because the sharpest focus occurs at the photochromic glass surface instead of in the center of the bulk material.

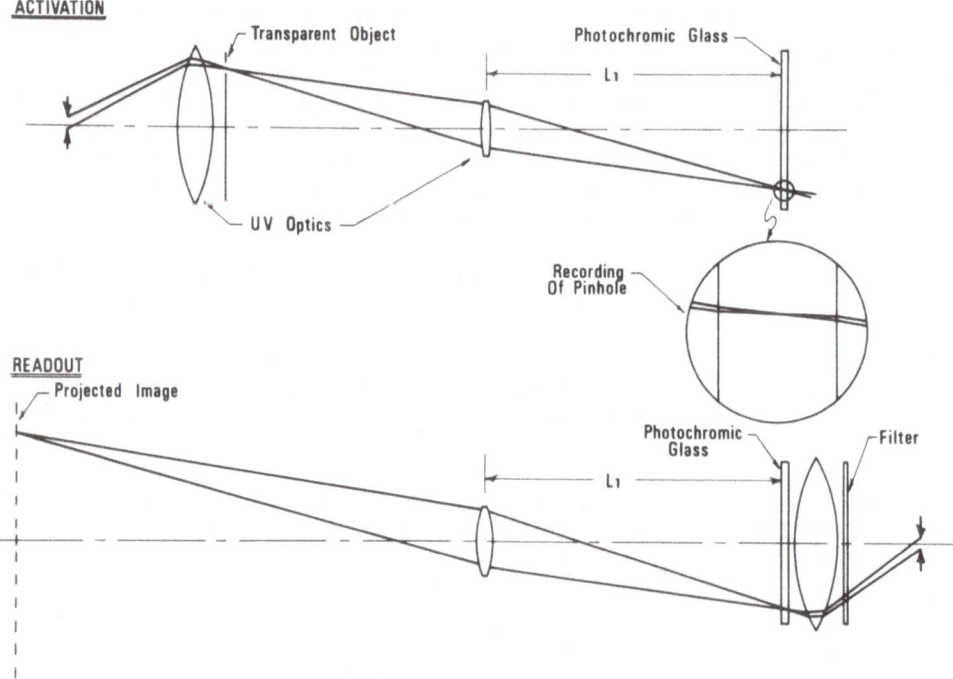

Fig. 14   Basic Scheme for Image Recording in Bulk Photochromic Glass

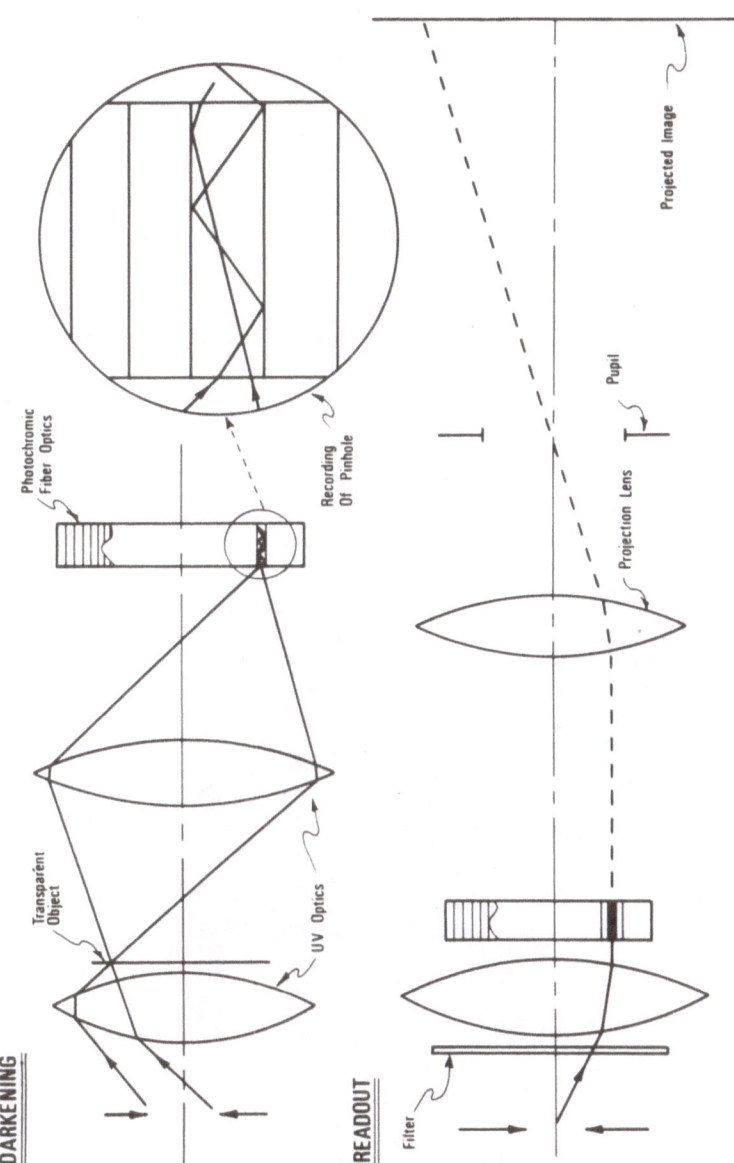

Fig. 15 Basic Scheme for Image Recording in Photochromic Fiber Optics

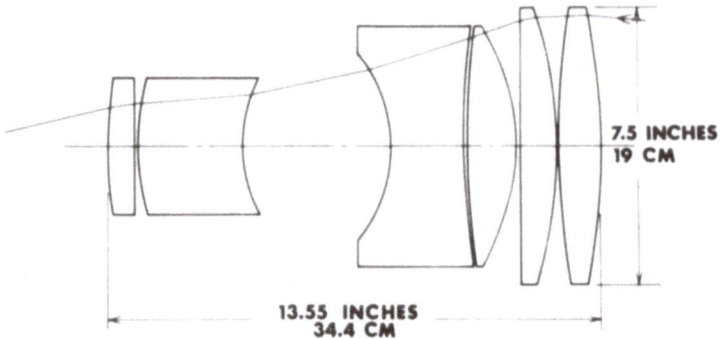

7.5 INCHES
19 CM

13.55 INCHES
34.4 CM

Fig. 16 Cross Section of Large Telecentric Lens

Fig. 17 Telecentric Lens

If doublepass readout is to be employed to increase the density available, then darkening and readout should be telecentric as far as the photochromic material is concerned.

For higher darkening speeds, photochromic fiber optic plates as described in Chapter: Photochromic Fiber Optic Plates, could be used. In that case, the resolution is determined by the fiber size but not by the f-numbers in the optical systems.

Figure 15 shows the basic scheme for image recording in photochromic fiber optic plates. The transparent object is imaged onto the surface of the fiber optics. Ultraviolet energy arriving on a fiber surface propagates down only that fiber, darkening its core. This trapping action is not dependent on the length of the fiber, allowing the thickness of the fiber plate to be chosen without affecting resolution. It is advisable to read out with a telecentric optical system such as shown in principle in Figure 15. The lamp source is placed in the focal plane of the condenser lens so that the light incident on the photochromic fiber optic plate is parallel to the optic axis (i.e., the light source appears to be at infinity). The light passes through the fibers with few reflections if the fibers are reasonably free from skew (the fibers in practically constructed plates are within 1° of parallel to the plate normal). The transmitted light passes through the projection lens which is provided with a pupil at its focal plane. This pupil appears to be at infinity and of constant angular size when viewed from the fiber optic plate so that the light intensity in the projected image is uniform. If uniformity is not critical, the pupil may be omitted, but the projection lens should be large enough to avoid serious vignetting at the maximum field height of the fibers. The cross section of a 7.5 inch diameter telecentric projection lens operating with a 4 inch square photochromic fiber optic plate is shown in Figure 16 and the constructed lens in Figure 17.

Fiber skew produces distortion between the image applied at one surface and the image seen at the other surface. To avoid this distortion for precision systems, the plate surface that receives the ultraviolet for the purposes of image forming should be arranged to face the projection lens for readout.

The following points are important for image storage in photochromic fiber optic plates:

a) The ultraviolet incident angle must be within the trapping angle of the fibers. The trapping angle is typically 35° or greater so that the lower limit of darkening f-number is determined by design of the lens needed.

b) Image darkening speed is proportional to the inverse square
of the f-number; thus, an f/2 fiber optic system is 300
times faster than an f/35 bulk plate system of comparable
resolution as mentioned earlier in this section.

c) Ultraviolet energy incident on the fiber cladding (about
20% of the total ultraviolet), rather than on the fiber
core, is untrapped and tends to reduce contrast and resolution somewhat.

APPLICATIONS

Hologram Storage in Photochromic Glass

Since the previous day of this convention was dedicated to
holography, the question of storing holograms in photochromic glass
will be discussed. Although such storage of holograms is experimentally possible, as shown by various laboratory demonstrations,
a more realistic appraisal of the practically possible applications
seems to be in order.

One of the major advantages of photochromic glass is its ability to change the recorded data without replacing the material or
without any chemical development process. However, since the mate
rial gives absorption primarily (phase holograms have never been
measured), the diffraction efficiency is only about 3.7% even with
very thin material [8] . The reconstruction diffraction efficiency of photochromic bulk glass is about ten times lower than thick
absorption type emulsions. Since for reconstruction the reference
angle and the wavelength must be the same as used for generating
the hologram, angle and waveband tolerances are very small for
thick photochromic glass. For a 3 millimeter thick plate, deviations of only 10 minutes of an arc and 100 Å in reconstruction
wavelength are allowable. Therefore, the reconstruction light
exposure either bleaches or darkens according to whether the hologram was originally bleached into darkened glass or was darkened
into bleached glass.

Although multiple recordings of up to 100 different holograms
in thick photochromic glass have been reported, the earlier recordings were seriously degraded by later recordings. In fact,
each recording uses some of the available dynamic range (Figure 1)
so that weak exposure for each recording must be used.All recordings are noise except for the one that is being reconstructed.
Therefore, only a limited number of holograms in photochromic
glass are practically possible because of the reduced modulation
depth and the increased noise.

Especially for this application, the large energy density
required for darkening of 30 mJ/cm^2 for 3 dB or for bleaching of
300 mJ/cm^2 for 2 dB reduction from 4 dB optical density constitutes
the major disadvantage. For comparison, even a very high resolution

photographic film such as 649F from Eastman Kodak has a sensitivity of about 17 mJ/cm^2/dB.

Having discussed the difficulties and disadvantages of using photochromic material for hologram storage, it may have become clear that the presently available photochromic glasses do not offer any real advantages for this application. This position may change if very thin photochromic material with, for instance, some electromagnetic development process (light or electrical field) become available.

Image Modification

In addition to storage of images, photochromic glass can be employed to modify images. One application might be to compress the dynamic range by projecting the image onto photochromic glass with ultraviolet energy and using a high f-number light source on the image side of the projection lens to retain good resolution. Another image modification is to achieve contrast enhancement. A low contrast object may be regarded as the superposition of a weak information object on top of a strong non-information containing background. Contrast enhancement is achieved by weakening the strong non-information background relative to the information object. The optics of a basic contrast enhancement system are shown in Figure 18. The same optical system is shown twice, once with the energy or source imaging rays shown and once with the information imaging rays shown. In both cases light, including some ultraviolet, is supplied through the pinhole and condenser lens to illuminate the transparency with parallel light.

The parallel light passing straight through the low contrast transparency corresponds to the strong non-information background and is brought to a focus at a single point in the photochromic glass. At this focus, the ultraviolet intensity is high and darkening of the glass occurs at this point. Some small portion of the non-information light passes through the dark spot in the photochromic glass to the second lens where it is made parallel and provides a low level of non-information background light at the image plane.

Information in the transparency image causes diffraction light scattering at all points of changing density. The scattered light is made parallel by the first lens so that the scattered light passes through the photochromic glass plate at full strength, bypassing the darkened spot. The ultraviolet content in the weak image is small and spreads over a large area of the photochromic glass plate so that the intensity is very low. The information image light passing through the photochromic glass plate is focused onto the image plane by the second lens. Consequently, the non-information light in the image is reduced relative to the information bearing light giving contrast enhancement.

Figure 19 is a photograph showing an example of the operation of this scheme. Figure 19a is the original image of nylon threads and air bubbles in plastic; Figure 19b shows the image after enhancement.

There are variations of this system possible, but the essence of the concept can be summarized as follows:

a) The light supply pinhole is focused onto the photochromic glass by the combination of the condenser and first lens (light need not be parallel at the transparency).

b) The transparency is imaged onto the image plane by the combined action of the first and second lens (light need not be parallel between these two lenses).

c) Any slight tendency for the photochromic glass, not at the pinhole image, to darken can be balanced out by bleaching the glass slightly with red light.

d) The extent of the portion of the spatial spectrum selectively reduced by the photochromic glass is determined by the size of the pinhole and f-number of the light at the photochromic glass.

Finally, it will be appreciated that the described system will increase the contrast of low contrast phase objects, as well as absorption objects, because phase objects scatter light.

Random-Access Memory

In order to assess the feasibility of photochromic glass to optically record digital information in the form of small darkened or bleached dots, the inherent resolution and the attainable two-dimensional storage density shall be discussed.

The definition capability of this material is far better than any other silver halide optical recording medium known, since the size and spacing of the silver halide crystals is very small as described in Chapter: Bulk Photochromic Glass.

The resolution of photochromic glass was determined by measuring its modulation transfer function using the two-dimentional hologram technique. After producing an interference pattern with two light beams (coming from the same laser) in the photochromic glass, one of the illuminating beams was blocked and the reconstruction (Bragg angle reflections of the stored fringe pattern) showed the image of the obscured light source. With this method and the

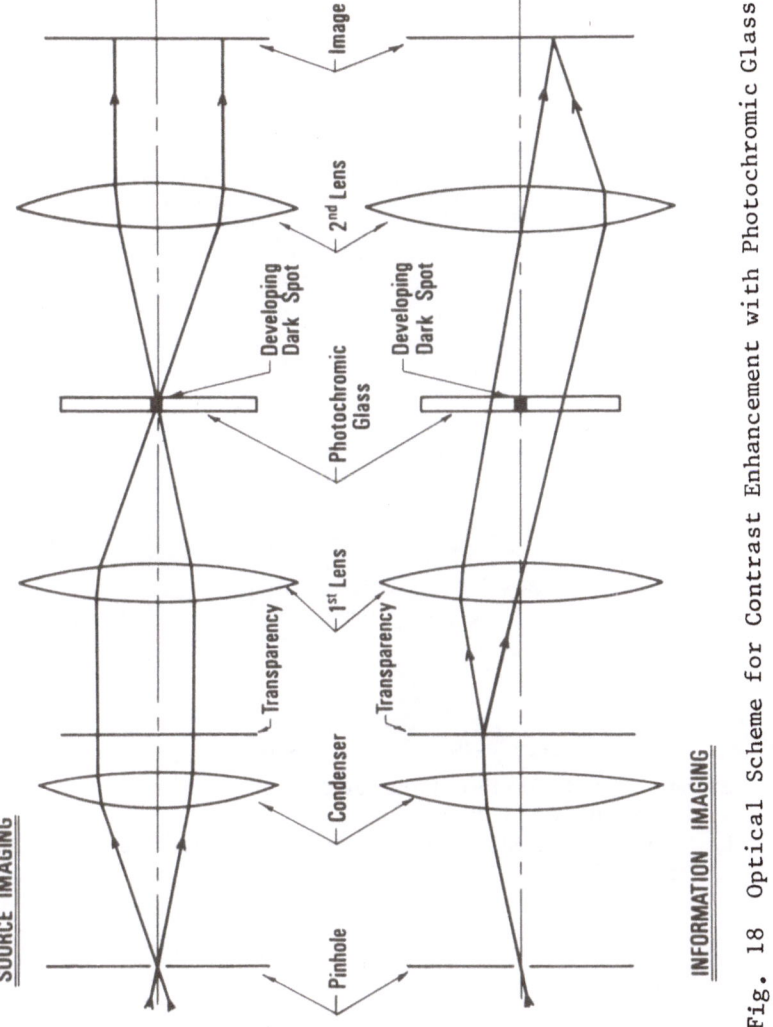

Fig. 18 Optical Scheme for Contrast Enhancement with Photochromic Glass

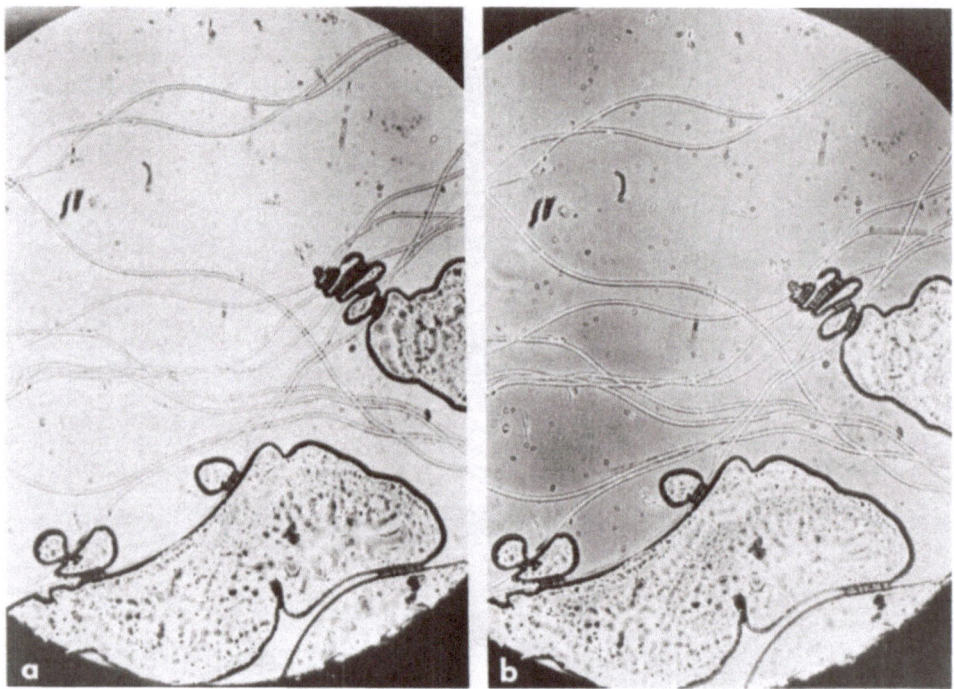

Fig. 19   Example of Contrast Enhancement
          a.   Conventional Photograph
          b.   Photographed with Contrast Enhancement Using Photochro
               mic Glass

relationship

$$f_s = (2/\lambda) (n_1/n_2) \sin \theta_1 \qquad\qquad 4.$$

(where θ_1 is the incident angle between the normal to the glass plate and the laser beams, n_1 is the refractive index of air, and n_2 is the refractive index of photochromic glass), the special frequency of the fringe pattern was measured to be $f_s = 2100$ cycles per millimeters for $\theta_1 = 80°$, $n_1 = 1.00$, $n_2 = 1.486$ and $\lambda = 633$ millimicrons [2].

However, this very high resolution cannot practically be exploited when information is stored digitally in photochromic glass, because of the necessary finite thickness of the glass and because of wave-optical limitations. Some details about the recording of digital information in photochromic glass are therefore discussed in the following paragraphs.

The optical power density obtainable from a laser light source is 4 to 5 orders of magnitude larger than that obtainable from the most intense conventional sources. Thus, in applications requiring large optical memories, the laser is the logical choise as a light source.

The transverse and axial power density distribution of laser radiation is determined by the normal modes of electromagnetic oscillation of the laser optical resonator. Most lasers are designed to oscillate in the fundamental TEM_{00} mode. Equation (5) gives the axial and transverse power density distribution (in air) of such a laser beam which is focused to obtain a small, intense spot.

$$I(u,v) = I_0 \frac{e^{-v^2/(1+u^2)}}{1 + u^2} \qquad\qquad 5.$$

where

$$U = \frac{K_0 Z}{8 F_0^2} \qquad\qquad V = \frac{K_0 r}{2\sqrt{2} F_0} \qquad\qquad K_0 = \frac{2\pi}{\lambda_0}$$

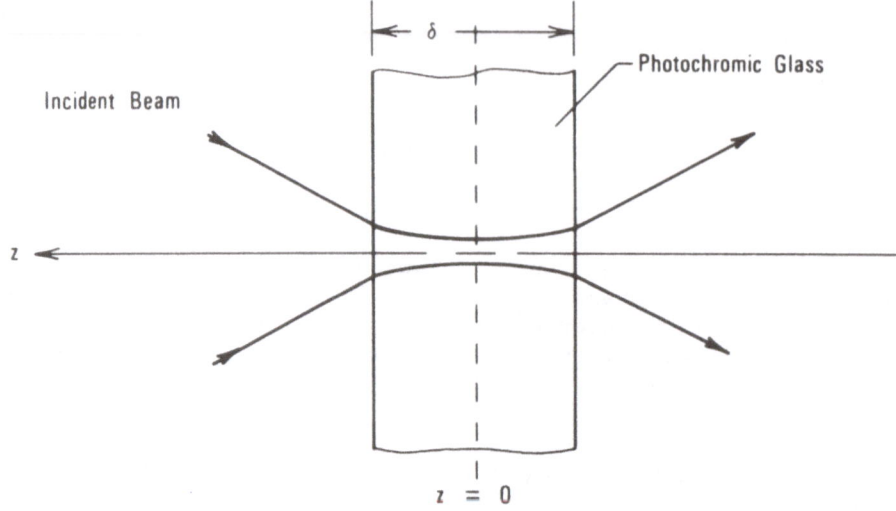

Fig. 20   Laser Beam Focused in Glass

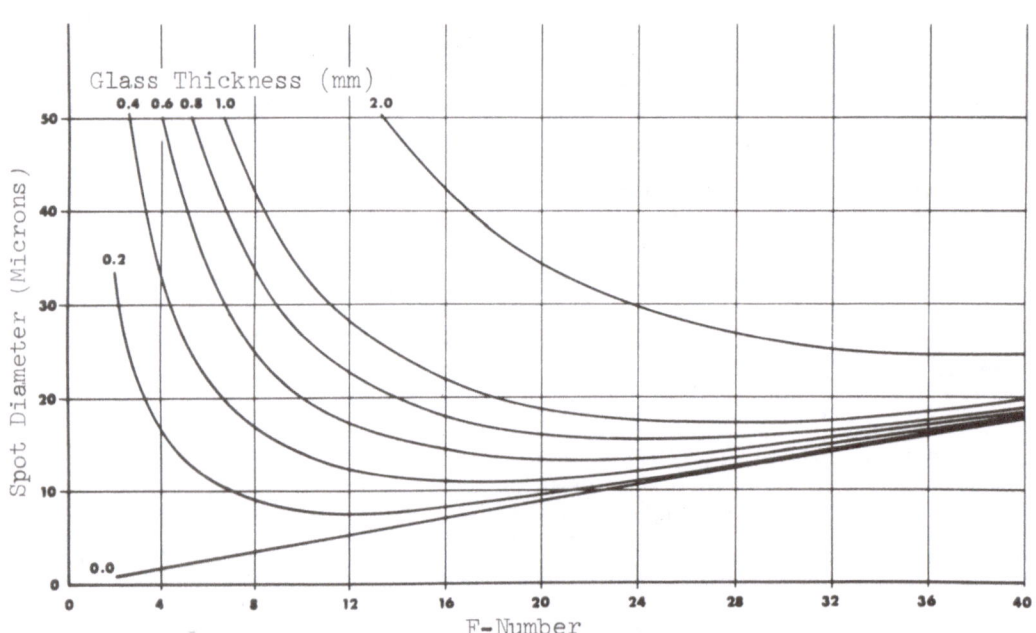

Fig. 21   Spot Diameter, d_0, vs. F_0 and Glass Thickness, with
          n = 1.5 and λ_0 = 0.35 Microns

F_O = f-number of the beam, and r,z are the usual cylindrical
coordinates, and all quantities are measured in air. From equation
(5), it follows that in the focal plane (z = 0) the e^{-2} diameter
of the Gaussian spot is 1.27 λ $_oF_O$ and contain 86.5% of the total
beam power.

In order to investigate the possibility of recording digital
information in bulk photochromic glass for a random-access optical
memory system, the irradiation of a plane parallel plate of photochromic glass of thickness δ by a focused laser beam has to be con
sidered. As indicated in Figure 20, the focal plane is located
at the midpoint of the glass plate (z= 0), so the irradiated volume
is minimized. The spot diameter d is defined as the e^{-2} diameter
of the beam at the surface of the glass. From equation (5), this
diameter is determined with

$$d = \frac{4 \lambda_o F_o}{\pi} \left(1 + \frac{\pi^2 \delta^2}{64 n \lambda_o^2 F_o^4} \right)^{\frac{1}{2}} \qquad\qquad 6.$$

where the refraction from air into the photochromic glass of refrac
tive index n is taken into account.

The spot diameter d is plotted in Figure 21 as a function of
F_O with δ as a parameter, using n = 1.5 and λ_o = 350 millimicrons
which corresponds to an available spectral laser line suitable for
activation of photochromic glass. Note, that in order to achieve
a spot diameter of 10 microns or less, δ must be less than 0.4
millimeters. Also, for fixed δ , there is an optimum value of F_O
for which the spot diameter is a minimum.

In order to allow sufficient darkening, most glasses require
a thickness of 0.5 to 1.0 millimeters; thus, the corresponding spot
diameters range from 12 to 17 microns. As mentioned in Chapter:
Thin Layer Photochromic Glass, thin photochromic glass with a thick
ness of 0.125 millimeters and with satisfactory saturation density
has been made. This would allow a spot diameter of about 6 microns.

If one takes 12 microns as an attainable spot diameter, then
the storage density for memory applications is about 7×10^5
bits/cm^2. Hence, for a 10 x 10 centimeter recording plane, the
total capacity is 35 megabits, allowing a factor of two decrease
for guard bands, etc. The capacity of present main high speed
internal computer memories is in the range of 1-10 megabits.

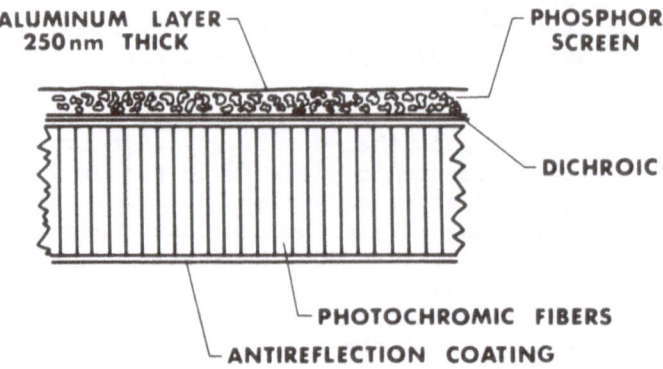

Fig. 22   Target Construction for Cathode Ray Tube

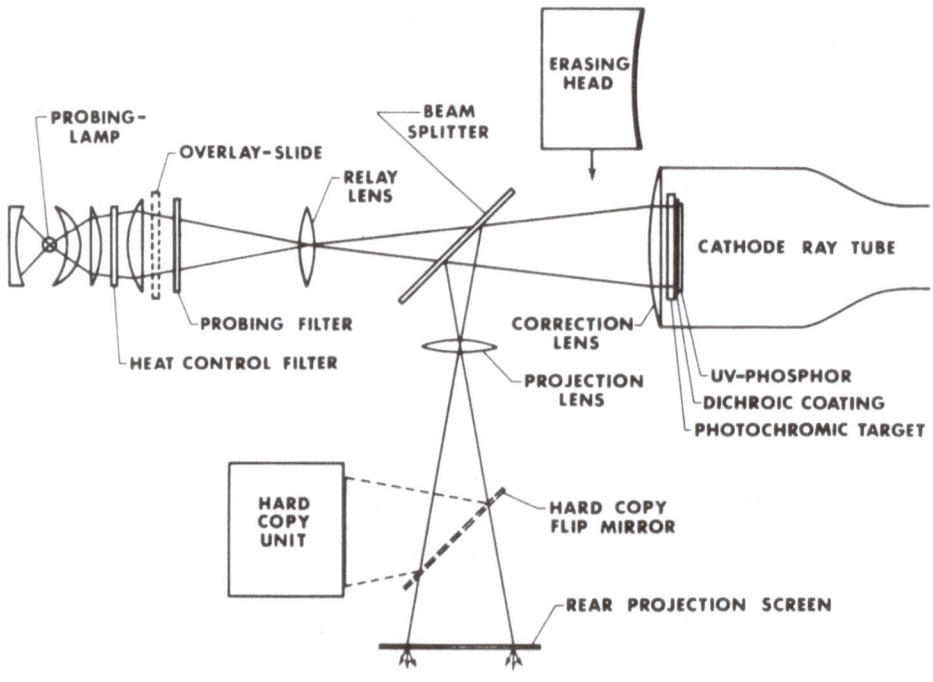

Fig. 23   Display Schematic of an Interactive Remote Terminal

In order to assess the feasibility of using bulk photochromic glass as the storage medium in a high speed random-access computer, it has to be determined if such a memory has performance characteristics that are comparable to present ferrite-core memories such as: capacity 3 x 10^6 bits, read-write cycle time 0.5 microseconds, and data rate 2 megaherts.

Considering the simple optical recording concept in which the storage medium is a plane parallel plate of photochromic glass, which is accessed by a two-dimentional digital deflector, each output position of the deflector corresponds to a bit location. It shall be assumed that the digital deflector consists of two cascaded Bragg cell acousto-optic deflectors, one for each deflection direction. No other deflector type is suitable for this application.

A photochromic glass of 100 microns layer thickness (Figure 21) shall be used. It will be fully optically bleachable, needing an energy density of 30 mJ/cm^2 for 3 dB of darkening and 600 mJ/cm^2 for 3 dB of bleaching. Applying maximum available laser power of 100 milliwatts at λ = 350 millimicrons and 2 watts at λ = 625 millicrons, and assuming an optical system efficiency of 50%, a write time per bit of 0.07 microsecond and an erase time per bit of 0.1 microsecond is available. The read-write cycle time of photochromic memory for "bit-at-a-time" organization with sequential erasing and writing is the sum of the access, erase, and the write times per bit. The read-only cycle time is the sum of the access and read times per bit and will be less than the read-write cycle time. Thus, the required deflector random-access time is given by τ = (0.5 - 0.07 - 0.1) = 0.33 microsecond or less, and the resolution of each of the one-dimensional deflectors must be N = $(3 \times 10^6)^{\frac{1}{2}}$ = 1730 or more

The characteristic equation for Bragg cell deflection of a Gaussian laser beam is

$$N = \frac{\Delta f \tau}{1.27} \qquad\qquad 7.$$

where Δf is the acoustic bandwidth. With equation (7), the required bandwidth is

$$\Delta f = \frac{(1.27)(1730)}{33 \times 10^{-8}} = 6.7 \text{ GHz}$$

A good fractional bandwidth for acoustic transducers is
$\Delta f/f_O = 0.5$, where f_O is the acoustic center frequency. Hence,
with $\Delta f = 6.7$ GHz, one gets $f_O = 13.4$ GHz as the required acoustic
center frequency for the Bragg deflector. Although such high
frequencies have been generated, transducers are difficult to
fabricate and bond, transducer conversion efficiency is low, and
acoustic attenuation is high in suitable acousto-optic materials.
Deflector limitations, then, make the simple concept of a fast
random-access memory presently unattractive.

Computer Driven Display

One of the major applications of photochromic glass has been
described as the storage display cathode ray tube in Chapter:
Photochromic and Cathodochromic Storage Tubes. If a dichroid
layer is applied on the photochromic fiber optic plate as indicated
in Figure 22, the image on the target of the cathode ray tube can
be projected onto large screens. Thus, a new concept for projection cathode ray tubes is offered because the brightness on the
display screen does not depend on the phosphor brightness; the
ultraviolet phosphor is only used for generating the image and an
auxiliary light source determines exclusively the screen brightness. The second advantage lies in the fact that the storage
cathode ray tube has to be addressed only once and it stores the
information until it is erased. Therefore, the large bandwidth
required for conventional cathode ray tubes, which is caused by
the fast phosphor decay time, is no longer necessary (9). A
flicker-free display without regenerating the image 30 to 60
times a second is therefore possible.

A storage cathode ray tube as an integral part of an optical
projection system is being used for an interactive computer-driven
terminal that can be used also in time-sharing systems. Its functional performance is explained with Figure 23. The characters
on the photochromic fiber optic plate are formed by a matrix of
5 x 7 dots and the projection light passes twice through the photo
chromic target, doubling the contrast. As indicated in Figure 23,
the faceplate of the storage cathode ray tube is a plano-convex
lens that is a part of the image projection system and directs all
the light from the probing source to impinge normally onto the
photochromic fiber optic target. The front surface of the target
is imaged through a beam splitter onto the rear-projection screen
of $8\frac{1}{2}$ x 11 inch size. Characters measuring 0.5 x 0.7 millimeters
on the photochromic fiber optic are enlarged 4.8 times by the
projection system and appear with 2.4 x 3.4 millimeter size on the
screen. A position pointer (cursor) provides accurate interaction
between the keyboard command switches and displayed data.

Information is erased from the photochromic fiber optic by
moving a bleaching head in front of the faceplate. The bleaching

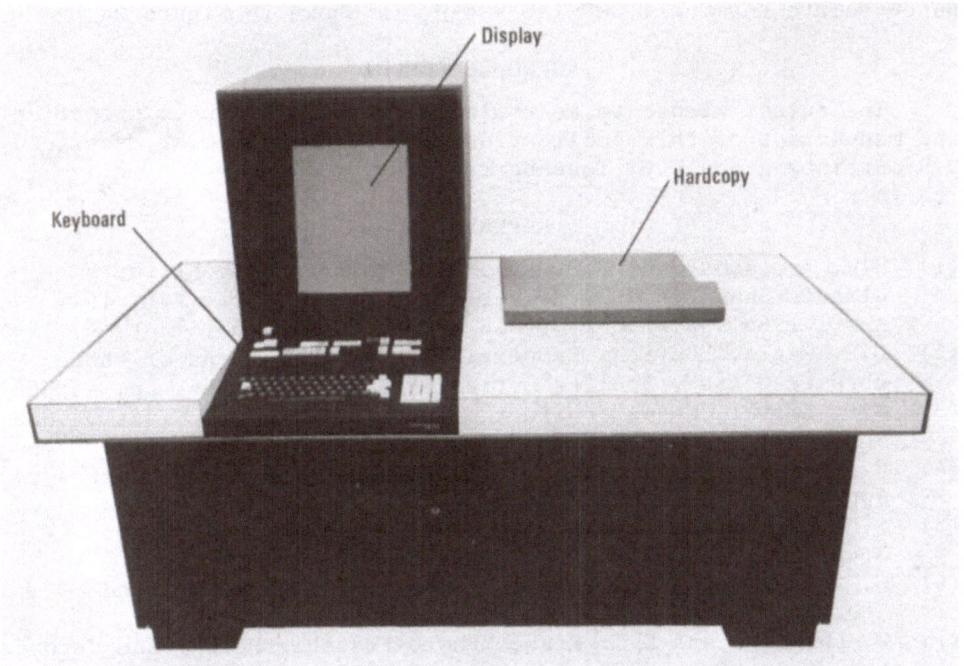

Fig. 24 The Corning 904^{TM} Remote Graphics Display Terminal

head is an optical cavity which contains three quartz iodide lamps and a broad band red filter.

In addition to the character generator that furnishes the signals for the alphanumeric display, a vector generator provides the system with line drawing capability.

Full-size hard copies are provided by reflecting the image into a built-in electrostatic copying machine. The complete computer remote terminal, CORNING 904^{TM}, is shown in Figure 24.

ACKNOWLEDGEMENT

The author wishes to acknowledge the help given in preparing the manuscript of this publication by R.F. Adrion, G.R. Mansfield, D.R. Steinberg and W.H. Touchberry.

REFERENCES

(1) W.H. Armistead and S.D. Stookey, "Photochromic Silicate Glasses Sensitized by Silver Halides, "Science, Vol. 144, April 1964, pp. 150-154; U.S. Patent 3,208,860, Sept. 8, 1965.

(2) G.K. Megla, "Optical Properties and Applications of Photochromic Glass", Applied Optics, Vol.5, June 1966, pp.945-960.

(3) G.P. Smith, "Chameleon in the Sun: Photochromic Glass", IEEE Spectrum, Vol. 3, No. 12, Dec. 1966, pp. 39-47.

(4) R.J. Araujo, "Kinetics of Bleaching of Photochromic Glass", Applied Optics, Vol. 7, No. 5, May 1968, pp. 781-786.

(5) S.D. Stookey, "Photochromic Glass Making", U.S. Patent 3,449,103, June 10, 1969.

(6) G.K. Megla and B.F. Ludovici, "Optical Information and Storage Device", Ital. Patent 845,551, July 1, 1969.

(7) W. Phillips and Z.J. Kiss, "Photo Erasable Dark Trace Cathode Ray Storage Tube", Proceedings of IEEE, November, 1968.

(8) A. Friesem and J.L. Walker, "Thick Absorption Recording Media In Holography", Applied Optics, Vol. 9, January 1970, pp. 201-214.

(9) G.K. Megla and D.R. Steinberg, "A New Time-Sharing Terminal, "Journal of the Society for Information Display, Vol. 7, 1970; Part I, Sept/Oct., pp.15-19; Part II, Nov/Dec, pp. 31-34.

HOLOGRAPHIC INFORMATION STORAGE

H. Kiemle

Siemens AG, Research Laboratories, Munich, West Germany

PRINCIPLES

In modern data processing, magnetic storage is widely applied since many years in various forms which have reached a high degree of perfection. Memories using cores, magnetic drums, discs, cards, and tapes are used, the capacities of which extend over 7 orders of magnitude.

We have therefore to ask ourselves why to look for new memory principles. New applications of computers not only require larger and larger storage capacities, but simultaneously shorter access times. As we will see later, optical and in particular holographic memories are promising in this sense.

This lecture will be restricted to digital information, because in modern computer techniques and time sharing systems all information is processed and stored in digital form.

1.1. Optical Storage Methods

It is well known that information can be stored in extremely high density in light sensitive media, since such media can reach a resolution limit near the wavelength of light.

Any optical memory working with a computer has to perform three main operations:

. Conversion of electrical pulses to optical patterns;

. Storage of these patterns on a light sensitive medium and

readout by projection of an image;

Conversion of the image back to electrical pulses.

The second step is analogous in its physical nature. For storing digital information, the input plane, where the pattern to be recorded has to be located, is divided into a regular raster which is two-dimensional in general since optical systems are able to transport and store information in the two spatial dimensions of the light beam cross section. The output plane, where the image occurs at readout, is of course divided into a corresponding raster. Every area element in these arrays can now signify one bit, and it can be agreed upon that light excitation above a given minimum level in it means the binary "1", no excitation the binary "0".

There are several different possibilities of constructing optical memories. The information stored may be either localized as in Fig. 1 (the input plane is imaged on the storage plane; at readout the latter is imaged on the output plane; every bit occupies a separated area element on the storage plane), or non localized, i.e. holographic as in Fig. 2 (imaging takes place only between the input and output plane; the storage medium is somewhere between, such that a number of bits is more or less equally distributed over it). An intermediate case is "carrier frequency photography", where storage is localized, but many different information patterns are superimposed on the storage medium. During exposure a grating is copied together with the pattern. Grating period and spatial orientation are different for each individual pattern, such that separation of the patterns at readout is possible by spatial filtering.

The two conversion operations may be divided in "non real time" methods where a certain amount of information is converted before storage and after readout - this kind of memories can be called "block - oriented" -, and in "real-time" methods where any bit is stored as it is converted, and converted back as it is read out. This latter method is analogous to magnetic recording in tracks on moving storage media.

1.2. Maximum Storage Density

In order to achieve maximum storage density, the array elements in the input and output plane are chosen as small as the resolution of the optical system permits; this resolution being chosen as high as possible within the limits set by our present technology.

In the case of localized storage, the minimum distance δ between two neighbouring storage cells (Fig.1) is given by the

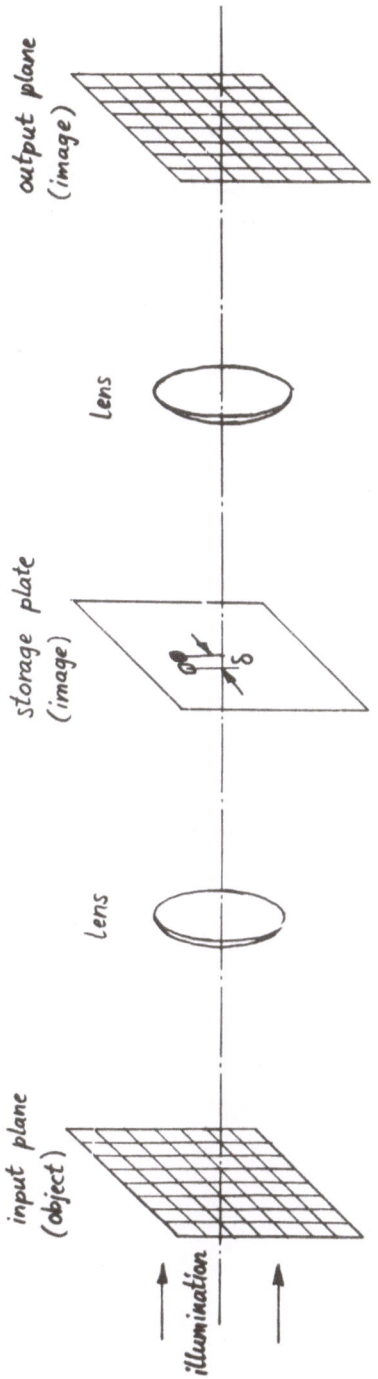

Fig. 1　The principle of localized optical storage.

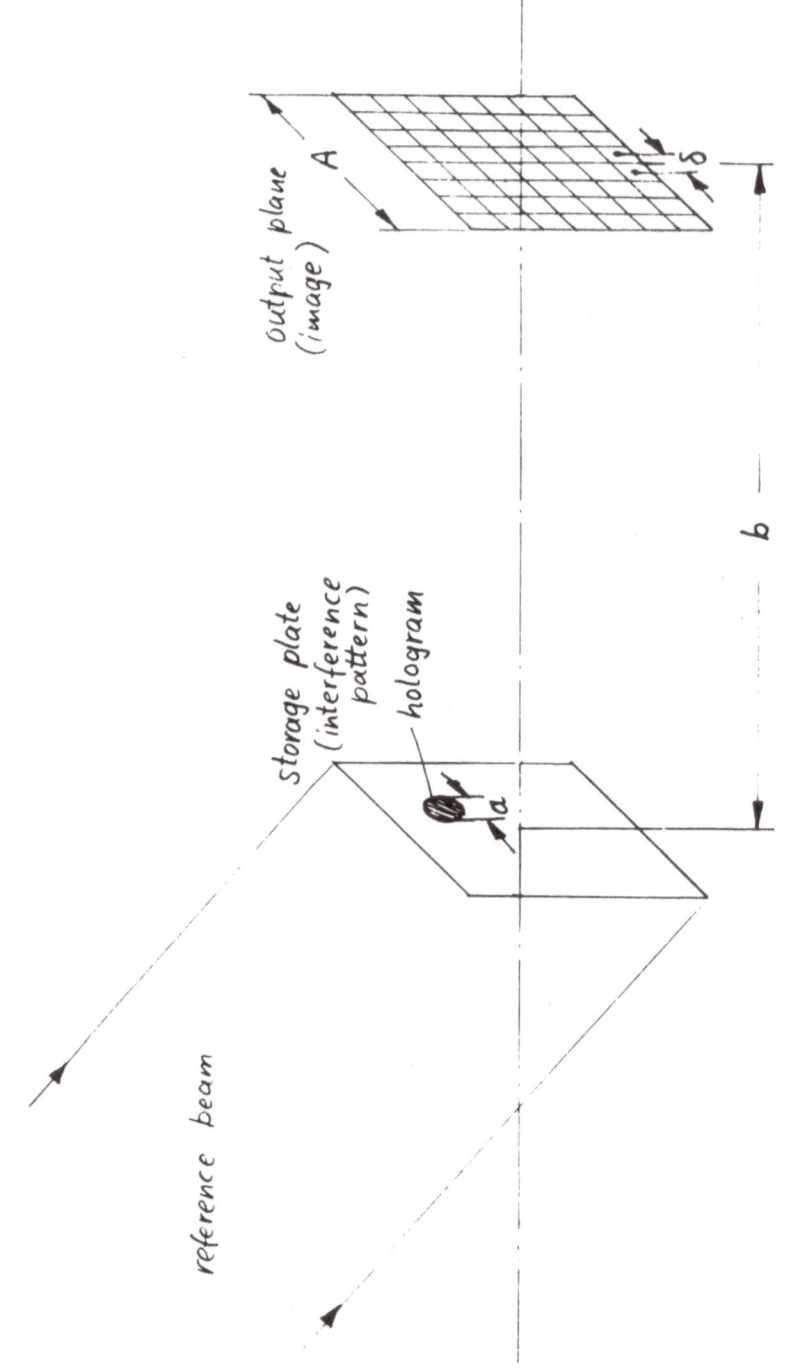

Fig. 2   The principle of holographic storage.

RAYLEIGH criterion   (1)   as

$$\delta = 1.22 \; \lambda/N \, ,$$

1.

where λ is the light wavelength and N the numeric aperture of the optical system. The maximum storage density S_ℓ therefore is

$$S_\ell = \frac{1}{\lambda^2} \left(\frac{N}{1.22} \right)^2 \; .$$

2.

Using $\lambda = 0.5 \, \mu m$ and N = 0.6 as typical values, we get $S_1 \sim 10^6$ bits/mm^2. This is several orders of magnitude higher than characteristic values for magnetic storage or integrated semiconductor memories.

In the case of holographic storage, diffraction at the boundary of the hologram determines how dense the storage cells may be positioned such that they are still separable in the reconstructed image. Analogous to eq.(1), their minimum distance δ (Fig.2) has to be

$$\delta = 1.22 \; \lambda/M \, ,$$

3.

where M = a/b is the numeric aperture of the hologram (assumed here to be circular), a its diameter and b its distance from the output plane. In a quadratic image field of side length A, a number $(A/\delta)^2$ bits can therefore be stored. The storage density S_h is

$$S_h = \frac{4}{\pi} \cdot \frac{1}{\lambda^2} \left(\frac{A}{1.22\ b}\right)^2 .$$ 4.

We see that with holographic storage, the ratio: width of the image field A/ image distance b, plays the same role as the numeric aperture N with localized storage. Comparing eqs.(2) and (4) shows that the theoretical maximum values of the storage density are practically the same in both cases.

Eq.(4) has been derived under the assumption that the conjugate image of the hologram is spatially separated from the direct image. To achieve this, no sacrifice of storage density is necessary since the angle of incidence of the reference beam is still a free parameter and can be chosen properly.

Our estimated storage density also applies to carrier frequency photography. This can be seen by considering the two-dimensional FOURIER transform of a carrier frequency photograph as the image plane and applying the same reasoning as for holographic storage.

1.3. Properties of Optical Storage Methods

Since with any of the optical storage methods mentioned the same theoretical storage density can be achieved, we have to ask for differences in their performances in order to see where their typical applications are. We will use two criteria: the influence of noise and differences in technical layout.

1.3.1. Influence of noise. We have to distinguish several sources of noise: "dynamic" noise which originates at the conversion of the optical information into electronic form and vice versa, and which appears as a time varying noise voltage at the output of the memory, and "static" noise which is due to inhomogeneities in the storage medium and other optical parts and gives rise to a time-indipendent uncertainty in the detected output level.

Theoretically, the dynamic noise can be made arbitrarily small relative to the static noise by increasing the light input into the optical part of the memory. Therefore we will consider here only the static noise.

In the ideal case, all optical parts in the ray paths are

clean and homogeneous, and only inhomogeneities of the storage medium give rise to static noise. Due to the state of the art of
photographic storage media, we find fundamental differences in the
influence of inhomogeneities on the information readout between
localized and holographic storage. In the first case, no silver
grains occur at binary zeros; fluctuations occur only in the binary ones. This does not introduce any error in readout. With ho
lographic storage, however, graininess gives rise to a granulary
background which is superimposed coherently on the image. As is
well known, the spatial intensity distribution of the background
is random with a mean grain size of the order of the resolution
limit. The grains alter the intensity distribution of the image.
We can easily estimate the probability that a grain exceeds half
of the intensity of a binary "1", thus falsifying the information.

After (2) the probability density of the intensity I in the
granulary pattern is

$$p(I) = \frac{1}{\bar{I}} \exp \left(- I \left/ \bar{I} \right. \right), \qquad\qquad 5.$$

$\bar{I}$ being the spatial mean value of the scattered intensity.
If in an image point representing the binary "1" the signal intensity is I_s, and if the decision threshold at the output converter
is set to $I_s/2$, the probability P that due to the granular background a "0" is interpreted as a "1" is given by

$$P = \int_{\frac{1}{2}I_s}^{\infty} p(I) \, dI = \exp \left(- I_s \left/ 2 \bar{I} \right. \right) . \qquad\qquad 6.$$

As an example, if the signal intensity is 20 dB above the
mean scattered intensity, in the average only one such error occurs in 10^{22} bits readout. At a ratio $I_s/\bar{I}$ of 10 dB, however, the
probability would already be 1 : 150.

Of course, the binary ones are also affected. The correspond
ing probability can be calculated in a similar way, but the random

phase variations in the scattered wave field have to be taken into
account since signal and noise are superimposed coherently.

These numerical examples show that for holographic storage recording media with low light scattering have to be chosen. The
stringent requirements of large capacity computer memories can be
met as far as light scattering due to the submicroscopic grainy na
ture of the materials is concerned, if careful processing is applied (3).

In practice, however, another kind of discontinuity occurs.
When coating large areas of a substrate with a light sensitive
layer such as a photographic emulsion or a photoresist, it is almost impossible to avoid local defects such as tiny holes, dust
particles, and scratches. Similar defects may occur later at hand
ling and storing the plates. Such defects are typically large com
pared with the wave-length, and therefore destroy completely certain amounts of information with localized storage. With holographic storage, however, they only reduce slightly the reconstructing hologram area, thus decreasing by the same amount I_s in eq.(6),
while their diffraction pattern adds to the background intensity.
This increases the error probability uniformly for the entire information, but no part of the information is lost completely.

The reason for this different benaviour is obviously, that on
the hologram every bit is distributed over an area which is large
compared with the dimensions of the defects. For localized storage one could think about using a redundant code to avoid the described information loss, but of course this would reduce the effective storage density. Holographic storage, on the other hand,
uses almost ideally the redundancy which consists in the difference
between the small error rates possible with defect-free recording
media and the error rates which can be technically tolerated, so
the storage density remains high.

1.3.2. Crosstalk at readout. Ideally, the light intensity in
any storage cell in the output plane should depend only from the
light excitation in the corresponding cell in the input plane, and
should be completely independent from the information content of
all the other cells. There are however several causes for crosstalk. Since high storage density is desired, the bits in the output plane are Airy diffraction patterns separated by δ as given
by eqs.(1) and (3). Their diffraction rings therefore fall into
neighbouring cells where they give rise to a background level (4)
Considering the well-known intensity distribution of the Airy pattern (1) we see that this crosstalk (which can be called linear
crosstalk to distinguish it from nonlinear crosstalk which will be
discussed later) leads to an uncertainty at readout which may be
much larger than that caused by noise. In principle, it could be
accounted for by electronic processing after readout, since it is

deterministic and not random.   But such a data processing would be
long and complicated.

Linear crosstalk is less problematic in localized storage than
in holography, since the exposure can be chosen so that due to the
curvature of the film characteristics the fringes of the Airy pattern are suppressed.

The cause of nonlinear crosstalk is the nonlinear relationship
between the light intensity at the exposure and the resulting density or phase shift in the developed record. Nonlinear crosstalk
does not occur in localized storage, but may be a problem in holographic storage. Nonlinear distortion of the interference field
leads to the generation of new spatial frequencies, as well harmo-
nics of the fundamental frequencies as combination frequencies (sum
and difference frequencies) (5). These new spatial frequencies
in turn lead to image points in cells in which no light should be.

In order to see which distortion products are most dangerous,
we consider two discrete frequencies out of the spatial spectrum
occupied by the stored information (Fig.3). The energy density W
at the recording due to these frequencies can be written as (5).

$$W = \bar{W} + W_1 \left[\cos(\vec{\omega}_1, \vec{x}) + \cos(\vec{\omega}_2, \vec{x}) \right] \qquad 7.$$

where $\bar{W}$ and W_1 are constants, $\vec{\omega}_1$ and $\vec{\omega}_2$ are frequency vectors in
the spectral plane characterizing the period and orientation of the
interference patterns considered, and x is a vector in the hologram plane from the origin to the hologram point in which W is determined.

With linear recording, the hologram would only contain the
original spectrum finely shaded in Fig.3. With quadratic distortion, we obtain a second order term

$$\left[\cos(\vec{\omega}_1, \vec{x}) + \cos(\vec{\omega}_2, \vec{x}) \right]^2$$

which yields new spectra

$$2\vec{\omega}_1, 2\vec{\omega}_2, \vec{\omega}_1 + \vec{\omega}_2, \text{ and } \vec{\omega}_1 - \vec{\omega}_2 .$$

They are coarsely shaded in Fig.3. As can be seen from this figure, they can be spatially separated from the object spectrum if we choose the incidence angle of the reference beam such that

$$2\,\omega_{x1} > \omega_{x2} \qquad\qquad\qquad 8.$$

This means that second order ghost points are eliminated from the output of the memory, if in the x-direction the bandwidth $\omega_{x2} - \omega_{x1}$ of the signal is smaller than 2/3 of its center frequency (5).

Cubic distortion yields a third-order term

$$\left[\cos\left(\vec{\omega}_1 , \vec{x}\right) + \cos\left(\vec{\omega}_2, \vec{x}\right)\right]^3$$

which generates sum spectra $3\vec{\omega}_1$, $3\vec{\omega}_2$, $2\vec{\omega}_1 + \vec{\omega}_2$, $2\vec{\omega}_2 + \vec{\omega}_1$, and difference spectra $2\vec{\omega}_1 - \vec{\omega}_2$, $2\vec{\omega}_2 - \vec{\omega}_1$. They are also indicated in Fig. 3. Here, unfortunately, we have no means to prevent overlapping of the difference spectrum with the original spectrum. The only possibility is to avoid the formation of the third order term by locating the operating point on the characteristic curve of the recording material at an exposure level where the third derivative of the curve vanishes.

Of course, there are still terms of higher order. They must be kept small by restricting the total exposure variation.

This is a difficult problem, if an appreciable number of bits is to be stored in the hologram. In the worst case, where all bits are binary "ones", the object of the hologram is a regular, coherently radiating point array with equal phases, which generates a far field exhibiting sharply concentrated, very intense spikes on the hologram, just as does a two-dimensional grating. If we choose the object to reference ratio and exposure such that these spikes remain in the linear portion of the response curve, most of the hologram area has a very low diffraction efficiency; if we expose however most of the hologram area correctly, the spikes are clipped by over- and under-exposure, and strong ghost points occur (Fig. 4a).

There are two ways around this problem. First, the hologram must not necessarily be recorded in the far field (i.e. in its FOURIER plane). A small distance apart the intensity spikes decrease considerably, and images like Fig. 4b are possible. Still better results can be obtained by varying the phase distribution in the object point array. This can be done by a phase mask, as proposed by BURCKHARDT (6).

1.4. Information Readout

In order to read out a block of information (a "page") with the localized storage method, a magnified image of the bit pattern has to be projected on the opto-electronic readout converter. Replacing one block by another requires either mechanical transport of the storage medium or a separate projection system for every "page" of the memory. The first possibility is unavoidably slow; access times amount at least to tens of milliseconds and because of the magnification, positioning requires extreme mechanical precision. The second way is expensive because it requires a large number of switchable light sources associated with a precise fly's eye lens of diffraction-limited resolution. A technical solution for such a memory has been proposed by SPITZ (7). He uses an array of holographically produced FRESNEL zone plates as the fly's eye lens.

With holographic storage, these problems do not exist. Holograms are capable of forming real images at an arbitrary scale without the need for separate projection systems. Since this projection system is "built-in", there is no magnification of the positioning error, regardless of the scale factor of the projected image. This facilitates and accelerates greatly positioning and adjustment of the storage plates if a change of plates is to be provided.

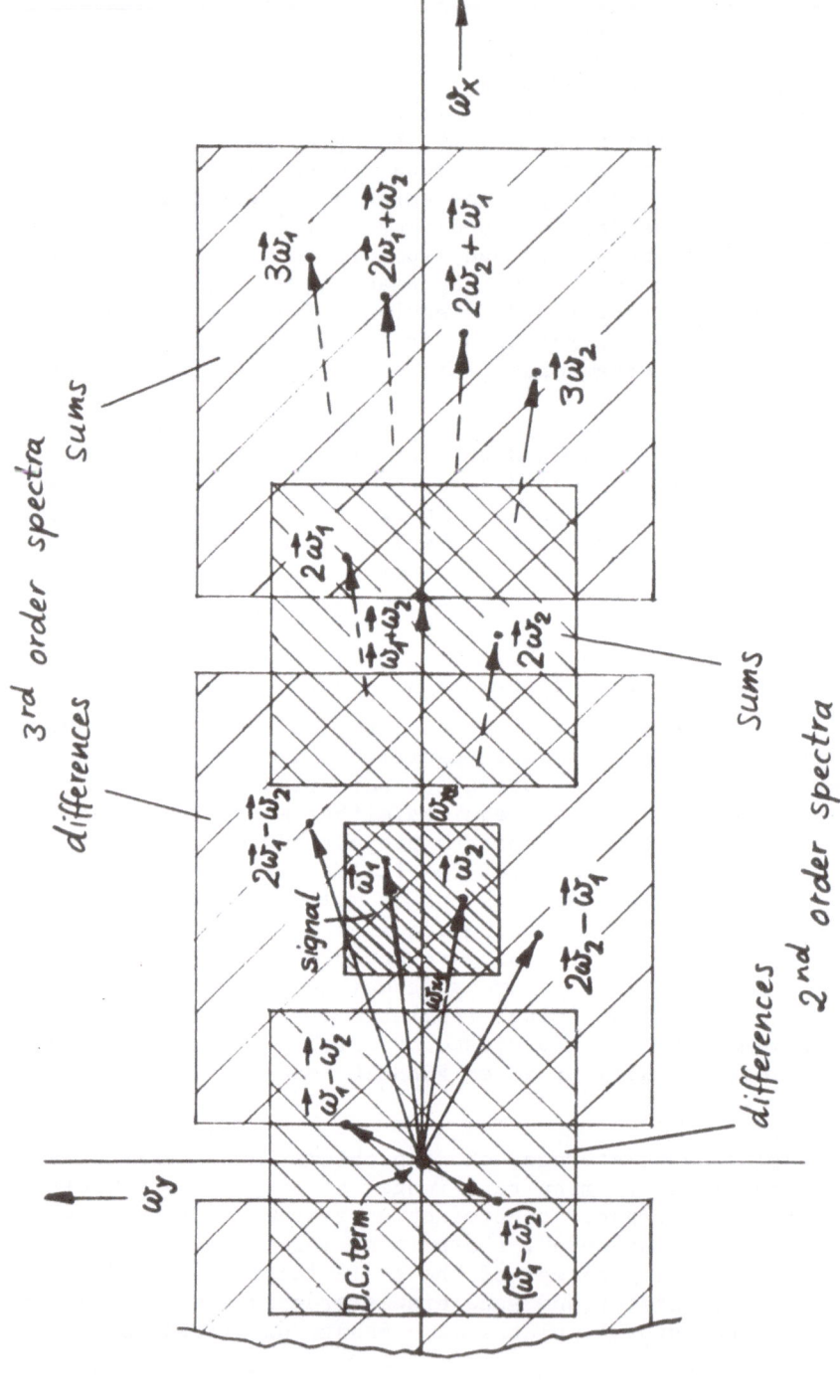

Fig. 3   Spatial frequency plane showing the signal spectrum and
the lower order spectra generated by nonlinear recording.

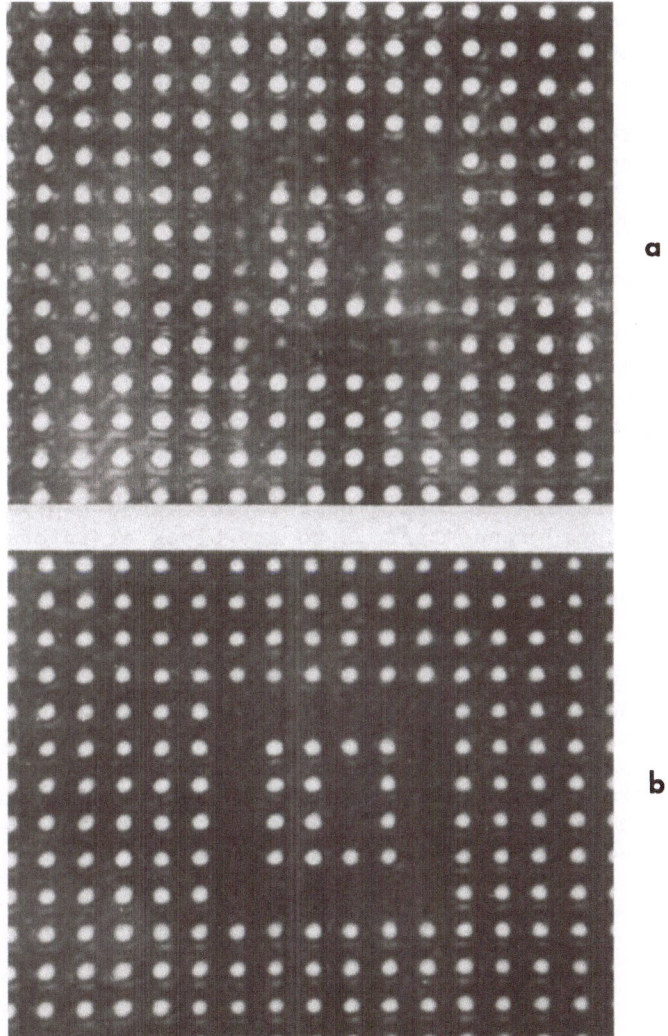

Fig. 4   a) Real image showing ghost image points generated by
         nonlinear distortion. b) Elimination of ghost images
         by recording slightly outside the FOURIER plane and
         careful exposure.   Courtesy of G. Goldmann.

1.5. Conclusions

Summarizing our basic results we can compare the method of lo calized and holographic storage. For all methods, the theoretical maximum storage density is equal. While in localized storage high densities cause the risk of blackouts due to local imperfections of the storage medium, carrier frequency storage is less sensitive to local defects because of the lower demagnification, and hologra phy is almost immune against such imperfections. Light scattering due to the grainy nature of the storage medium is no problem in lo calized storage, while it has to be taken into account in hologra phy; modern recording media and processing procedures however assure a highly sufficient signal – to-noise ratio.

An important problem is crosstalk which may occur in holography and carrier frequency photography. In holography, method have been worked out with which nonlinear crosstalk can almost be elimi nated. Linear crosstalk due to diffraction depends on the light wavelength, since it is a diffraction effect, and suggests use of short wavelengths, while static noise decreases with increasing wavelength; the choice of the optimum wavelength for holographic sto rage is therefore rather determined by other criteria such as spec tral sensitivity of the recording material used. While holography permits very rapid random access to any block of information of large storage plates, this requires mechanical transport of the plate with localized storage. These results are summarized in Table 1.

This shows that for digital information, holography offers significant advantages over localized storage for block-oriented memories. For tape-like recording, where access to entire blocks is not necessary, the advantage of immunity against local defects is till sufficient to make holography the most promising optical approach. On the other hand, localized storage methods seem to be better adapted to the storage of analog data.

Table 1

Advantages and problems of localized and holographic digital
storage at very high storage densities and capacities

	STORAGE	
	localized	holographic
static noise	no problem	can be kept sufficiently low
local defects	loss of in-formation	no problem
linear crosstalk	avoidable	problematic
nonlinear crosstalk	no problem	avoidable
block access	slow	fast
technical layout	complicated	relatively simple

DEVICES

In the following, storage devices on the basis of holography and their state of the art will be described. We will concentrate our discussion on the block-organized memory (8, 9), because most of the international development activity has been devoted to this type.

The storage plate of the block-organized holographic memory is divided into an array of individual, independent holograms ("sub-holograms"). In every hologram, a block ("page") of information is stored and transferred in parallel to the opto-electronic output converter. The holograms are addressed by deflecting the recon-structing beam by means of an electro-optic or acousto-optic light deflector. We will now consider the various elements necessary for data recording and reading.

2.1. Data Recording

The simplest way to convert the electrical information to be recorded into optical form is to produce a transparent mask on film either by scanning with a modulated light beam or by photo-graphing a suitably addressed matrix of light sources. This me-thod presents no particular problems and is commonly used in the laboratory. Since it is an indirect method, however, it is slow and may be useful in practice only for fixed memories or cases where the information changes only in long time intervals. For all other cases, especially erasable memories, a more satisfactory so-lution would be a two-dimensional light modulator. This modulator has to be capable of storing the incoming signals until the entire input format is filled, in order that the complete object wave can be recorded. Modulators consisting of an array of electro-optic or acousto-optic modulators would in principle be possible but dif-ficult to realize; furthermore, they do not have inherent storage capability. Devices such as the TITUS tube modulator (10) could be combined with electrostatic storage; but they are still in an early stage of development.

Another possible approach is to use a nematic liquid crystal matrix (11). Nematic liquids have the property to be transpa-rent without an applied electric field, but to scatter light if a field of suitable strength is applied. This is due to scattering centers which are in a fast random movement. This causes the scat-tered light to become incoherent, which means that it is not re-corded on the hologram. A thin layer of such a liquid is put bet-ween transparent plates covered with segmented transparent electro-des forming a matrix (Fig.5). Matrix elements which are not ad-dressed do not alter the coherence in the object wave and are thus recorded, while addressed elements destroy the coherence and there-fore miss in the reconstructed image. This is shown in Fig. 6.

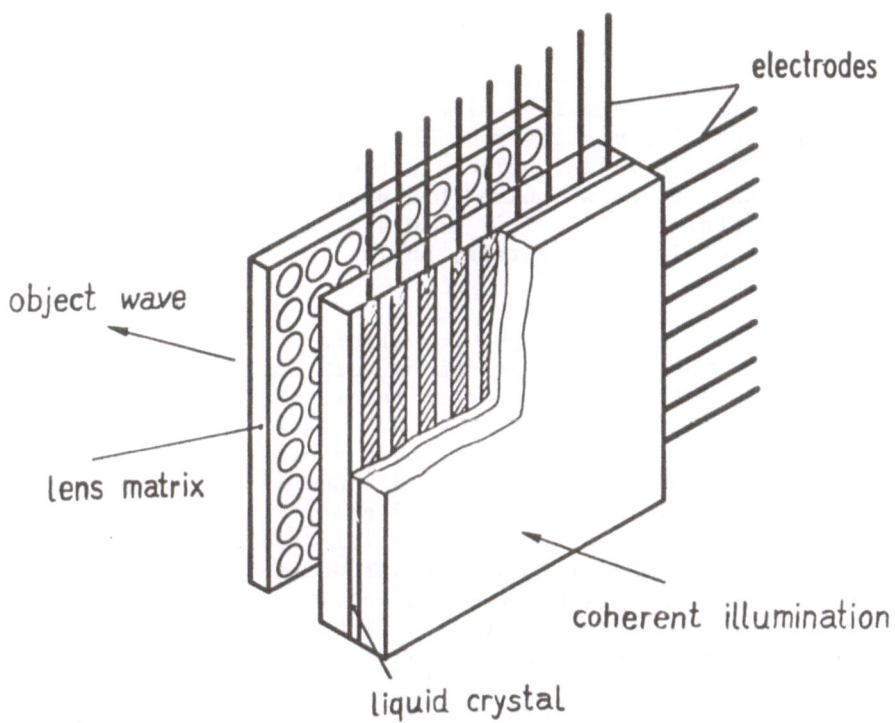

Fig. 5 Liquid crystal light modulator for the conversion of electrical signals into optical form.

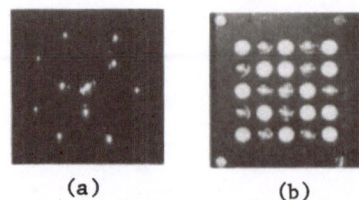

(a)   (b)

Fig. 6   Holographic image (a) and conventional photograph (b) of
         a simplified model of a matrix after Fig. 5. Part of the
         matrix elements are scattering light due to electrical
         excitation and are therefore not recorded on the hologram.

     Storage of the electrical signals can easily be achieved by
using a thin ferroceramic layer in series with the liquid crystal,
which stores the electrical pulses in the form of a polarization
pattern. By an activating pulse this pattern is transferred to
the liquid crystal (12) . This modulator is slow because of the
time constants of liquid crystals, but at least 1 hologram per s
can be recorded.

     None of these dynamic input converters is ready for immediate
technical use, and further development work has still to be done.
An alternative to this parallel data recording is sequential record
ing, where a single modulated light beam scans the input plane by
means of a light deflector. It has been shown (13) that if the
phases of the object waves are altered randomly by a phase mask as
described in 1.3.2., in order to avoid ghost images, the individual
interference patterns get a random spatial displacement relative to
each other, and because of their incoherent superposition the contrast and hence the diffraction efficiency decreases inversely
with the number of bits per hologram. Therefore sequential record
ing is inferior to parallel recording.

     There are two basic geometric requirements which have to be
fulfilled at the recording: In order that the reconstructed image
is a complete replica of the input plane, all the light leaving
the input transparency or modulator must be concentrated on the
sub-hologram surface. In addition, the real images of all sub-ho
lograms have to occur at the same location of the output plane in
order that the opto-electronic readout converter has not to be moved. Both requirements lead to a preferable recording position of
the hologram at or near the optical FOURIER transform plane of the
object. To achieve this, a converging spherical illumination beam
is used and the hologram placed near its focus. The FOURIER trans
form lens has to be well corrected in order not to decrease the
possible storage density.

     Of course, the object illumination beam as well as the refe-
rence beam have to be moved when going from one sub-hologram to
another, while the data input window has to keep its position relative to the storage plate.

2.2. Data Readout

Besides the storage plate and the laser, the principal parts
of the readout setup as depicted by Fig.7 are the beam deflector
and the optoelectronic output converter. Two fast deflector principles are in use: electro-optic and acousto-optic deflection. We
will not discuss them here in detail but only briefly review how
they work and which performances are achieved today.

An electro-optic deflector contains an electro-optic polarization switch such as a Kerr cell or an electro-optic crystal. If a
suitable electric field of typically several kV/cm is applied, the
polarization direction of the incident linearily polarized laser
beam is rotated by 90°. Behind the polarization switch there is a
polarization selector, e.g. a WOLLASTON prism, which directs the
beam in one of two possible directions, depending on its polarization. Such a unit is a binary deflector. In order to obtain many
different directions, many stages of this kind have to be mounted
in series. With n stages, 2^n directions can be obtained. With
two sets of 7 stages each, for instance, we obtain a two-dimensional array of 128 x 128 directions. Electro-optic deflectors can
be operated with typical access times between 100 ns and 1 μs .
However, in continuous operation the access time is limited ,because
capacities have to be charged and discharged very rapidly, which
leads to high power dissipation in the electronics, as well as heating due to losses in the electro-optic material.

Acousto-optic deflectors use BRAGG reflection of laser beams
by transverse sound waves in transparent media. Such a wave acts
as a three-dimensional phase grating. The deflection angle depends
on the sound wavelength. Switching the light beam to a new direction is therefore possible by changing the sound wavelength. The
time which is required for switching has a lower limit given by the
time the new sound wave takes to travel through the whole light
beam cross section. Since on the other hand the number of resolvable beam directions is proportional to the beam diameter, we have
a trade-off between short access times and the number of resolvable directions. With present day technology and materials access
times of a few μs for 100 resolvable directions can be achieved
(14) .

A fundamental difference of acousto-optic deflection as compared with the electro-optic method is that with one single stage
many directions can be addressed; for two-dimensional deflection
only two stages are necessary which are oriented perpendicular to
each other. This simplification makes acousto-optic deflection
attractive if the number of resolvable directions per stage has
not to exceed about 100, while above this number digital deflec-

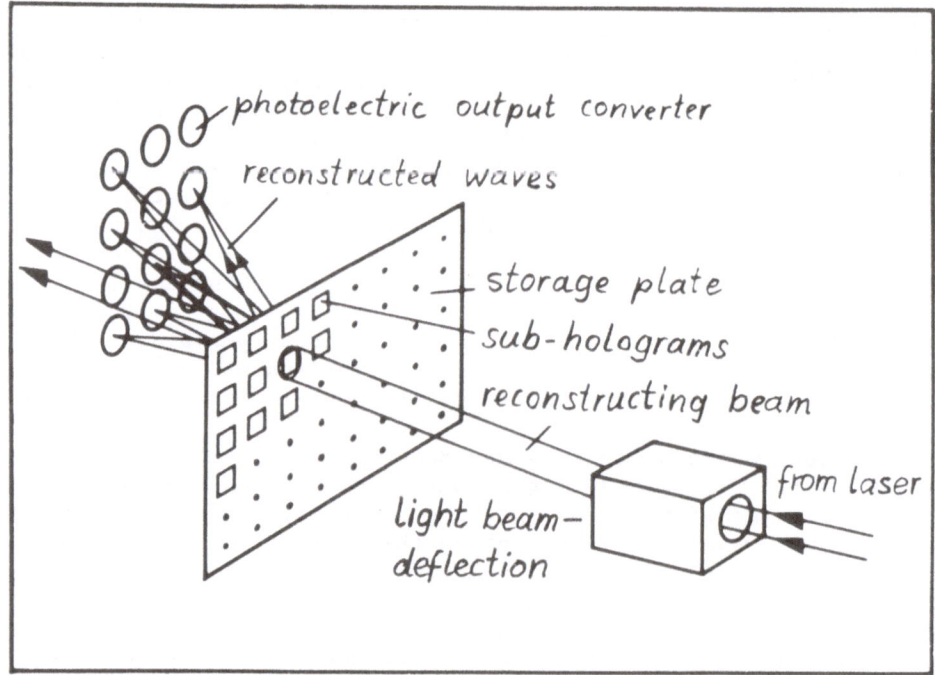

Fig. 7 Readout part of a block-organized holographic memory.

tors may appear better despite their complexity, since they permit
shorter access times (14) .

The opto-electronic output converter has to convert the intensity distribution in the output plane into electrical pulse trains
which can be fed into the computer. Various known devices seem to
be suited at first sight: television camera tubes such as vidicons
and arrays of phototransistors or photodiodes. But besides the generally slow response of camera tube targets, linearity and stability of the electron beam sweep is a problem. Further, such tubes
have only one electrical output; a readout in parallel channels is
not possible.

A better solution would be an array of light sensitive semiconductor elements fabricated in LSI technique. But everybody who
is familiar with LSI production knows that it is a formidable task
to realize a matrix of e.g. 100 x 100 elements without any defecti
ve element. A promising approach seems to be the composition of a
large matrix from smaller chips which can be produced more easily.
But this raises new technologic problems related to the intercon-
nection of the chips, and of course reduces the information content of the sub-holograms due to the loss of image space between
the chips.

Thus, satisfactory readout converters are not yet available
and will still need a good deal of development.

2.3. Information Capacity of Block-Organized Holographic Memories

The information capacity C of a block-organized memory (14)
is obviously the number of bits which can be readout in parallel
– it is given by the number l_x . l_y of detector elements in the
readout converter – times the number of sub-holograms, which is
given by the number m_x . m_y of directions which the beam deflec
tor is able to resolve in the x and y dimension:

$$C = l_x \, l_y \, m_x \, m_y \, .$$
9.

With reasonable values l_i and m_i of the order of 100, the
capacity of a storage plate reaches $C = 10^8$ bits. This is an attractive order of magnitude if we remind that a random access is
possible within a few µs or less. It is interesting to ask if and

how much this number could be increased if we were able to construct light deflectors and readout matrices with still more positions. Let δ be the distance between neighbouring detectors. The
necessary sub-hologram diameter a is then given by eq.(3) as

$$a = 1.22 \, \frac{b \, \lambda \, k_1}{\delta \, \cos^2 \psi} \, , \qquad\qquad 10.$$

where $k_1 > 1$ is a security factor necessary to avoid excessive linear crosstalk, and ψ is the maximum angle which the reconstructed object beams form with the normal of the output plane (see Fig.
8). The side lengths H_x, H_y of the hologram storage plate are
given by

$$H_{x,y} = 2b \, \tan \psi - I_{x,y} \, \delta \, . \qquad\qquad 11.$$

The number of sub-holograms in the x- and y- direction are
given by $m_{x,y} = H_{x,y}/ak_2$, where $k_2 > 1$ is another security factor
to avoid crosstalk between neighbouring sub-holograms.

From esq. (10) and (11) we obtain

$$m_{x,y} = \frac{(2b \, \tan \psi - I_{x,y} \delta) \, \delta \, \cos^2 \psi}{1.22 \, b \, \lambda \, k_1 k_2} \, . \qquad\qquad 12.$$

The design parameter b is optimized if we choose it to be
large compared with the linear dimensions of the detector matrix:

$$b \gg \frac{I_{x,y} \, \delta}{2 \tan \psi} \, . \qquad\qquad 13.$$

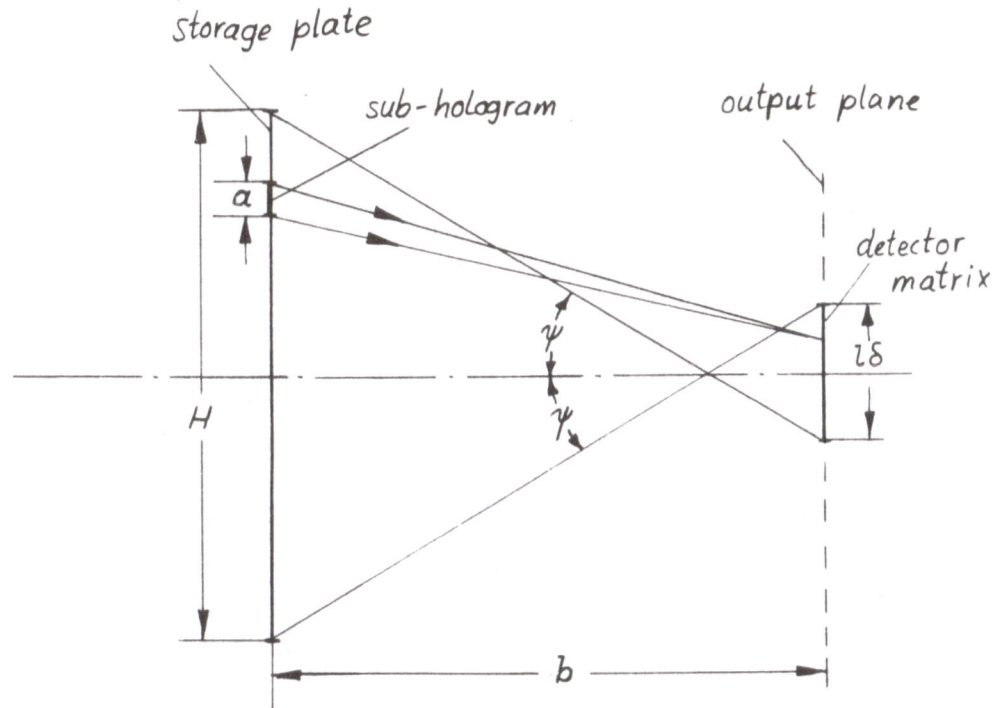

Fig. 8   Geometry at readout.

In this limit the number of sub-holograms in one dimension becomes

$$m_x = m_y \sim \frac{2\delta \sin\psi \cos\psi}{1.22 \; \lambda \; k_1 k_2} \; ; \qquad\qquad 14.$$

which attains its maximum value if the design parameter ψ is chosen to be 45°:

$$m_{max} \sim \frac{\delta}{1.22 \; \lambda \; k_1 \; k_2}. \qquad\qquad 15.$$

The capacity at optimum geometric parameters therefore becomes

$$C_{max} \sim \frac{\delta\ell_x \cdot \delta\ell_y}{\left(1.22 \; \lambda \; k_1 k_2\right)^2} \qquad\qquad 16.$$

It is inversely proportional to the square of the wavelength; short wavelengths are favorable in accordance with eq.(4).

A remarkable result is that C_{max} does not depend on the number of resolution elements of the detector matrix but only on the detector area. However, when designing a memory, we cannot consider eq.(16) alone, but have also to take eq.(15) into account which reminds us that any saving in readout elements, i.e. any increase of δ at a constant detector area, requires of course a corresponding increase in the number of resolvable beam directions of the deflector. The consequence is that we can hope to reach C_{max} after eq.(16) only if we succeed in adapting the detector matrix and the beam deflector to each other. Reminding the properties

of acousto-optic deflectors we see that at constant storage capaci
ty C_{max}, an increase in the number of detector elements leads to a
decrease of the number of beam directions and thus to a faster ran
dom access.

In order to illustrate this analysis, suppose we have a detec
tor matrix of $l_x = l_y = 100$ elements per dimension, and a detec
tor spacing of $\delta = 200\,\mu$m. Using He - Ne - laser light with
$\lambda = 0.633\,\mu$m and security factors $K_1 = K2 = 1.5$, we obtain
$C_{max} \sim 1.3 \cdot 10^8$ bits. After eq.(15), this requires a deflector
with 115 x 115 positions.

The results of the above analysis are not the only limitations
of the storage capacity. An important boundary condition is also
that the zero diffraction order of the reconstructing beam never
hits the detector, and that the angles it forms with the object wa
ves always remain within the limits set by the recording materials;
this makes it difficult to optimize the geometric layout in the a-
bove sense. Further limitations are due to the recording geometry
which has already been discussed. Finally, increasing the number
of bits stored in the sub-holograms reduces the light power availa
ble on the detector elements, because the diffraction efficiency
of the holograms is limited.

Of course, arbitrary capacities can be achieved by exchanging
the storage plates mechanically. But in this way the short access
times would become illusory. For further increase of the information capacity per storage plate, we have two principal ways. The
first is to use volume holograms, in which many sub-holograms could
be superimposed. But even if we neglect the problem of light power which is still more stringent here, we run into problems. To
address the sub-holograms separately, we would have to vary either
the angle of incidence of the reference and reconstructing beam at
constant wavelength - together with the scanning of the storage
plate this would lead to a very complicated deflection system -,or
to vary the wavelength at constant incidence angle, which requires
a tunable laser. Such lasers indeed exist, but since they use non
linear frequency conversion by parametric effects, their amplitude,
mode, and frequency stability is still rather poor. A second way
to increase the storage capacity would be to leave the concept of
binary storage and to use more than two amplitude levels. But if
we remember the problem of crosstalk suppression and take into account the nonuniformity of even the most advanced storage media
available today, this way is also not likely to be promising. We
have therefore to be cautious when predicting storage capacities
of future holographic memories.

ACKNOWLEDGEMENT

The author gratefully acknowledges the work of many collea-

gues which contributed to the matter presented here. In particular, thanks are due to Dr.M. Lang and G. Goldmann for providing
the manuscripts cited prior to publication as well as the photographs in Fig.4.

REFERENCES

(1) M. Born and E. Wolf, Principles of Optics. 3rd ed., Pergamon
      Press Ltd., London/New York (1964/65).
(2) W. Martienssen and E. Spiller, Holographic Reconstruction
      without Granulation. Phys. Letters 24A, No.2 (1967), 126 -
      128.
(3) H. Kiemle, Phase Holograms in Photographic Emulsions for Di-
      gital Data Storage. Optics Techn. 1 (1969), 146 - 148.
(4) G. Goldmann, Recording of Digital Data Masks in Quasi Fourier
      Holograms. To be published in Optik.
(5) H. Kiemle and D. Roess, Einführung in die Technik der Holo-
      graphie. Akad. Verlagsgesellschaft Frankfurt (1969).
(6) C.B. Burckhardt, Use of a Random Phase Mask for the Record-
      ing of Fourier Transform Holograms of Data Masks. Appl.Optics
      9 (1970), 695 - 700.
(7) E. Spitz, Mémoire Optique à Grande Capacité avec Accès Aléa-
      toire et Rapide. AGARD Conference Proc. "Opto-Electronics
      Signal Processing Techniques", Toensberg/Norway (1969).
(8) V.A. Vitols, Hologram Memory for Storing Digital Data. IBM
      Tech. Discl. Bull. 8, No.11 (1966), 1581 - 1583.
(9) L.K. Anderson, Holographic Optical Memory for Bulk Data
      Storage. Bell Lab. Rec. 46, No.10 (1968), 318 - 325.
(10) G. Marie, Projection d'images de télévision sur grand écran.
      Nouveau tube relais optique utilisant l'effet Pockels. Les
      laboratoires d'électronique et de Physique appliquée, extrait
      de Techniques Philips No.1 (1969).
(11) H. Kiemle and U. Wolff, Application de cristaux liquides en
      holographie optique. Nouv.Rev. d'Optique Appl., Suppl. au
      tome 1, No.2 (1970), 19; Optics Comm. 3, No.1 (1971), 26 -
      28.
 12) J.G. Grabmaier, W.F. Greubel and H.H. Krüger, Liquid Crystal
      Matrix Displays Using Additional Solid Layers for Suppression
      of Parasite Currents. Mol. Crystals and Liquid Crystals 12
      (1971). (to appear soon)
(13) M. Lang, On the Diffraction Efficiency of Transmittance Sto-
      rage Holograms. To be published in Optics Comm.
      D. Graf, Holographic Memories. Proc. of the 1970 IEEE Compu-
        · Group Conference, Washington, D.C.

HOLOGRAPHY THROUGH THE ATMOSPHERIC TURBULENCE

A. Consortini

Istituto di Ricerca sulle Onde Elettromagnetiche del

C.N.R., Firenze (Italy)

INTRODUCTION

It is well known that the velocity of propagation of an e.m. wave in the vacuum is c=3 10^8 m/sec.

When the radiation propagates through a medium, the velocity changes by a factor 1/n, where n represents the refractive index of medium. If V denotes the velocity in the medium, one has

$$V = \frac{c}{n}$$ 1.1.

The quantity n depends on the characteristics of the medium. For instance if the medium is a fluid n depends on the density, temperature and pressure.

If the characteristics of the medium vary, for instance as a function of the point P, they give rise to a consequently varying refractive index.

A turbulent medium is a medium whose characteristics vary in a random way both in space and time, around a mean value.

In order to have an idea of random functions, let us refer to a measure of the atmospheric temperature (Fig.1). If a sufficient ly sensitive thermometer is placed at a given point in the atmosphere, we can see that the value of the measured temperature in a short time interval (e.g. 1 sec) is not constant. On the contrary

235

it fluctuates presenting small irregular and random variations
about a mean value. The temperature at a given time may be therefore considered as a sum of two parts, $T=T_0+ \delta T$. T_0 is a constant
quantity corresponding to the mean value, δT is an additional
small quantity, whose value cannot be predictable. This is an e-
xample of a random function. Note that T is not a given function
of time but it can be treated only in terms of probability. One
can only know the probability that the function has a given value
T. The quantity T_0 is the mean value evaluated by taking into account the probability.

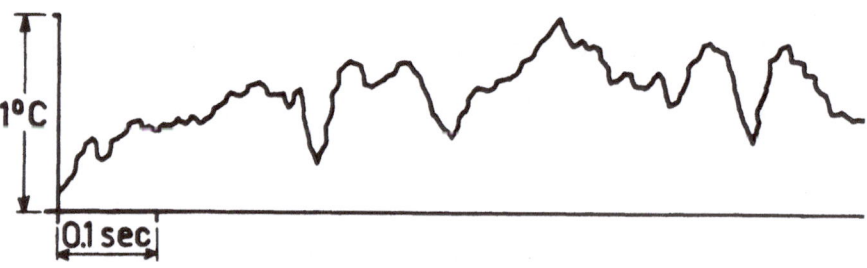

Fig. 1 - An example of temperature recording

In the above example we have examined the temperature as a
function of time. However the same thing can be said for the temperature as a function of space, for instance of the point P along
a given direction or in a given volume.

The characteristic quantities of a turbulent medium are space
and time dependent.

For random processes generally the hypothesis of hergodicity
is made. Under this hypothesis the mean value of the function
$f(P,t)$ taken at a given point over a long period of time (time ave
rage), coincides with the ensemble average taken at a given time.
Under this hypothesis one can treat the problems only in terms of
time averages.

The process is called stationary if the mean value does not
varies in time.

From the point of view of electromagnetic propagation, a random medium is characterized by a refractive index which fluctuates
in a random way about its mean value.

The fluctuations of the refractive index on their turn, cause
amplitude and phase fluctuations of the electromagnetic waves, tra

versing the medium. As a consequence these fluctuations produce
unavoidable effects on an imaging system. All of us, for instance,
know the effect of scintillation of stars, or of remote terrestrial
sources, which is due to the atmospheric turbulence. In this case
the imaging system is the eye.

The problem of imaging through turbulent media, and especially
the atmosphere, is an old problem in the field of astronomy (1,2,
3) and has been extensively investigated in the recent years (see
for example Refs.4,5,6,7,8,9).

There is an important distinction to do: the case of imaging
with incoherent illumination and the case of coherent illumination.
Here we are interested in the second type of illumination. This
problem which was important for some type of microscope, gained im
portance, with the advent of the laser, in connection with the ima
ging through the turbulent atmosphere.

As an example of the influence of atmospheric turbulence on
laser radiation, in Fig.2., the diffraction pattern is shown, of a
circular aperture illuminated by a laser beam having traversed a
layer of turbulent atmosphere 55m thick at about 1.5m above the
ground. Fig. 2 a) refers to a pupil aperture R=0.5 cm, b) R=1 cm
and c) R=2 cm. The exposure time was 1/5000 sec. As one can see
the phase and amplitude fluctuations, induced by the atmosphere,
give rise to a deterioration of the image, which in general causes
a loss of resolution.

With the advent of holography the problem arose of the influence of the turbulence on the holographic images. The aim was essentially to see if holographic images can give a better resolution
than conventional images. Alternatively one could use holography
as a means for investigating the turbulence.

Two different methods of making holography through turbulent
media can be examined. In the first one, only the beam from the
object traverses the medium. This scheme has been proposed to the
purpose of obtaining high resolution in the presence of aberrations
(10, 11, 12) due to random inhomogeneous media, not varying in
time.

In the second case, which is of interest for us, both the
beams from the object and from the reference traverse the turbulent medium. This method, was proposed by Goodman (13) and seems
particularly suitable for having the holographic images as free as
possible from turbulence effects.

The basic concept is the following: if the object of interest,
coherently illuminated, and the mutual coherent reference source,

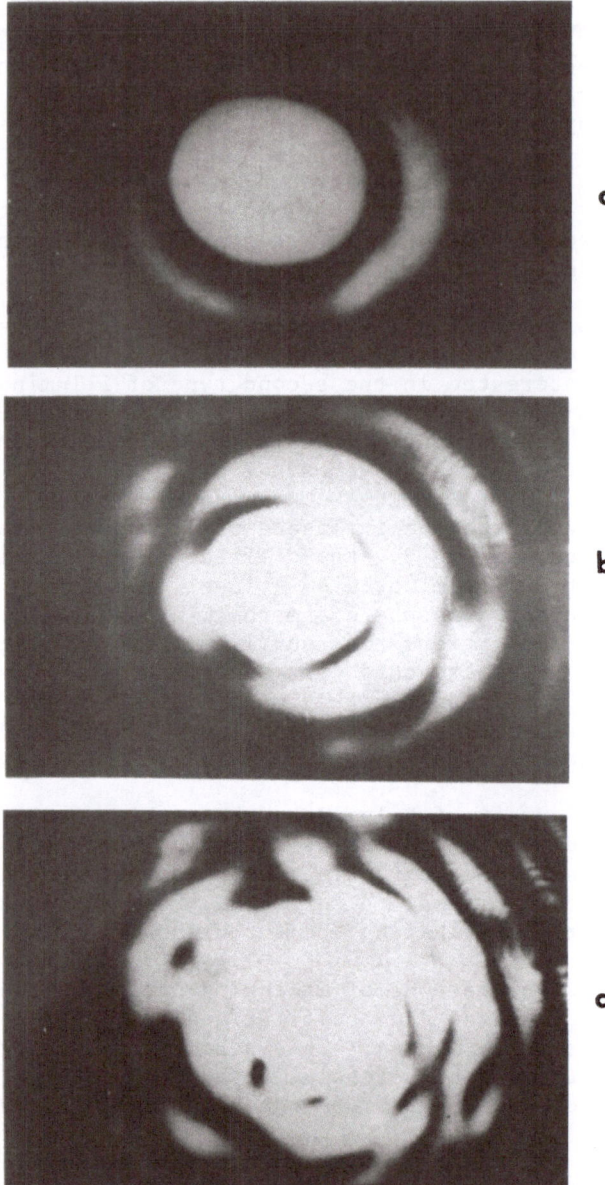

Fig. 2 - The diffraction pattern of a cyrcular aperture alluminated
by a laser beam after a path in the atmosphere for pupil
radius: a) R = 0.5 cm), b) R = 1 cm, c) R = 2 cm

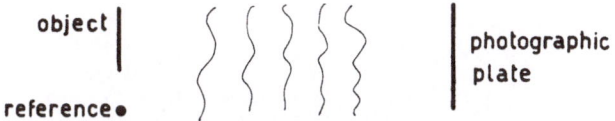

Fig. 3 - Hologram recording in the lens-less Fourier transform
holography through a turbulent medium

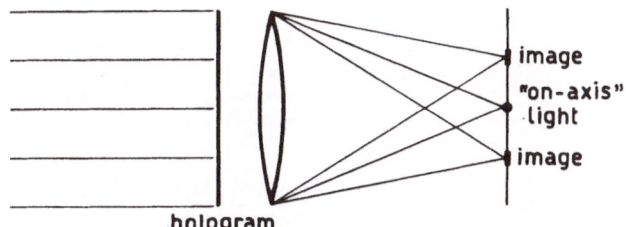

Fig. 4 - Image reconstruction in the lens-less Fourier transform
holography

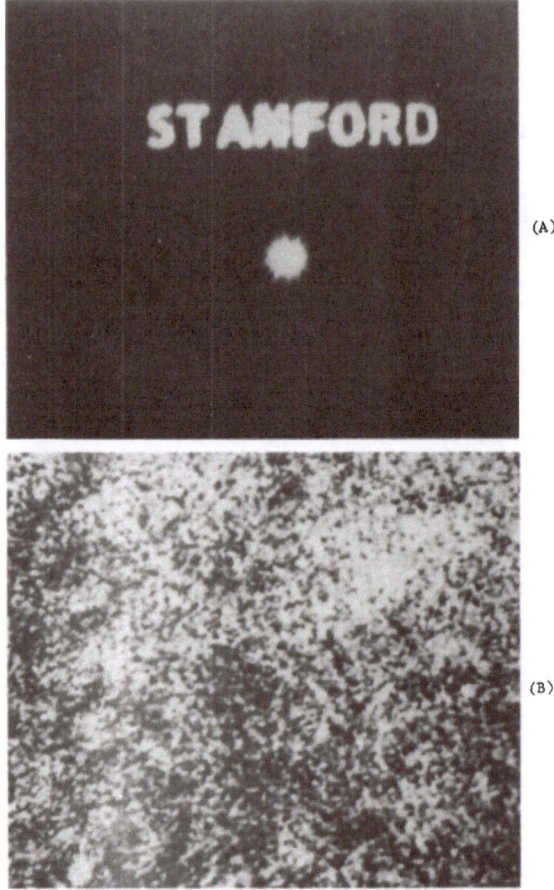

Fig. 5 - Conventional image: a) in the absence of random medium,
b) in the presence of a random medium constituted by a
shower glass (from Goodman J.V.: Introduction to Wave
Optics - Mc.Graw Hill Book Co., N.Y., 1968 p.267-268)

are located sufficiently close together, both the waves from object
and reference are subjected to equal phase and amplitude fluctuations by turbulence. The similarly aberrated wavefronts will inter
fere giving rise to a undistorted hologram, from which, an image
free from the effects of turbulence can be obtained.

The technique used is lens-less Fourier-transform holography.
The object and reference source are located equidistant from the
hologram-recording plane, where a photographic plate is located,
Fig. 3.

The interference pattern between reference wave and the waves
from object is directly recorded. Recall that the wave from the
source must be must more intense than that from the object.

The reconstructed image is the Fourier transform of the ampli
tude transmittance of the developed hologram. It is obtained by
using the Fourier transform properties of a lens. The hologram
(Fig.4) is placed in front of a positive lens and illuminated by a
collimated coherent beam. Twin images are obtained in the focal
plane of the lens, symmetrically located about a bright image of
the reference beam. This method has been successfully applied by
Goodman itself to the holography through a thin stationary phase-
perturbing plate (shower glass). In Fig.5 and 6 an example from
Goodman is reported. Fig. 5 refers to conventional imaging: a) in
the absence of random medium and b) in the presence of random medium.

Fig. 6 refers to wavefront-reconstruction imaging in the presence of random medium. As one can see, the image resolution in
the presence of random medium can be markedly improved by the holographic technique.

The method has been subsequently extended to holography
through turbulence and investigated both theoretically and experimentally by Goodman and by Gaskill (14, 15).

DESCRIPTION OF THE TURBULENCE

In order to analyze the effect of the turbulence on a hologram let us recall some features of the turbulence, with particular reference to the atmosphere. Recall that the refractive index
of a turbulent atmosphere is a random fluctuating quantity. In or
der to describe the statistical behaviour of a random fuction

$$f(P) = \langle f(P) \rangle + \delta f(P) \qquad\qquad 2.1.$$

(where brackets denote infinite time average) one usually introduces the correlation function of the fluctuations, B_f

$$B_f(P_1,P_2) = \left\langle \delta f(P_1)\ \delta f(P_2) \right\rangle \qquad\qquad 2.2.$$

and the structure function D_f

$$D_f(P_1,P_2) = \left\langle \left[\delta f(P_1) - \delta f(P_2) \right]^2 \right\rangle \qquad\qquad 2.3.$$

The function B_f gives a measure of how much the fluctuations at points P_1 and P_2 are correlated. If they are totally uncorrela ted B_f vanishes. On the contrary it reaches its maximum value, when they are totally correlated. The structure function D_f is a more general function which was introduced by Kolmogorov in order to treat processes with stationary first increments.

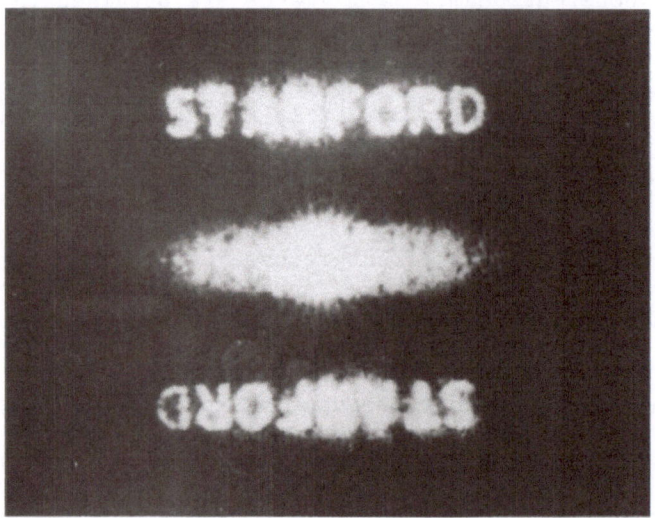

Fig. 6 - Holographic image in the presence of the same random
medium as in Fig. 5 (from Goodman J.V.: Introduction to
Wave Optics- Mc Graw Hill Co. N.Y. 1968 p. 268)

The process is called homogeneous and isotropic if B_f and D_f, depends only on the distance r apart between the two points. Here we will made as usual this hypothesis. In this case:

$$D_f(r) = 2\left[B_f(0) - B_f(r) \right] \qquad\qquad 2.4.$$

and the function B_f or D_f can be used indifferently.

Sometimes it is useful to introduce the so called spatial spectral density $\phi_f(x)$, representing the three dimensional Fourier transform of the function $B_f(r)$. Here we will limit ourselves to give some relations between $\phi_f(x)$ and $B_f(r)$:

$$\phi_f(x) = \frac{1}{2\pi x} \int_0^\infty r\, B_f(r)\, \sin(xr)\, dr \qquad 2.5.$$

$$B_f(r) = \frac{4\pi}{r} \int_0^\infty x\, \phi_f(x)\, \sin(xr)\, dx \qquad 2.6.$$

The description of the turbulence is generally made by means of the correlation B_T or the structure function D_T of the atmospheric temperature. Since at optical frequencies the refractive index of the atmosphere is related to the temperature by the equation:

$$n = 1 + \delta\mu = \frac{80\ 10^{-6}p}{\langle T \rangle^2} T \qquad 2.7.$$

where p is the pressure in millibars, the functions B_n and D_n of the refractive index are related to $B_T(r)$ and $D_T(r)$, respectively.

A given form for B_T (ore D_T) and consequently for B_n (and D_n) is called a model of the turbulence. A model of the turbulence was given by Kolmogorov on the basis of the physical processes involved in the turbulence itself. This model has been extensively used in the past, even out of the limits of its applicability.

Following the model of Kolmogorov one has:

$$D_n(r) = c_n^2\, l_o^{2/3}\, \left(\frac{r}{l_o}\right)^2 \qquad \text{for } r < l_o \qquad 2.8.$$

$$D_n(r) = c_n^2\, r^{2/3} \qquad l_o < r < L_o \qquad 2.9.$$

where c_n^2 is a constant, called "structure constant" of the refractive index; the quantity l_o, called inner scale of the turbulence, is of the order of the millimeter and depends on the kinematic viscosity of the medium, and on the rate of conversion of atmospheric energy into heat. L_o, called the outer scale, is related to l_o through the Reynold's number R_e:

$$l_o = \frac{L_o}{R_e^{3/4}}$$ 2.10.

As an order of magnitude some Authors assumed L_o of the order of the height from the ground (16,17). However this value is not confirmed by recent experimental results (15, 18, 19, 20).

For $r > L_o$ the function D_T, and consequently D_n are not well defined in the sense that an universal expression for D_T cannot be given: since the turbulence cannot be considered as homogeneous and isotropic. However there is an important physical fact which gives some indications. When the distance between the points is sufficiently large, $r \gg L_o$, the fluctuations at point P_1 are completely uncorrelated from those at point P_2, so that for $r \gg L_o$ the function B_n (P_1, P_2) vanished. This indicates (see also eq.2.3) that D_n tends to saturate. Although there is no universal form for D_n when $r > L_o$, however for the application to the study of the electromagnetic wave propagation it is necessary to know D_n also for $r > L_o$ (21, 22).

Another model, mentioned by Tatarsky (3), and recently used by Strobehen, (23), is the Von Karman model:

$$D_n(r) = 2a^2 \left[1 - \frac{2^{2/3}}{\Gamma(1/3)} \left(\frac{r}{L_o}\right)^{1/3} K_{1/3}\left(\frac{r}{L_o}\right) \right]$$ 2.11.

Here $K_{1/3}$ (x) denotes the MacDonald function of order 1/3, Γ (1/3) the Gamma function. The parameters of this model are 'a' and L_o. This model, however, while taking into account the saturation, and giving the correct dependence, 2/3, when r is sufficiently small, does not have the proper behaviour for $r < l_o$.

Here we will refer to a more practical model "Gaussian model", which has been extensively used in the past by many Authors (1, 2, 24, 25) and appeared particularly suitable at near ground levels (15, 19, 20):

$$D_n(r) = 2 \langle \delta^2 \mu \rangle \left[1 - \exp(r^2/r_o^2) \right]$$ 2.12

$$B_n(r) = \left\langle \delta^2 \mu \right\rangle e^{-r^2/r_o^2}$$

2.13.

where $\left\langle \delta^2 \mu \right\rangle$ represents the mean square fluctuation of the refracti
ve index. The quantity r_o, called 'microscale', gives an idea of
the distance at which the fluctuations are no more correlated and
can be assumed of the order of L_o. As an approximation:

$$C_n^2 \sim \left\langle \delta^2 \mu \right\rangle / L_o^{2/3}.$$

In Fig.7 the quantity $B_n(r)/\left\langle \delta^2 \mu \right\rangle$ is represented versus r/r_o.

At present there is a great deal of interest in knowing the
models of the atmosphere under different meteorological conditions.

Another problem is to derive experimentally the parameters of
a given model, for instance in the Gaussian case the quantities
$\left\langle \delta^2 \mu \right\rangle$ and r_o. These quantities are generally obtained by measuring the parameters of a wave propagated through the turbulent me-
dium. Holography can be useful way of making these measurements.

WAVE PROPAGATION THROUGH A TURBULENT MEDIUM

Let us consider now the effect of the turbulence on a wave pro
pagating through it.

As is well known, the problems in the field of optics are sol
ved by considering only one of the six components of the electromagnetic field and by treating the problem as if this component
was the only one describing the optical phenomenon. With reference to an orthogonal reference system let u denote such a component, which in the absence of turbulence satisfies the well known
wave equation:

$$\nabla^2 u_o = \frac{1}{V^2} \frac{\partial^2 u_o}{\partial t^2} \qquad V = \frac{c}{n}$$

3.1.

where ∇^2 is the Laplace operator.

The method for solving this equation consists in introducing
the so called complex amplitude v_o and searching for a monochromatic solution

$$u_o = v_o e^{-i\omega t}$$

3.2.

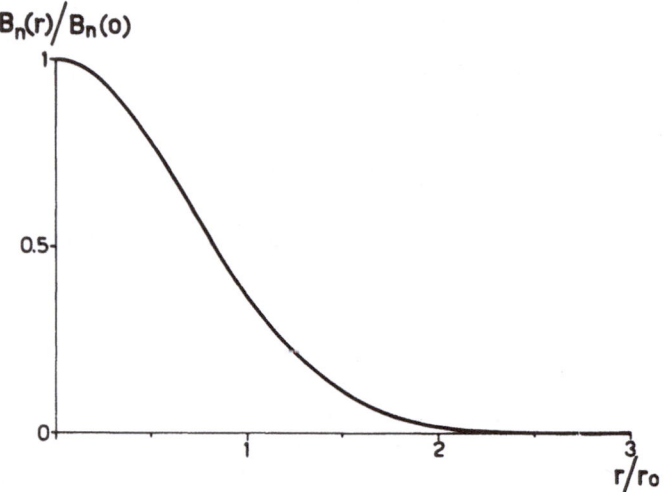

Fig. 7 - The normalised correlation function of the refractive index
          fluctuations for a Gaussian model of turbulence, plotted
          versus r/r_o

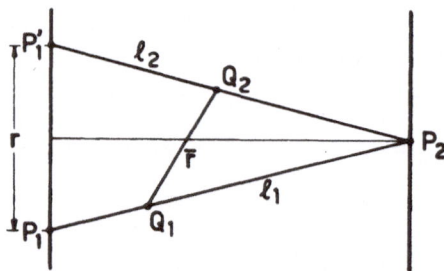

Fig. 8 - Geometry   of interest, from the holographic point of
          view, for the computation of the correlation function of
          phase fluctuations. The particular case is considered of
          P_1 and P_1' symmetrically located with respect to P_2

of equation (3.1).

The quantity v_0 can be written in the usual form

$$v_0 = A_0(x,y,z) \, e^{i\phi_0(x,y,z)}$$

3.3.

Obviously a complex quantity does not have a physical meaning, but its real or imaginary part do. A_0 is called amplitude and ϕ_0 phase. Symbols with an index zero have been used, because this quantity refers to the unperturbed wave. We will use the same symbols without index referring to a wave propagating through a turbulent medium. In our case the equations (3.1), (3.2) and (3.3) read

$$\nabla^2 u = \frac{(1+\delta\mu)^2}{c^2} \frac{\partial^2 \mu}{\partial t^2}$$

3.4.

$$u = v(x,y,z) \, e^{-i\omega t}$$

3.5.

$$v(x,y,z) = A(x,y,z) \, e^{i\phi(x,y,z)}$$

3.6.

respectively.

Due to the fact that $\delta\mu$ fluctuates, A and ϕ will fluctuate about their unperturbed values.

For our purposes it is useful to rewrite v, by introducing the unperturbed amplitude A_0.

$$v(x,y,z) = A_0 e^{i\phi+X}$$

3.7.

$$X = \ln A/A_0$$

3.8.

In order to study the holography through a turbulent medium one should know ϕ and A as solutions of the equation (3.4). Up to now only approximate solutions have been given. The simplest one is obtained by the geometrical optics and can be usefully applied to waves having a wavelength much less than the dimensions of the inhomogeneities.

assume that only phase fluctuations are present, or better that the amplitude fluctuations can be neglected with respect to the phase fluctuations. With reference to Fig. 8 let us breefly examine the solution for the phase given by the geometrical optics.

In the absence of turbulence the phase difference between points P_2 and P_1, given by the geometric optics approximation, is

$$\phi_o = k \int_{1_1} ds = ks_{12} \qquad\qquad 3.9.$$

where s_{12} is the distance from P_1 to P_2.

In the presence of turbulence the instantaneous phase is given by

$$\phi_1 = k \int_{1_1} (1+\delta\mu)\, ds_1 = \phi_o + k \int_{1_1} \delta\mu\, ds_1 \qquad\qquad 3.10.$$

The phase fluctuation from the unperturbed value at P is therefore given by

$$\delta\phi = \phi-\phi_o = k \int_{1_1} \delta\mu(Q_1)\, ds_1 \qquad\qquad 3.11.$$

where $\delta\mu(Q_1)$ are the fluctuations of the refractive index at point Q_1 on the ray-path 1_1, and ds_1 an element of 1_1.

As $\delta\mu$ fluctuates, $\delta\phi$ fluctuates. It is now an easy matter to evaluate correlation or structure functions of the fluctuations along different rays. As we will see in the sequel we are interested in knowing the structure function at point P_2 of the phase fluctuations $\delta\phi_1$ and $\delta\phi_2$ along two rays 1_1 and 1_2 starting from two mutually coherents points P_1, P_1' respectively, Fig.8.

Let us start from the correlation function from which the structure function will be derived.

From (3.11) one has

$$\delta\phi_1\, \delta\phi_2 = k^2 \int_{1_1} \int_{1_2} \delta\mu_1\, \delta\mu_2\, ds_1\, ds_2 \qquad\qquad 3.12.$$

where the abbreviate notations $\delta\mu_1 = \delta\mu(Q_1)$ and $\delta\mu_2 = \delta\mu(Q_2)$

are used. In the hypothesis of homogeneous turbulence by taking
the mean values and recalling eq.(2.2) one obtains

$$B_\phi(r) = k^2 \int_{l_1} \int_{l_2} B_n(\bar{r}) ds_1 \, ds_2 \qquad 3.13.$$

where $\bar{r}$ denote the distance between Q_1 and Q_2.

The structure function can be derived from eq.(3.13) by recalling eq. (2.4).

The correlation function of phase fluctuations of a wave propagating in a atmosphere having a gaussian model of turbulence can be obtained by introducing eq.(2.13) into (3.13). One has

$$B_\phi(r) = k^2 \langle \delta^2 \mu \rangle \int_{l_1} \int_{l_2} e^{-r^2/r_o^2} ds_1 ds_2 \qquad 3.14.$$

The evaluation of this integral in the hypothesis of a wave
path $z \gg r_o$, gives

$$B_\phi(r) = \langle \delta^2 \mu \rangle \, k^2 \, \pi \, r_o^2 \, z \, \frac{1}{2r} \, \mathrm{erf} \, (\frac{r}{r_o}) \qquad 3.15.$$

where

$$\mathrm{erf}(y) = \frac{2}{\sqrt{\pi}} \int_0^y e^{-x^2} dx \qquad 3.16$$

and consequently, by recalling Eq.(2.4)

$$D_\phi(r) = 2 \langle \delta^2 \mu \rangle \, k^2 z \, \sqrt{\pi} r_o \left[1 - \frac{r_o \sqrt{\pi}}{2r} \, \mathrm{erf}(\frac{r}{r_o}) \right] \qquad 3.17.$$

In Fig. 9 a plot is given of B_ϕ (r), normalized to its maximum value, versus r/r_o. From this graph it appears that, when r
is sufficiently small (r $<$ r_o) the phase fluctuations are correlated: when r increases the correlation goes down and vanishes for
r $\gg$ r_o.

From the application point of view a comparison is interesting
between figure 7 and figure 9. From Fig. 7 it appears that the
fluctuations of the refractive index are correlated up to a distance smaller than that of the phase fluctuations. The curves of

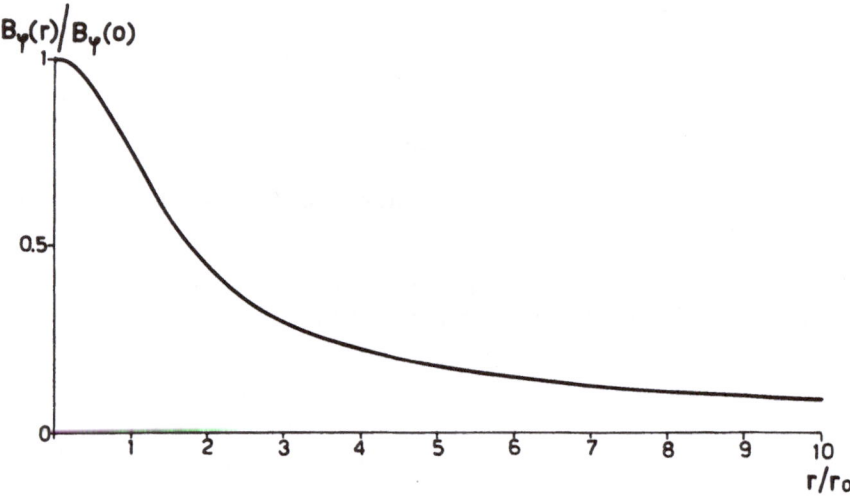

Fig. 9 - Normalized correlation function of the phase fluctuations,
plotted versus r/r_o, in the particular geometry of Fig.8

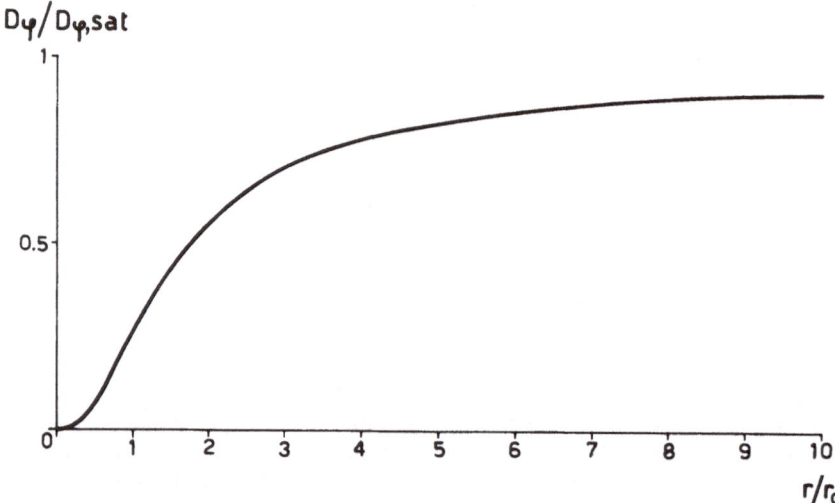

Fig.10 - Normalized phase structure function plotted versus r/r_o,
in the particular geometry of Fig.8

Figs.7 and 9 reduce to a value $\frac{1}{2}$ of their maxima for $r = r_0$, and for $r = 2.5 \, r_0$, respectively. This feature is characteristic of the geometry in which we are interested (see Fig. 8), and is due to the fact that the two rays are converging and travel a portion of the path, (near P_2) at a distance less than r_0. The same does not happen, if the two rays are parallel.

From (3.17) one can see that when $r \gg r_0$ D_ϕ tends to saturate. The saturation value D_ϕ ,sat is given by:

$$D_\phi , \text{sat} = 2 \langle \delta^2 \mu \rangle k^2 \sqrt{\pi} \, r_0 z \qquad\qquad 3.18.$$

In Fig.10 a plot of $D_\phi \left(\frac{r}{r_0} \right) / D_\phi$,sat is represented.

The geometrical optics requires the condition

$$r_0 \gg \sqrt{\lambda L} \qquad\qquad 3.19.$$

to be valid.

HOLOGRAPHY THROUGH A RANDOM MEDIUM

Let us now pass to examine the problem of the holography through a random medium. Two types of exposures can be made: long and short exposures. In the first case the time exposure of the hologram is large with respect to the characteristics times of the turbulence. In this case the quantities of interest are time averages. Short exposures refer to the opposite case, in which the turbulence can be considered frozen-in, and each hologram refers to instantaneous values of the quantities of interest. Each image will be different from the others, and one can speak only in terms of "mean image", which is obtained by taking mean values over a collection of images.

Here the long exposure case will be examined and some results for the short exposure case will be presented.

The object is assumed planar and located in the plane x_1, y_1 . The hologram is recorded at a distance z in a plane x_2, y_2 . The distance z is assumed large with respect to the other distances of interest (Fraunhofer case). Without loss of generality, we will limit ourselves to a case of cylindrical symmetry along y, so that we consider only functions of one coordinates, x. An application of the Kirchhof-Fresnel diffraction theory gives the complex amplitude $v_2 (x_2)$ in the recording plane, starting from the knowledge of the complex amplitude $v_1 (x_1)$ in the object plane. In the absence of turbulence one has:

$$v_2(x_2) = \int_{-\infty}^{+\infty} v_1(x_1) \, K(x_1,x_2) \, dx_1 \qquad\qquad 4.1.$$

where

$$K(x_1,x_2) \propto \frac{e^{ik\varrho}}{\sqrt{\rho}} \qquad\qquad 4.2.$$

where ρ denotes the distance from the points x_1 and x_2.

In the usual approximations of the diffraction theory

$$K(x_1,x_2) \propto \exp\left\{ i\left[kz + \frac{k}{2z} \, (x_2 - x_1)^2 \right] \right\} \qquad\qquad 4.3.$$

The presence of the turbulence introduces an additional phase fluctuation $\delta\phi$, so that K must be replaced by

$$K_1 = K \, e^{i\delta\phi} \qquad\qquad 4.4.$$

We are interested in the intensity in the recording plane:

$$I(x_2) = v_2(x_2) \, v_2^*(x_2) \qquad\qquad 4.5.$$

Due to the fact that the fluctuations due to the turbulence are rapid, with respect to the exposure time, during an exposure the phase fluctuation will take an ensemble of random values, so that one has to deal with the mean value $\langle I(x_2) \rangle$ of $I(x_2)$, given by:

$$\langle I(x_2) \rangle = \langle v_2(x_1) v_2^*(x_2) \rangle \qquad\qquad 4.6.$$

By using Eq.(4.1) and (4.4) the equation (4.6) becomes, apart from constant factors:

$$\langle I(x_2) \rangle = \int_{-\infty}^{\infty} \int_{-\infty}^{\infty} v_1(x_1) v_1^*(x_1') \frac{e^{ik(\rho_1 - \rho_2)}}{\sqrt{\rho_1 \, \rho_2}} \left\langle e^{i(\delta\phi_1 - \delta\phi_1')} \right\rangle dx_1 \, dx_1' \qquad 4.7.$$

The factor

$$m(r) = \left\langle e^{i(\delta\phi_1 - \delta\phi_1')} \right\rangle \qquad\qquad 4.8.$$

represents the effect due to the turbulence. We recall that $\delta\phi_1$ denotes the phase fluctuation at then point x_2 over the recording plane, suffered by a ray starting from the point x_1 on the object plane (see Fig.8), and $\delta\phi_1'$ that is corresponding to the point x_1'.

It has been shown by several Authors (5,6) that, in the hypothesis, generally accepted, that the statistics of the phase fluctuations be Gaussian, one has:

$$m(r) = \left\langle e^{i(\delta\phi_1 - \delta\phi_1')} \right\rangle = \exp\left[-\frac{1}{2} \left\langle (\delta\phi_1 - \delta\phi_1')^2 \right\rangle \right]$$

where $\left\langle (\delta\phi_1 - \delta\phi_1')^2 \right\rangle$ represents the structure function $D\phi$ of the phase at the point x_2 due to the rays starting from x_1 and x_1'.

For the Gaussian model of turbulence $D\phi(r)$ has been evaluated in section 3, Eq.(3.16) in the particular case that x_1 and x_1' are symmetric with respect to x_2. One has:

$$m(r) = \exp\left[-\frac{1}{2}D\phi(r) \right] \qquad\qquad 4.9.$$

By introducing eq.(4.9) into (4.7) and with the standard approximations (4.3) one obtains (see Appendix 1):

$$\left\langle I(x_2) \right\rangle = F_2 \left\{ m(r) \int_{-\infty}^{\infty} w_2(x_1)w_2^*(x_1-r) \, dx_1 \right\} \qquad 4.10.$$

where $F_2 = F(\frac{x_2}{\lambda z})$ denotes the Fourier transform at the spatial frequency $\frac{x_2}{\lambda z}$ and

$$w_2(x) = v_1(x) \exp(ikx^2/(2z)) \qquad\qquad 4.10'.$$

In the absence of turbulence $D\phi = 0$ and $m(r) = 1$, consequently the Fourier transform of the integral of Eq. (4.10) represents $I(x_2) = \left\langle I(x_2) \right\rangle$ in the absence of turbulence.

The intensity is now recorded on the hologram. Precisely according to the photographic process, the exposure $E(x_2)$ of the photographic plate is given by

$$E(x_2) = \tau \left\langle I(x_2) \right\rangle \qquad \qquad 4.11.$$

where τ denotes the exposure time.

It has been observed by Kozma (26) that, in the case of high contrast films, used for the holography, it is convenient to operate in the linear region of the curve of the amplitude transmittance t, versus the exposure E, rather than in the linear region of the more usual curve density D, versus $\log E$. Accordingly the amplitude transparency of the plate is given by:

$$t(x_2) = t_o - \sigma \tau \left\langle I(x_2) \right\rangle \qquad \qquad 4.12.$$

where σ is a coefficient different from the γ of the plate.

Let us now examine the reconstruction process, (Fig.4). The plate is located in front of a positive lens, of focal length f, and illuminated by a plane monochromatic wave. The wavelength is generally the same as for the recording process.

In the absence of turbulence the reconstructed image consists of a point, due to the reference plane wave, and a pair of images of the object, symmetrically located, with respect to the spot (in Appendix 2 a simple demonstration of this assertion is reported.

In order to analize the problem in the presence of turbulence let us neglect the limiting effect of the lens (that is, we will assume that the diffraction effect of the lens be small).

In this hypothesis, the modulus of the complex amplitude $u(x_3)$ in the focal plane, is proportional to the modulus of the Fourier transform of the transparency. One has

$$\left| u_3(x_3) \right| = \left| F\left[t_o - \sigma \tau \left\langle I(x_2) \right\rangle \right] \right| \qquad \qquad 4.13.$$

The term containing t_o is the on axis image of the reference plane beam and does not give information. By neglecting also constant factors, which are not of interest, and recalling (4.10) Eq. (4.13) can be written as:

$$\left| u_3(x_3) \right| = \left| F_3 \left\{ F_2 \left[m(r) \int_{-\infty}^{\infty} w_2(x_1) \, w_2^*(x_1 - r) \, dx_1 \right] \right\} \right| \qquad 4.14.$$

where F_3 denotes the Fourier transform at the frequency $x_3/(\lambda f)$.

We recall here that a Fourier transform of a function $f(x)$ does not represents the function, but $f(-x)$. This means that in our case one has to deal with an inverted image.

In conclusion one has, apart from constant factors:

$$\left| u_3(x_3) \right| = \left| m(-\frac{x_3 z}{f}) \int_{-\infty}^{\infty} w_2(x_1) \, w_2^* \left(x_1 - \frac{x_3 z}{f} \right) dx_1 \right| \qquad 4.15.$$

We are interested in the intensity $I_3 = u_3 u_3^*$. By observing that $m(r)$ is a real even function, I_3 can be written as:

$$I_3(x_3) = \left[m \left(\frac{x_3 z}{f} \right) \right]^2 \left| \int_{-\infty}^{\infty} w_2(x_1) w_2^* \left(x_1 - \frac{x_3 z}{f} \right) dx_1 \right|^2 = m^2 \left(\frac{x_3 z}{f} \right) I_0 (-\frac{x_3 z}{f}) \qquad 4.16.$$

As the factor m^2 represents the effect due to the turbulence and tends to 1 in the absence of turbulence, clearly I_0 represents the image in the absence of turbulence. It is worth noting that this effect appears as a factor in the intensity. Therefore the intensity, at a point P in the image, depends on the object only through the intensity in the corresponding point of the object, as it happens in the absence of turbulence. This conclusion would be not valid if, due to the turbulence, each point of the object was imaged in a spot. More precisely, each point of the object gives a spot in the image plane, however this is due to the finite dimen sions of the lens, here neglected, and not to the turbulence (14). In other words there is not blurring of the image due to turbulence.

The function $m^2(r)$ has its maximum when $r=0$, then decreases. In fig. 11 a plot is given of the quantity

$$m_1^2(r) = \exp \left\{ - \frac{2 \, D_\phi(r)}{D_\phi, sat} \right\}$$

in the case of a Gaussian model of turbulence in the particular geometric situation analized in section 3. Obviously, in order to obtain $m2(r)$ one has to take the $D_{\phi,sat}$-th power of $m_1^2(r)$.

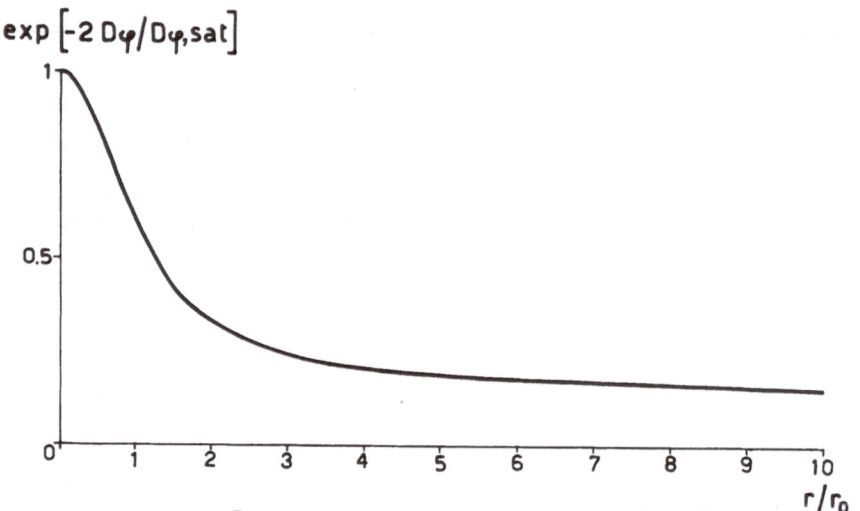

Fig.11 - A plot of m_1^2 (r) = exp $[-2D_\varphi$ (r) $/D_{\varphi,sat}]$ for a Gaussian
        model of turbulence versus r/r_o

The quantity $D_{\phi,sat}$ (Eq.3.19) depends on the parameters of
the turbulence $r_o, \langle \delta^2 \mu \rangle$, as well as on the wavelength and on the
path-length in the medium.

If, for instance $D_{\phi,sat}$ if of the order of one radiant m^2(r)
reduces to 1/5 of its maximum value for $r=4.5r_o$. If $D_{\phi,sat}$ is se
veral radiants, as it has been found in some practical situations,
(15,19) then m^2(r) goes more rapidly down. For instance, if
$D_{\phi,sat}$ = 5 rad, m^2(r) reduces to $3,10^{-4}$ of its maximum for the sa
me value of r. In this case the effect of the factors m^2(r) is a
reduction of the effective field of view. Within the field there
is no blurring of the image.

These results have been qualitatively confirmed by experimental holographic images taken through an artificial turbulence (14).
Figs.12 and 13 show some results from Ref.14.

In Fig. 12 are represented: a) conventional image and b) reconstructed holographic image, without turbulence are represented.
Fig. 13 shows an experimental result with the artificial turbulence. One clearly see the reduced dimensions of the holographic i-
mage.

The dimensions of the visible part are related to the micro-

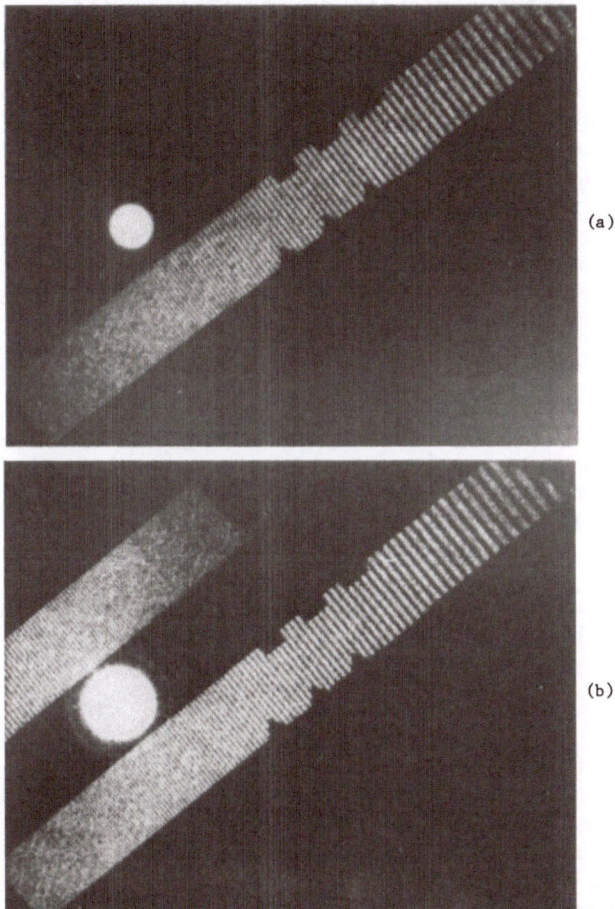

Fig.12 - Long exposure: a) Conventional and b) holographic images
in the absence of turbulence (from J.Opt.Soc.Am., 58,
pag. 605, May 1968)

scale of the turbulence. This result suggests that holographic i-
mages can be used as a means for investigating the turbulence.

As to the problem of short exposures, here we will limit ourselves to describe the results due to Gaskill (14). In this case
"short time" means an exposure time short compared with the fluctuation time of the medium. In practical situations an exposure
time of 1/200 sec can be considered short. Here one has to deal
with an instantaneous value for the intensity at the image plane,
then one has to average over a collection of images. In Ref.14
it has been shown that, in this case, there is no limitation of
the field of view, due to the turbulence, differently from the case of long exposure. Moreover there is a limited region of good
resolution. In this region a turbulence dependent background is
superimposed on the image that one could obtain in the absence of
turbulence. Outside, this region, the resolution decreases down
to a value 2/3 of that of an usual long exposure conventional image.

At present in the literature no result is found of short exposure experiments made in the presence of turbulence, even artifi
cial. The turbulence is generally simulated by means of a shower
glass plate. This is possible because in this case the situation
is frozen-in and the medium does not varies during the exposure.

An example of holography through a glass plate, made by Goodman has been shown in Fig.6: the corresponding conventional image
in Fig. 5 b).

REFERENCES

(1) S. Chandrasekhar, Mon.Not.R.Astr.Soc.112,475, 1952.
(2) L.A. Chernov, Wave propagation in a random medium, McGraw-
     Hill, New York 1960.
(3) V.I. Tatarski, Wave propagation in a turbulent medium, McGraw-
     Hill Book Co. - New York 1961.
(4) J.V. Ramsay, Optica Acta 6, 344, 1959.
(5) A. Consortini, L. Ronchi, A.M.Scheggi, G. Toraldo di Francia,
     Influence of the atmospheric turbulence on the space coheren-
     ce of a laser beam. Alta Frequenza 33, 714-154E (1963).
(6) R.E. Hufnagel and N.R. Stanley, Modulation transfer function
     associated with image transmission through turbulent media.
     J.Opt.Soc.Am., 54, 52, Jan. 1964.
(7) G.O. Reynolds and T.J. Skinner, Mutual coherence function ap-
     plied to imaging through a random medium J. Opt. Soc. Am. 54,
     1302 (1964).
(8) C.E. Coulman, Dependence of image quality on horizontal range
     in a turbulent atmosphere. J. Opt.Soc.Am. 56, 1232, Sept.1966.
(9) D.L. Fried, Optical resolution through randomly inhomogeneous
     media for very long and very short exposures J. Opt. Soc. Am.

56, 1372, Oct.1966.

(10) N.E. Leith and J. Upatnieks, Holographic imaging through diffusing media. J.Opt.Soc.Am.56, 523 (1966).

(11) H. Kogelnik, Holographic image projection through inhomogeneous media, Bell System Techn. J. 44 2451 (1965).

(12) J. Upatnieks, A. Vander Lugh, E. Leith, Correction of lens aberrations by means of holograms. Appl. Optics 5, 589 (1966).

(13) J.W. Goodman, W.H. Huntley jr., D.W. Jackson and M. Lehmann, Wavefront-reconstruction imaging through random media. Applied Physics Letters 8, 311, 1966.

(14) J.D. Gaskill, Imaging through a randomly inhomogeneous medium by wave front reconstruction. J. Opt. Soc. Am. 58, 600, 1968.

(15) J.D. Gaskill, Atmospheric degradation of holographic images. J. Opt. Soc. Am. 59, 308, 1969.

(16) D.L. Fried and J.C. Cloud, Radio Science 1, 405 (1966).

(17) V.I. Tatarski, Wave propagation in a turbulent atmosphere. Nayka, Moscow 1967 (in Russian).

(18) M. Bertolotti, M. Carnevale, L. Muzii, D. Sette, Applied Optics 7, 11 (1968).

(19) P. Burlamacchi, A. Consortini, L. Ronchi, G. Toraldo di Francia: Alta Frequenza (Special Issue) 38, 149 (1969).

(20) A. Consortini, L. Ronchi, L. Stefanutti, Applied Optics 9 2543 (1970).

(21) A. Consortini, L. Ronchi, Lettere al Nuovo Cimento 2, 683 (1969).

(22) R.F. Lutomirski and H.T. Youra, J. Opt. Soc. Am. 61, 482 (1971)

(23) J.W. Strohbehn, J. Geophys. Res.71, 5793 (1966).

(24) P. Beckmann, Radio Sci. 69D, 699 (1965).

(25) A. Consortini, L. Ronchi, A.M. Scheggi, G. Toraldo di Francia, Radio Sci. J. Res. 1, 523 (1966).

(26) A. Kozma, J. Opt. Soc. Am. 56, 428 (1966).

(27) G. Toraldo di Francia, Ottica 8, 3, (1943).

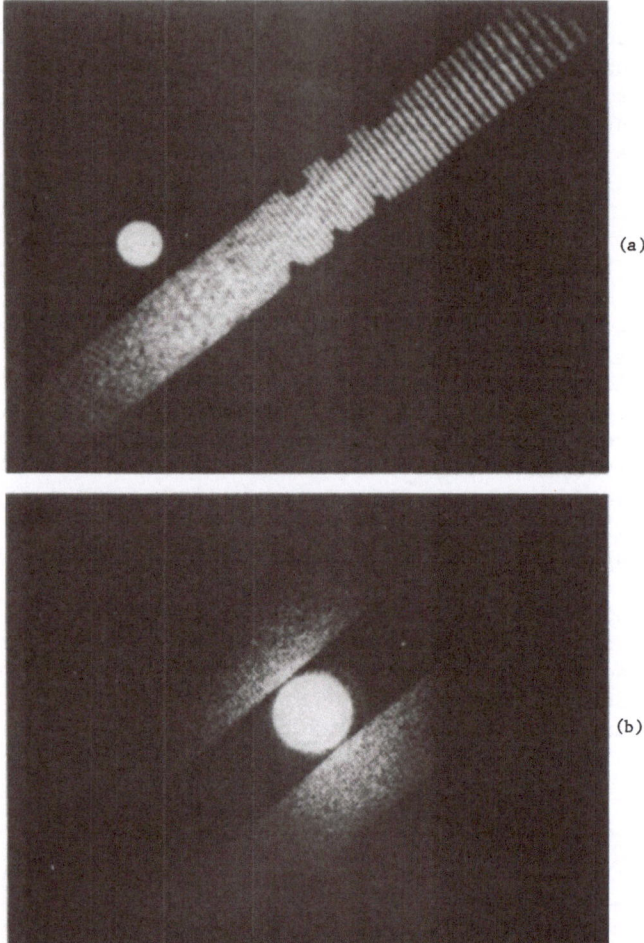

Fig.13 - Long exposure: a) Conventional and b) holographic images
          in the presence of turbulence (from J.Opt.Soc.Am., 58,
          pg. 606, May 1968)

APPENDIX 1

Starting from eq. (4.7) let us introduce the approximation $x_1, x_2 \ll z$. The quantity $\rho_1 - \rho_2$ results

$$
\begin{aligned}
\rho_1 - \rho_2 &= \frac{1}{2z}(x_2 - x_1)^2 - \frac{1}{2z}(x_2 - x_1')^2 = \\
&= \frac{1}{2z}(x_2^2 - 2x_2 x_1 + x_1^2) - \frac{1}{2z}(x_2^2 - 2x_1' x_2 + x_1') \\
&= \frac{1}{2z}(x_1^2 - x_1'^2) - 2x_2(x_1 - x_1')
\end{aligned} \qquad \text{A.1-1.}
$$

With the position

$$
x_1 - x_1' = r
$$

$$
x_1 = x_1
$$

Eq. (4.7) on account of (A.1.1), becomes

$$
\langle I(x_2) \rangle = \int_{-\infty}^{\infty} \left\{ \int_{-\infty}^{\infty} v_1(x_1) e^{ik\frac{x_1^2}{2z}} v_1^*(x_1 - r) e^{\frac{ik}{2z}(x_1 - r)^2} m(r) dx_1 \right\} e^{-ik\frac{x_2}{z}r} dr
$$

The external integral represents the Fourier transform of the quantity in parenthesis, taken at the frequency $x_2/\lambda z$. With the position (4.10') Eq. (4.10) is immediately derived.

APPENDIX 2

In order to analyze the reconstruction process let us start from the simple case when both object and reference are point sources, located at the two points x_1 and x_1' respectively.

The amplitude of the source, a_2, is assumed larger than that of the object, here assumed unity. For simplicity without loss of generality the source and object are supposed in phase. In this case, apart from constant factors, the complex amplitude on the recording plane is given by

$$v_2(x_2) = a_r e^{ik\rho_1} + e^{ik\rho_2}$$
A.2-1.

and consequently

$$v_2(x_2)v_2^*(x_2) = (a_r^2 + 1) +$$

$$+ a_r e^{ik(\rho_1 - \rho_2)} + a_r e^{-ik(\rho_1 - \rho_2)}$$
A.2-2.

In the approximations of Appendix 1 one can write

$$I(x_2) = v_2(x_2)v_2^x(x_2) = (a_r^2+1)+a_r \exp\left[\frac{ik}{2z}(x_1^2-x_1'^2)\right] \times$$

$$\times e^{-\frac{ik}{z}(x_1-x_2')x_2} + a_r \exp\left[-\frac{ik}{2z}(x_1^2-x_1'^2)\right] e^{\frac{ik}{z}(x_1-x_1')x_2}$$
A.2-3.

Obviously this quantity is real, in fact the sum of the second and third term is a cosine function.

In the reconstruction process the amplitude trasparency $t(x_2)$ is proportional to $I(x_2)$ so that apart from constant factors the complex amplitude $u(x_2)$ on the plane π of the plate in front of the lens is still given by (A.2-3) which can be written as:

$$u(x_2) = u_o + u_1 e^{ikax_2} + u_1^* e^{ikax_2}$$
A.2-4.

where

$$u_o = a_r^2 + 1$$

$$u_1 = a_r \, \exp\left[\frac{ik}{2z}(x_1^2 - x_1'^2)\right] \qquad \text{A.2-5.}$$

and

$$a = \frac{1}{z}(x_1 - x_1') \qquad \text{A.2-6.}$$

Recall now the principle of the inverse interference (27). Let Σ be a material surface with amplitude transmittance $t(P)$, separating the space in two regions S_1, S_2 Fig. A1, on which a wave impinges propagating from S_1. Let $A(P)$ and $\phi(P)$ be the amplitude and phase of the field on the emergence surface of Σ (on the side of S_2). The inverse interference principle states that: if a set of waves 0_1, 0_2, 0_n is found, propagating in S_2 away from Σ, which by interfering on Σ, give rise to a complex amplitude distribution $A(P) \exp\left[i\phi(P)\right]$, the actual field at any point Q of S_2 is given by $0_1(Q)+0_2(Q)+\ldots+0_n(Q)$.

In our case Σ is the plate. The terms of eq. (4.4) can be treated separately. Let us start from the first term. The field which reproduces the real quantity u_o is a plane wave propagating normally to Σ :

$$A_s = u_o e^{ik\xi} \qquad \text{A.2-7.}$$

where ξ is a coordinate normal to Σ . The effect of the lens is simply to focus this wave into its focus.

Let us now pass to the second term of (A.2-2)

$$u_1 e^{ikax_2} \qquad \text{A.2-8.}$$

Due to the linear dependence on x_2, this term can be consider ed as the field on Σ of the plane wave

$$u_1 e^{ikax_2 + ik\sqrt{1-a^2}\,\xi} \qquad \text{A.2-9.}$$

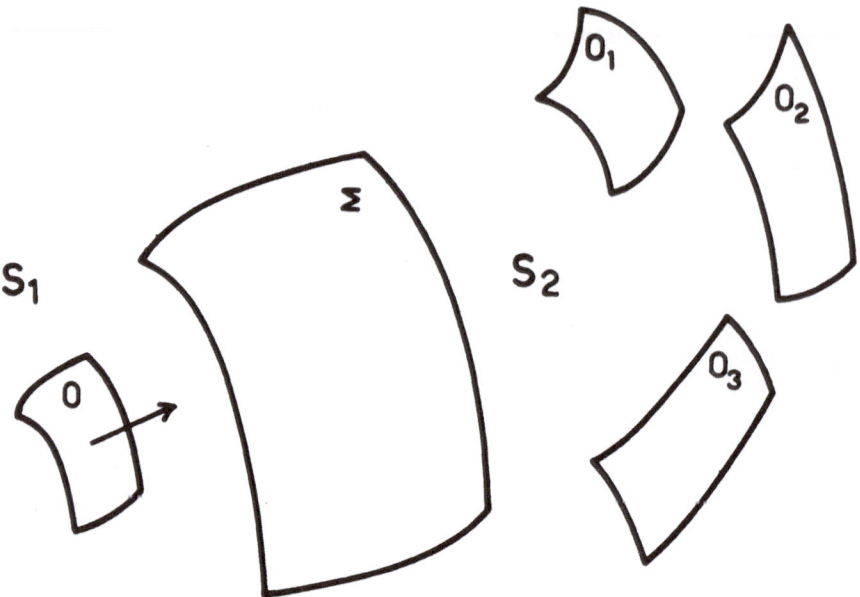

Fig.A.1- Inverse interference principle: O incident wave, Σ material
surface, O_1, O_2, O_3 diffracted Waves.

This plane wave propagates in the direction making with the
axis x on Σ an angle whose cosine is given by (A.2-6).

This wave is focused by the lens in a point of the focal plane at a distance h from the axis given by

$$\frac{f}{h} = \tan \alpha \qquad\qquad A.2\text{-}10.$$

as it can be immediately seen by considering the ray passing through
the optical center of the lens.

In our hypothesis $z \gg x_1, x_1'$, by using eq.(A.2-6), one obtains

$$\frac{f}{h} \simeq \frac{z}{(x_1 - x_1')}$$

that is

$$h = \frac{(x_1 - x_1')f}{z} \qquad\qquad A.2\text{-}11.$$

The same analysis can be made for the third term of eq.(A.2-4). Here the wave makes an angle $\pi - \alpha$ with respect to the x_2 axis, that is it is symetrically directed with respect to the axis of the lens. This wave is focused in the focal plane at a distance $- h$ from the axis.

ADAPTING HOLOGRAM INTERFEROMETRY TO INDUSTRIAL CONDITIONS

Nils H. Abramson

Laser Research Group Division of Production Engineering

Royal Institute of Technology Stockholm (Sweden)

In 1967 a "Laser research group" was formed at the division
of Production Engineering, Royal institute of Technology, Stockholm, Sweden. The aims of the group was to study, adapt and develop methods using the laser in industrial applications. The work
was very soon split into two categories: Machining and measurement.
Measurement was split into: Aligment, surface inspection, ordinary
laser interferometry and hologram interferometry. Hologram interferometry was used for the measurement of dimension, deformation
and vibration. It was soon found that the methods described in
the literature were not always sufficient for the practical problems the group had to face as it offered hologram service to the
Swedish industry.

The methods described in the following are attempts to better
adapt hologram interferometry to workshop conditions. The methods
are based on the idea that they should be easy to use and to under
stand. The customer who is often a mechanical engineer should be
able to check the results without having to learn complicated optical theories and equations. The research work was made by the
author and sponsored by the Swedish Board for Technical Development.

INTRODUCTION

Optical interferometers are very important scientific tools.
The standard unit of length is based on interferometric measurement
using the wavelength of krypton 86 and our view of the whole world
of Physics is greatly dependent on the unit of length. Einstein's
relativity theories were constructed to explain the results of some very important interferometric experiments made by Michelson
and Moorly. The invention of holographic interferometry has in the

last years to a great extent widened the possibilities of interferometric measurement.

The application of interferometry to industrial metrology has
however not reached the magnitude that could be expected. Even if
there are some very sensitive and very accurate measuring and inspection interferometers commercially available, and even if the
uses of optical metrology has widended since the invention of the
laser, the possibilities of interferometry for engineering uses
are still only utilized to a relatively small extent.

With these opinions in mind the laser research group has worked out new methods with which the applications of ordinary and holographic interferometry for visual inspection can be better adapted to workshop conditions. It is the hope of the group that these
methods will widen the uses of interferometry so that it will become just as an important tool to Engineering as it already has
been to Science.

ADVANTAGES AND DISADVANTAGES OF HOLOGRAM INTERFEROMETRY

Hologram interferometry can be used for measurement of dimension, deformation, vibration and changes in refractive index. It
is a general agreement that hologram interferometry has following
advantages to ordinary interferometry.

1)   Displacement of non optical rough surfaces that scatters light
in a diffuse way can be measured with interferometric accuracy because each small surface defect is compared with it's own image.

2)   An object can be compared with itself as it was during earlier
exposure. Thus the dimensions of the object during two or more
conditions can be compared.

3)   The amplitude of vibration can be measured by means of a time
averaging method.

4)   When the measuring object is to be compared with a master, the
latter can be substituted by it's holographic image which might
make possible a saving of time and space.

Some properties of hologram interferometry that have been
disadvantages to its engineering uses are:

1)   The complicated and timeconsuming evaluation of the fringe
pattern. The sensitivity is influenced by both the angles of illumination and that of observation in respect to the displacement
to be measured and these angles may in a hologram vary independently over the surface of the object.

2)   The often too high sensitivity of the technique which makes
necessary a high stability and a good isolation from vibrations of
the system used during exposure of the hologram.

3)   The limited size of the holographic image defined by the cohe
rence length of the light.

INTRODUCTION OF A NEW DIAGRAM, THE "HOLO-DIAGRAM"

The simplest possible interferometer consists of just a flat
place of glass that is placed in contact with the surface to be
studied. Light illuminating this device is divided into two components as it is partly reflected by, partly passing through the
glass surface nearest to the object. The latter component is reflected by the object surface and at the glass surface recombined
with the first.

If the difference in path-length between those two components
is shorter than the coherence length of the light, interference
fringes are formed that like the level-lines of a map reveal the
macro- and microscopic topography of the object surface. Each
fringe corresponds to a difference in path length between the two
components of one wavelength. The distance between two adjacent
fringes corresponds to a level difference of half the wavelength
if the object is illuminated and studied in a direction that is
normal to the surface. In all other cases the level differences
is greater.

The level difference between two points on the studied surface
thus can be calculated:

$$h = k \cdot n \cdot \frac{\lambda}{2}$$

where h is the level difference

n is the number of fringes between the two points

λ is the wavelength of the light in use

k is a constant, introduced by the author, that is a function of the angles of illumination and of observation

When illumination and observation are normal to the studied
surface (as in ordinary interferometers) the k-value is 1. For all

other angles the k-value is greater than 1. The flat glass plate
(the reference surface) which was placed in contact with the object
surface might be substituted by it's mirror image as for instance
in the Michelson interferometer, or by it's holographic image as in
Hologram Interferometry. The fringe formations dependence of the
k-value and of the coherence length of the light can be calculated
in exactly the same way whether the real reference surface is used
of if it is substituted by it's image.

     The work by the group has been concentrated on following
items:

1)   A method to optimize the utilization of the coherence length
of the light.

2)   A study of how the k-value can be varied in ordinary and holo
graphic interferometry.

3)   A method to simplify the evaluation of the k-value.

4)   A study of the advantages that can be gained by a planned variation of the k-value.

     A diagram has been constructed named the Holo-diagram (short
for hologram diagram) which has turned out to be a useful device
for the use of both ordinary and holographic interferometry (see
reference 1).
The diagram (Fig. 1) is based on a set of ellipses with the two
common focalpoints at the origin of the illumination (A) respective
ly the point of observation (B). The lightpath from A to B via an
object (C) is constant if C is moved along one ellipse. The diffe
rence in light path when the object is moved from one ellipse to
an adjacent is constant for the whole set of ellipses. The k-value
is a direct function of the distance between adjacent ellipses. The
position lines of constant k-value are formed by arcs of circles
which are drawn in the diagram with respective k-value indicated.

     Optimizing the use of the coherence-length available and eva-
luation of the fringes is made in a quick and easy way using this
diagram and the displacement thus evaluated is the projection of
the real displacement on to the normal of the ellipse that inter-
sects the object surface. The positioning of the object to get a
fringe formation of wanted resolution and wanted direction of sensitivity is also easily performed. It is also in this way possible
to minimize the destructive influence of an unwanted movement or
vibration of the object during exposure (see reference 2).

THE HOLO-DIAGRAM APPLIED TO FRINGE EVALUATION

     By utilization of it's threedimensional properties it is possi

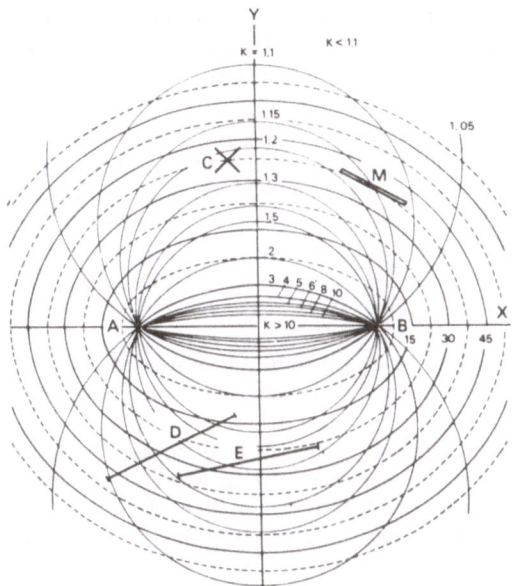

Fig. 1 The Holo-Diagram

ble to get much more information from an interference hologram than
from an ordinary photo of an interference pattern (see reference 3).

Using three different points of observation during reconstruc
tion of a hologram it is possible to find not only one projection
of the displacement but to evaluate the amplitude and direction of
the real displacement in all three coordinates. It is also possible to calculate the displacement even if the fringes between the
point of observation and the fix point are not in sight or does
not even exist. All these calculations are made in a simple way
by the use of the Holo-diagram and as the trigonometric calculations are already built into this diagram all that is needed is a
holo-diagram, a slide rule, a measuring rod and a pencil.

The holo-diagram can also be used the other way round. It
can be used for the prediction of fringepatterns caused by simple
basic object motions (e.g. translation or rotation). There exists
many, more or less complicated mathematical methods for the predic
tion of fringepatterns but the use of the holo-diagram appears to
be the most direct method that often with no calculations at all
quickly produces an approximate view of the patterns.

Many complicated patterns have been predicted by the use of
the diagram and verified by experiments. Some patterns are sur-
prisingly complex because we have all the time been looking for the
most extreme cases which give most information of the main structure of the patterns. At large the result corresponds to results
presented by other workers in this field but there does exist some
interesting differences which mainly have been observed because of
the generality of the holo-diagram.

It is found that the three most important points for the fringe
formation are:

  The point source of illumination (A)
  The point of observation (B)
  The tangential point of the object surface and an ellipsoid
  with the focal-points A and B

The theory of the holo-diagram is relevant both for diffuse
and specular reflection. Special diagrams for pure translation of
the object in the plane of the holo-diagram have been presented.
There exists in the vicinity of A and B an area of nearly constant
sentivity to translations perpendicular to AB. Just one observation of the patterns in some cases might be sufficient for an ab-
solute measure of the object displacement even if no fix point of
the object is in sight.

Finally we believe it will never be possible in a general way to interpret a complex interference pattern using just one static study of its apparent localization in space. Instead we believe in a dynamic study of the parallax by a continous study of the fringes as the point of observation is moved around.

The technique to change the k-value in order to gather more information from a pattern of interference fringes can also be utilized in ordinary interferometers. This way it is possible to construct interferometers with continuously variable sensitivity, which produce fringes even on very rough surfaces (see references 5,6).

Finally it has recently been found that there exists an analogy between hologram interference fringe patterns and the moiré patterns formed by two sets of the holo-diagram. Using this analogy the hologram interference fringes caused by any sort of in-plane translation or rotation can be simulated and studied in a simple way (see reference 10).

SOME METHODS TO LOWER THE SENSITIVITY TO UNWANTED MOVEMENTS

If holograms of large objects are to be made in workshop conditions it is often found that unwanted movements and vibrations destroy the formation of the holographic image. To solve this problem we have tried three ways.

First, as already told, we have used the holo-diagram to find out where to position the object in relation to the point of illumination and that of observation so that the influence of movements is minimized.

The second method which does not lower the resolution is to fix the reference mirror to the object (see reference 9).

A third way that we have found very useful is to substitute hologram interferometry by the moiré patterns of projected fringes (see reference 7, 8). With that method it is possible to make a picture of the object to be measured with its surface covered by fringes very similar to those of an interference hologram. These fringes can be used to measure dimension, deformation or vibration but the sensitivity to fringeformation can be changed from that of a hologram down to any value, for instance one fringe per centimeter of displacement.

With the methods described here we have tried to build a bridge between the very sensitive, very delicate laboratory interferometric measurement methods that already exist and the robust engineering measuring tools that are needed in, or very close to, the production line.

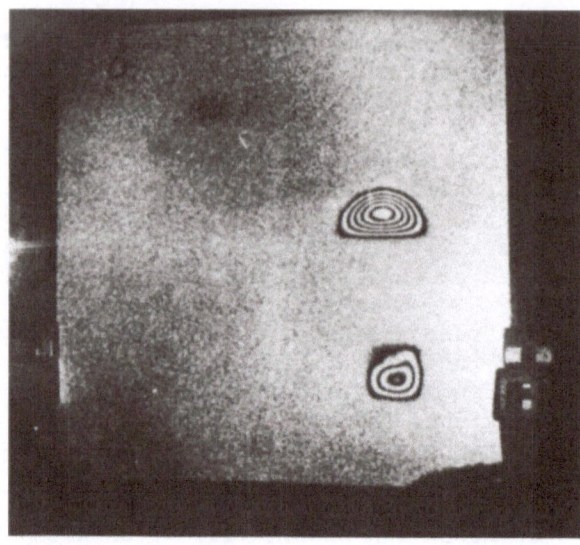

Fig. 2   Non Distructive Hologram Test of the bonding of a sandwich
         contsruction for airplanes. A pressure difference of 27 mbar
         between the two exposures reveals two errors.

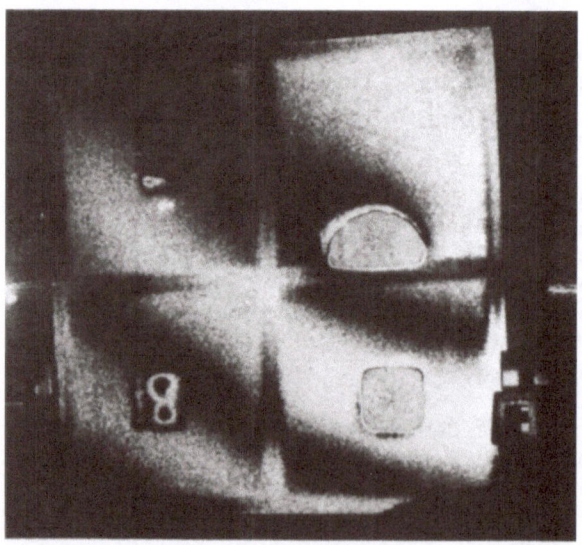

Fig. 3   When the pressure difference was rised to 200 mbar more
         errors are found

REFERENCES

(1) Nils Abramson: The "Holo-diagram" a Practical Device for Making
      and Evaluating of Holograms, Applied Optics, 6. 1235 (1969).
(2) R.L. Kurtz, H.Y. Loh: A holographic Technique for Recording a
      Hypervelocity Projectile with Front Surface Resolution. Applied
      Optics, 5. 1040 (1970).
(3) Nils Abramson: The "Holo-diagram" II: A Practical Device for
      Information Retrieval in Hologram Interferometry. Applied
      Optics, I. 97 (1970)
(4) Nils Abramson: The Holo-diagram III: A Practical Device for
      Predicting Fringe Patterns in Hologram. Interferometry. Appli-
      ed Optics, IO. 2311 (1970)
(5) Nils Abramson: The "Interferoscope" a new type of interfero-
      meter with Variable Fringe Separation. Optik, Zeltschrift für
      Licht- und Elektronenoptik, I. 56 (1969).
(6) J.D. Bries: Interferometric Flatness Testing of Nonoptical
      Surfaces. Applied Optics, I3. 519 (1971)
(7) Nils Abramson: "Interferometric Holography" Without Holograms.
      Laser Focus, 26. December 1968.
(8) Robert E. Brooks, Lee o. Heflinger: Moiré-Gauging Using Optical
      Interference Patterns. Applied Optics, 5. 935 (1969).
(9) Nils Abramson: A single Arrangement for Making Holographic
      Interferometry. Discussion. Proceedings of the Symposium on
      "Lasers and the Mechanical Engineer". The Inst. Mech. Engrs.,
      1968 - 1969, pp. 85.
(10) Nils Abramson: Moiré Patterns and Hologram Interferometry.
      Nature, May I7, 1971, page 65.

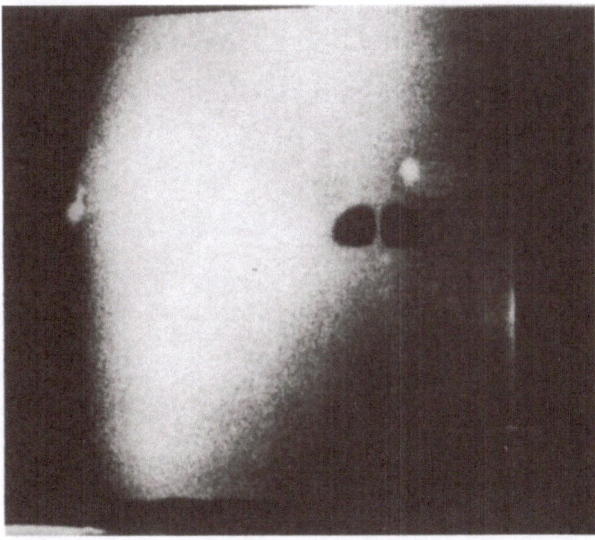

Fig. 4 A vibration of 4800 Hz reveals two of the errors. This work
        was made in cooperation with SAAB, the Swedish Aircraft
        Industry.

ADVANCED CONCEPTS OF HOLOGRAPHIC NONDESTRUCTIVE TESTING

Leonard A. Kersch

G C O, Inc.

Ann Arbor, Michigan (U.S.A.)

INTRODUCTION

In three short years holographic interferometry has gone from a curiosity looking for a problem to solve to a viable tool for use in nondestructive testing.

Dr. Donald Wells of GCO, Inc., Ann Arbor, Michigan, at the National Spring Conference of the American Society for Nondestructive Testing in 1969 (1) cited a few prominent examples of the use of holographic interferometry for nondestructive testing in the air frame industry as well as a brief introduction to the technical aspects of holography.

Since that introduction, much headway has been made to encompass new ideas and applications in HNDT as well as to improve on many of the existing techniques. Some of those new ideas and improvements will be discussed herein.

FRINGE CONTROL IN REAL-TIME HOLOGRAPHY

Perhaps the most exciting and outstanding progress that has been made is in the ability of HNDT to truly function as a mass production nondestructive evaluation technique. Whereas inspection rates of ten square meters per minute are realizable and are within the "state of the holographic art".

It became apparent in the earlier days of HNDT that the maximum displacement of a flaw or anomaly in a laminated structure would usually occur during the time that the maximum temperature gradients would occur in the part during the thermal stressing.

The difficulty in holographic viewing at this time, however, was that the maximum spatial frequency in the holographic fringe system would simultaneously exist because of the large gross body deformation of the total object. Since it was this fringe system that would be used to evaluate the part, the ability to analyze would break down.

The most one could hope for was that after the part had cooled down sufficiently, and the gross body displacement returned to near zero, the flaw would have some hysteresis and would still be visible with the reduced spatial frequency.

This hysteresis condition was found to exist in the class of relatively thin skins on laminate and honeycomb structures <.070 cm, but in the thicker sections most information was lost after the object had sufficiently cooled down and the fringe frequency reduced. A little over a year and a half ago, it was found that by properly changing the apparent position of the point light source being used to illuminate the object in the holographic system (2), we could, in fact, artificially reduce the spatial frequency of the fringe system on the object during the maximum temperature gradient. Earl ier attempts had been made with only partial success to reduce the spatial frequency of the fringe system by slight changes in the lo cation of the reconstruction hologram. However, because of non-uniqueness of individual points on the object with all points on the object surface, several serious side effects would occur when the hologram plate itself was moved; the two most serious being the loss in visibility of the fringe system and the inability to control the location of the fringes in space. None of these short-comings were found to exist with the modulation of the wavefront.

The optical elements necessary to accomplish this new type of fringe control are shown in Fig. 1. The hologram schematic is typical except for the addition of a lens between the object and the element used to expand the laser beam which illuminates the object and a tiltable mirror which can rotate about its vertical and horizontal axes. The lens is not necessary if the position of the beam expander element can be moved; however, in cases where optical spatial filtering is desired, it is difficult to move this element. The second lens is placed at least five lengths from the beam expander so that its aerial image of the beam expander's focal spot moves when the lens is moved.

Using fringe control in a real-time holographic recording system is straightforward. Initially a hologram of the ambient object is recorded, processed, and replaced for real-time viewing. Next the object is thermally stressed for a time necessary to bring out the detail. (Usually 10 seconds is sufficient for most materials). This thermal stressing brings about a high spatial frequency fringe system which is not usable for flaw detection. The fringe control

is now used to reduce the fringe frequency. First, the tiltable
mirror will move the center of the fringe system either up and
down or back and forth. Secondly, moving the lens along its optical axis will increase the spacing between fringes (i.e., lower
the spatial frequency); at this point the flaws become detectable.

One example of the affects of reducing the spatial frequency
on flaw detection is shown in Figs. 2 and 3. An adhesive bonded
honeycomb sandwich structure with .089 cm aluminum face sheets was
thermally stressed increasing the skin temperature 15° F. above
ambient. With this temperature change the fringe frequency is so
high that flaw detection is virtually impossible. Changing the
lens and mirror in the object illumination beam immediately reduces the spatial frequency. In Fig. 2, a 1.27 cm diameter unbond
was not visible with the high spatial frequency while in Fig. 3,
the fringes have been reduced and the location of the anomalous
region is unmistakable.

Figure 4 is taken from a typical real-time hologram showing an
advanced composite fan blade immediately after thermal stressing.
On the lower left side of the blade a disbond line extends along
the left edge. Although this disbond runs along the entire left
edge, all information is lost in the upper left region due to the
high spatial frequency. The ability to move the fringes at will
easily allows inspection of the entire edge with just one hologram.

With the ability then to control the spatial frequency immediately after the temperature has been raised, the holographic
technique can extract the maximum amount of detail concerning the
part's structural integrity.

FRINGE CONTROL IN DOUBLE-EXPOSURE HOLOGRAPHY

Further consideration of the optical wavefront construction in
a holographic system revealed that if a one-to-one relationship
could be made between points on the object's surface and points on
the hologram, then the fringe control could just as well be accomplished by manipulation of the reference illumination field instead of the object field. This relationship was established by
using an optical imaging system (i.e., a lens used to image the
object onto the hologram plate) as shown in Fig. 5.

The ability to control the spatial frequency in a hologram by
adjusting the reference beam essentially eliminates the need for
real-time holography. Since the object field in a real-time system
can be replaced by one exposure in a double-exposure hologram, the
same information could be obtained in a double-exposure hologram.
The optical arrangement for recording a double-exposure hologram
with fringe control is shown in Fig. 6.

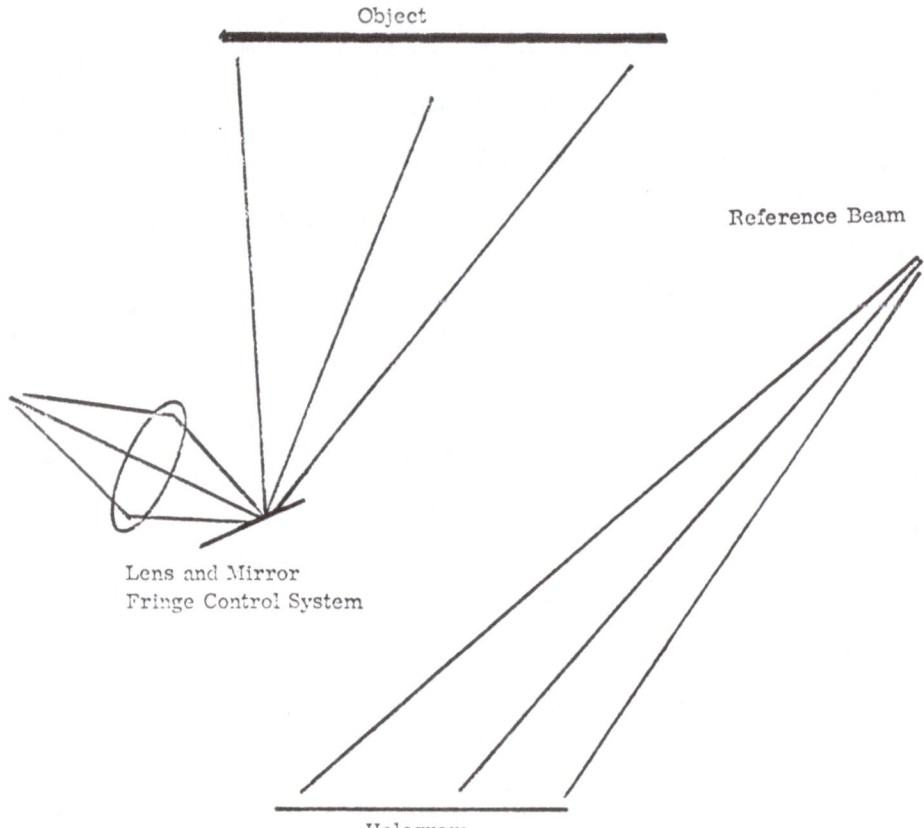

Fig. 1   Real-time fringe control.

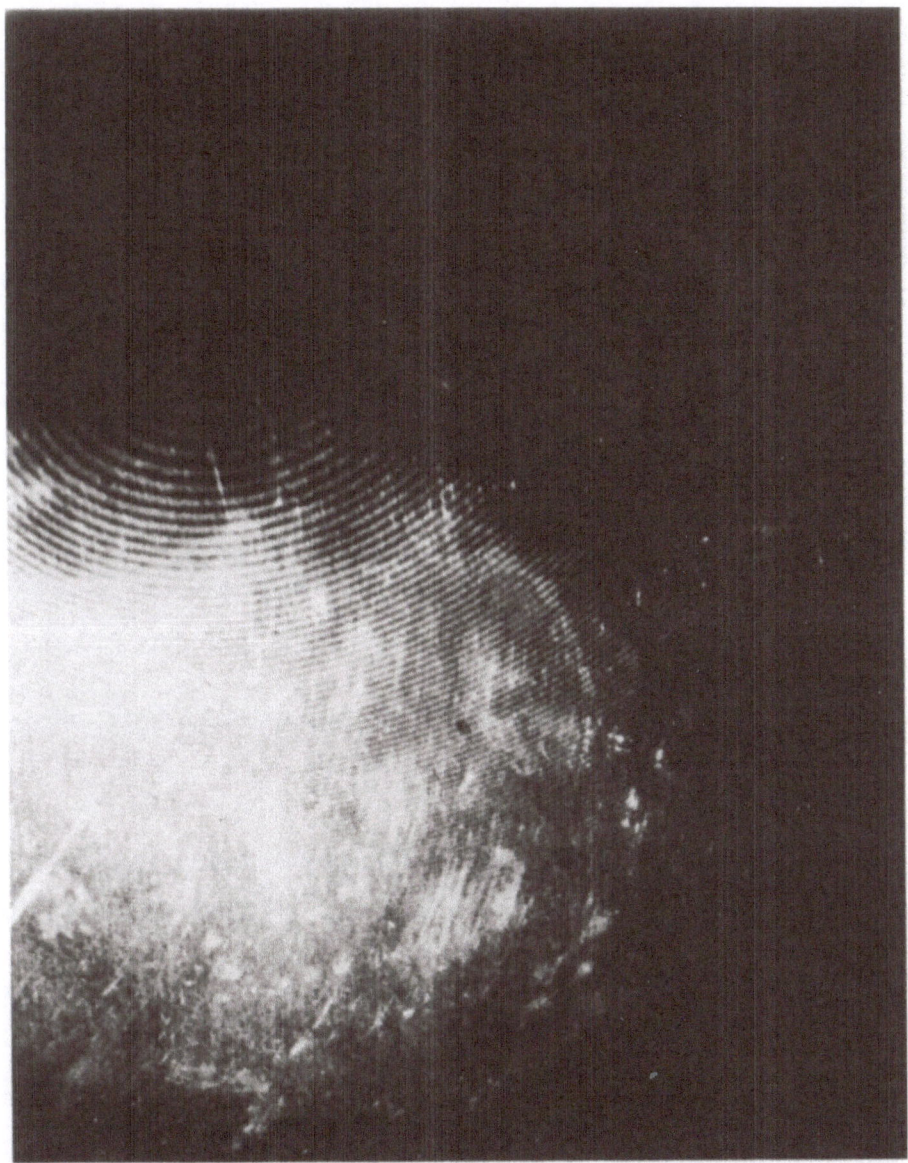

Fig. 2. Real-time holography on Aluminum Honeycomb Structure (no fringe control).

Fig. 3. Real-time holography on same aluminum structure as Fig. 2
         (fringe control).

Fig. 4. Advanced composite turbine blade; flaws seen with fringe
         control.

The first exposure is made of the object in its ambient state with reference beam number 1 on and number 2 off. The object is then thermally stressed and the second hologram is made with reference beam number 2 on and number 1 off. The hologram is processed and viewed in a readout station as shown in Fig. 7. With both beams on, the holographic fringe system is entirely adjustable by manipulation of either reference beam.

The advantages of such a system are obvious. First, the system is not dynamic as in real time, consequently, fringe control correction is much more easily achieved. Secondly, the inspection rates are significantly increased as the part is not held up while the new hologram reconstruction is being examined for flaws. The readout stations are much lower in cost than the recording station making it financially feasible for several readers to analyze results while the recording station is holographing new parts coming down the line.

Using these new fringe control techniques, the variety of parts inspected with HNDT has significantly increased. To date, we have examined successfully multilaminate structures at all depths with overall thicknesses in excess of .78 cm (see Fig. 8). On the other end of the scale, very thin laminated structures with overall thicknesses of less than .02 cm have been successfully examined.

Other types of material configurations such as rubber-to-metal laminates, plastic laminates, plastic-honeycomb, and paper-to-metal laminates are now being successfully tested (see Fig. 9 and 10).

The use of these various fringe scanning techniques in real time are now available on several of the industrial systems which GCO manufacturers. Two such systems employing these techniques are shown in Figs. 11 and 12. Figure 11 is of a Mark I system which weighs 7,000 pounds and can analyze a specimen 120 cm on a side. Figure 12 is of a Mark III system weighing 24,000 pounds. This in dustrial analyzer can inspect a laminated specimen up to 2 meters by 6 meters.

Another laminated structure which has lent itself to readily being inspected by HNDT has been the automobile, truck, and aircraft tire. Recent developments in holographic machine design and in some new innovative optical configurations have led to the development and production of a holographic tire analyzer which can inspect laminated tires for carcass ply separations; belt-edge, tread and sidewall separations; multiple cord fracture; porosity; and voids with production levels of over 20 tires per hour.

The operations sequence for obtaining the holographic data is

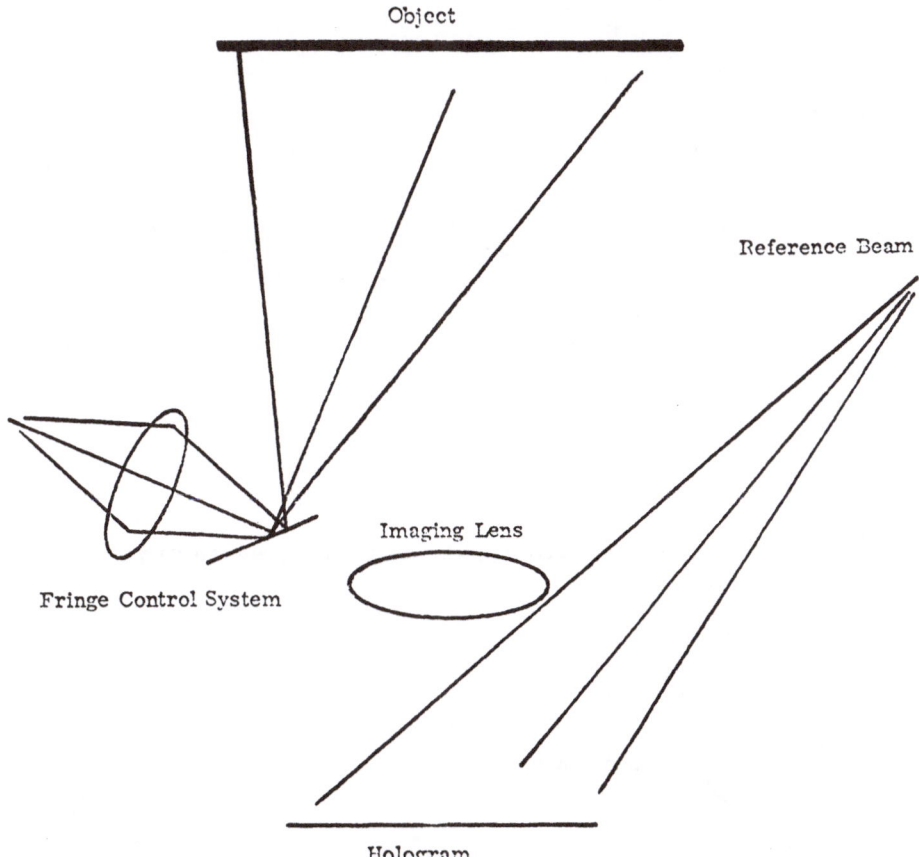

Fig. 5 Imaging holographic system.

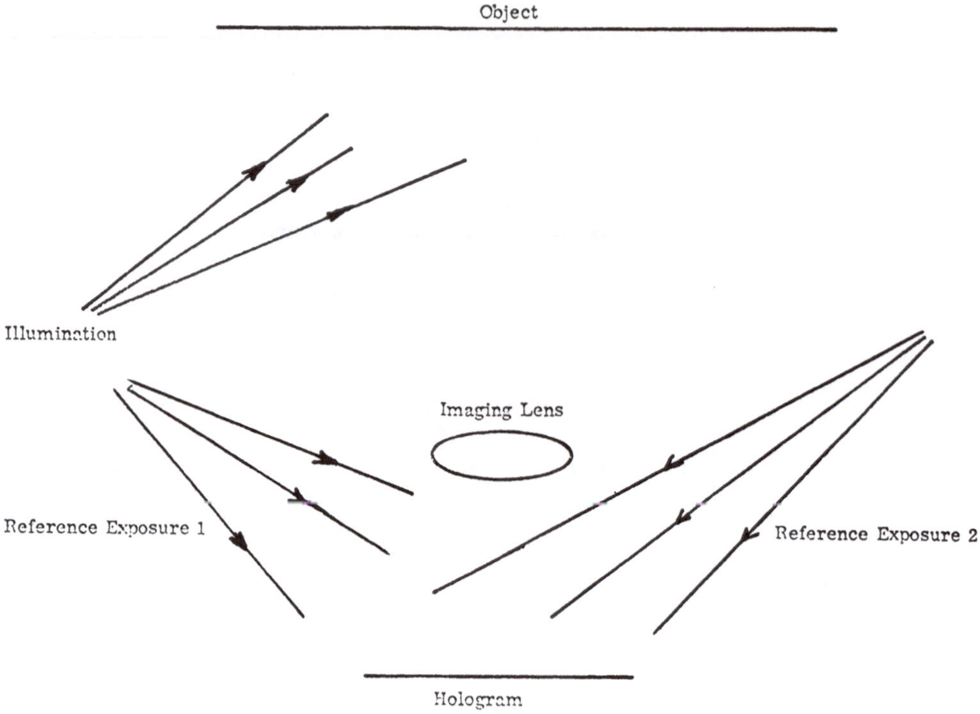

Object

Illumination

Imaging Lens

Reference Exposure 1

Reference Exposure 2

Hologram

Fig. 6 Double-exposure holographic recording system with fringe control.

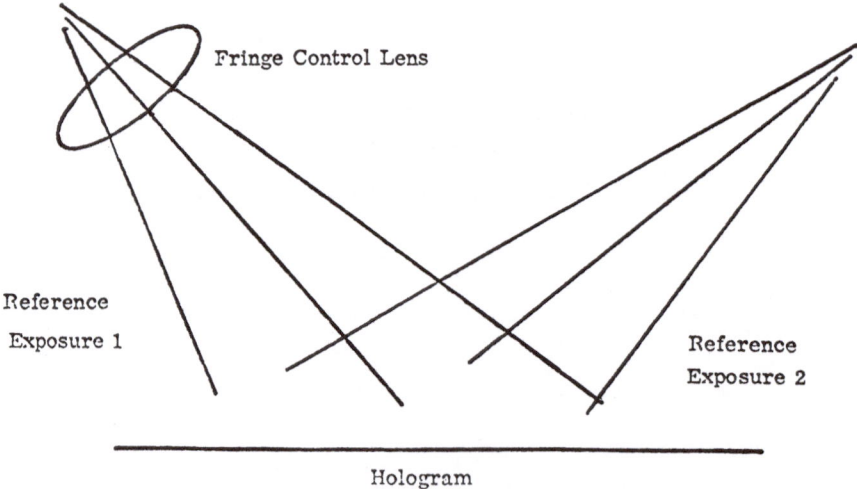

Fringe Control Lens

Reference Exposure 1

Reference Exposure 2

Hologram

Fig. 7 Double-exposure fringe control readout system.

Fig. 8. Multilayer laminated structure with two 1.27 cm diameter
flaws .38 cm deep; taken from double-reference-beam,
double-exposure hologram.

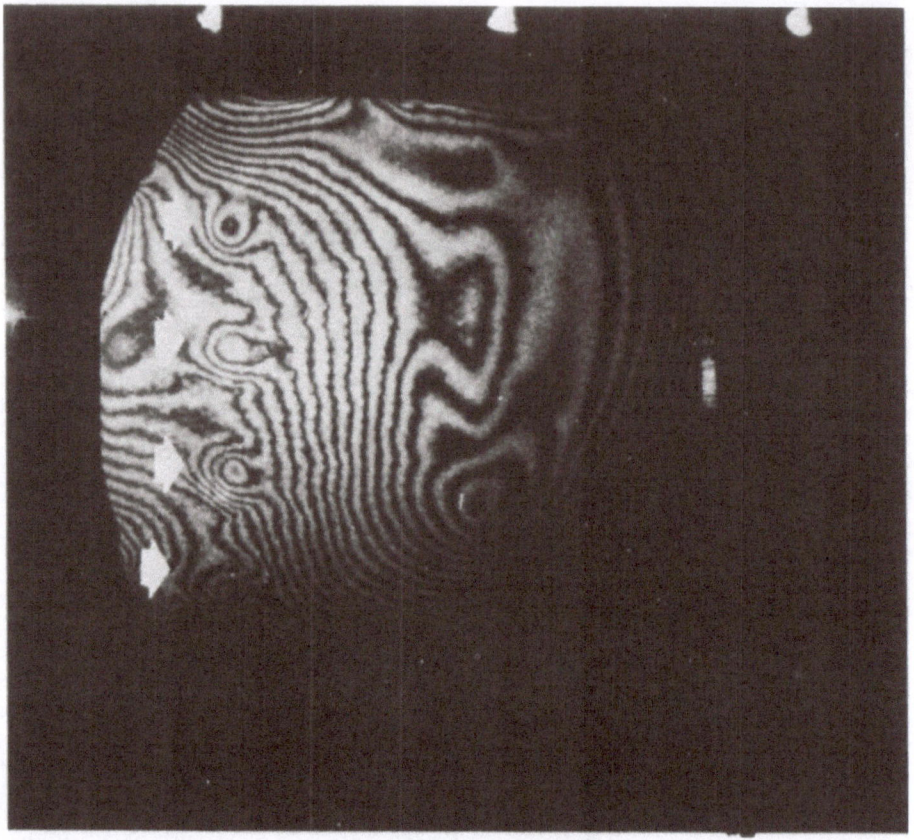

Fig. 9.   Plastic, phenolic impregnated honeycomb-sandwich struc-
          ture; four 1.27 cm diameter disbonds .076 cm deep.

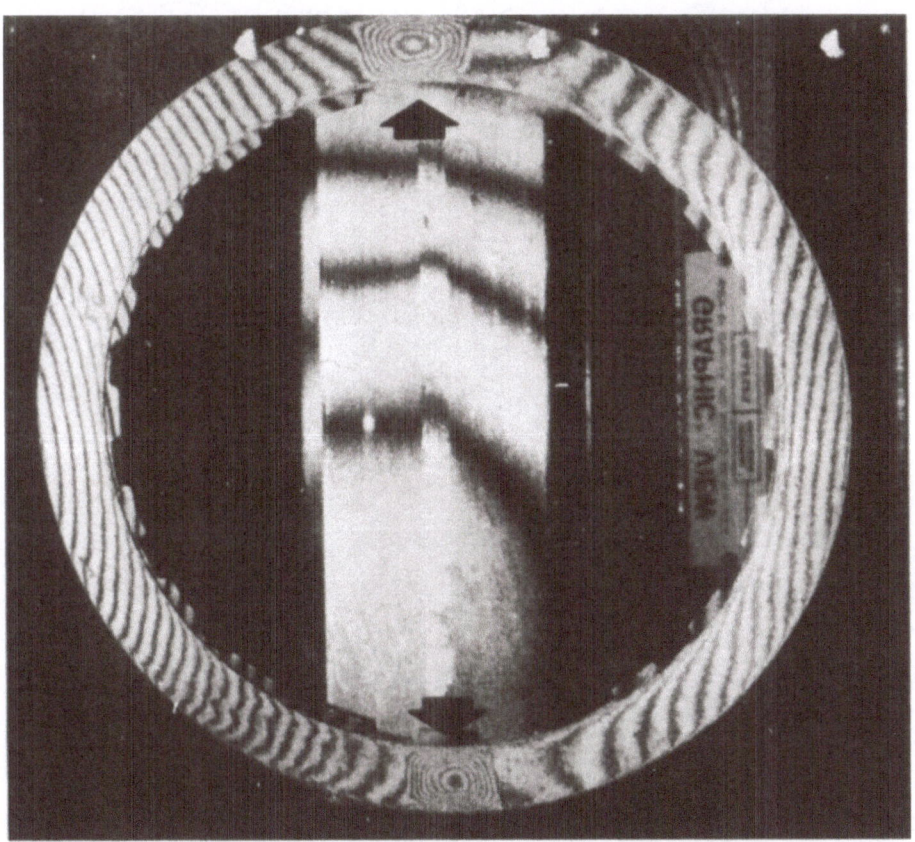

Fig. 10. Paper-to-metal laminate (clutch disc); two disbonds
          at paper-to-metal interface.

Fig. 11   Mark I system weighing 7,000 lbs.   Using fringe scanning
          techniques.

Fig. 12   Mark III system weighting 24,000 lbs.   Using fringe
          scanning techniques.

Fig. 13a. Hologram of a fiberglass tire showing 1" separation in
           shoulder.

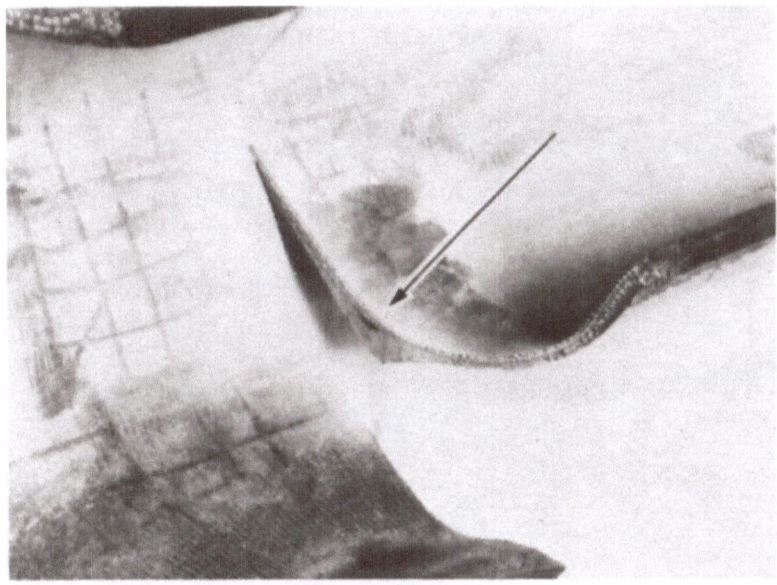

Fig. 13b. Destructive results of same tire showing flaw which
           was detected holographically.

Fig. 14a. Hologram of a rayon tire showing two 1/2" dia.
Flaws detected holographically.

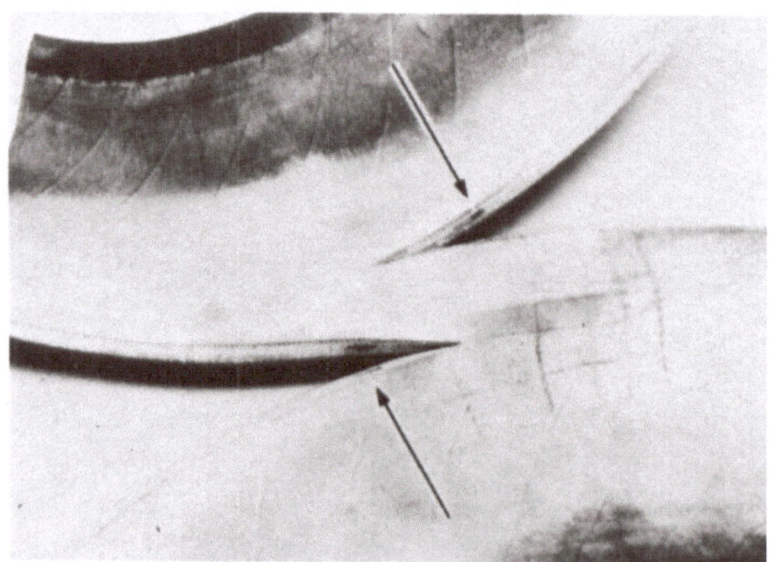

Fig. 14b. Destructive tests confirmed holographic results.

to first record a hologram of the tire's inner surface; second, stress the tire by slightly changing the enveloping air pressure; and thirdly, record the second exposure of the tire's inner surface at the second pressure. The resultant double-exposure hologram when reconstructed indicates all but the most minute structural anomalies.

Some typical results taken from double-exposure holograms of vacuum stressed tires are shown in Figs. 13 through 15.

The automatic HNDT tire analyzer, which recorded these holograms, is seen in Fig. 16. This particular machine is capable of inspection rates of up to 20 tires per hour. Tire machines now on the drawing board will be capable of inspection rates of 75 to 100 tires per hour. These machines are completely automatic and require only the loading and unloading of the tire, developing, and read-out of the film. The entire holographic process and stressing process is performed automatically.

SMALL AMPLITUDE VIBRATION DETECTION

A second area of application that holographic interferometry has found itself to be very useful is in the field of vibration analysis.

The ability of optical holography to record the vibrational patterns of a vibrating object and to measure their surface displacements down to several millionths of a centimeter has greatly assisted engineers in the design of structures from high power mechanical transducers to turbine blades.

There are many situations, however, where the surface displacement of an object under acoustical excitation is much less than millionths of a centimeter and consequently, holography has been of no use. This condition exists in essentially all structures where the excitation frequency is greater than 100 kHz. The only systems that have heretofore been able to measure these very small displacements have done so on a point-by-point basis using piezoelectric transducers.

Recent work in holographic vibrational analysis has shown that if the phase or frequency of the reference wave in a holographic recording system is shifted such that all points on a surface not moving reconstruct in a dark field rather than a light field, as is the usual case with standard vibrational holography, then the motion sensitivity of the system is significantly increased.

Two possible yet different techniques that can be used to modulate the reference field enhancing the motion sensitivity of the holographic recording system are the following:

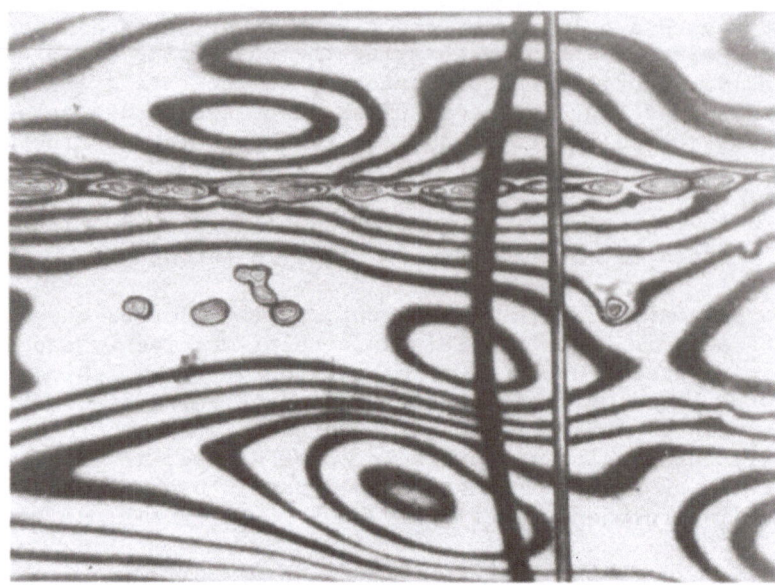

Fig. 15a. Holographic results indicate tread separation from
carcass in shoulder region.

Fig. 15b. Destructive results confirm holographic results.

Fig. 16   Cutaway sketch of automatic HNDT tire analyzer used to
          record above holographic results.

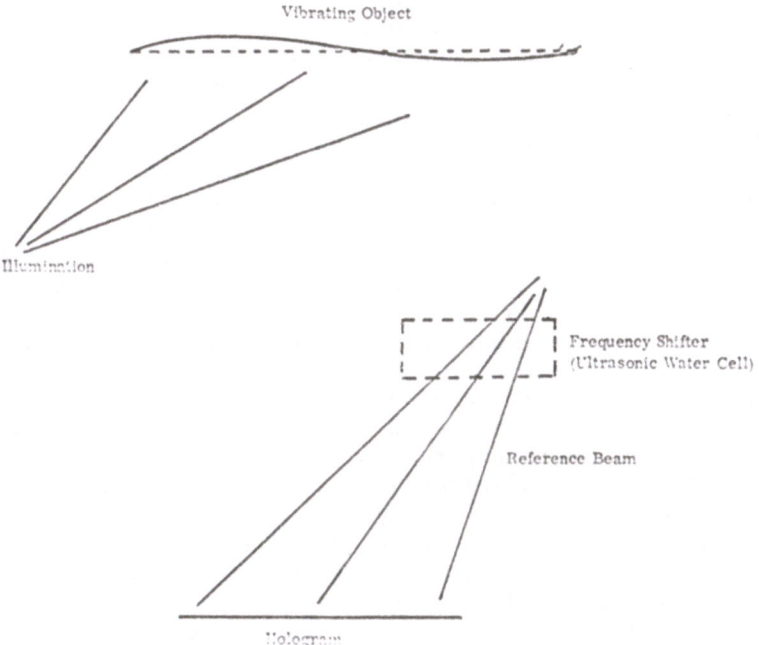

Fig. 17   Time-average holographic recording system with frequency
          shifted reference beam.

Phase Modulated Reference Field Holography (3): This technique
was used by the author and is useful to approximately 1 megahertz.
The reference wave is phase modulated by vibrating a transducer dri
ven mirror at the same frequency as the object under study and with
an amplitude equal to .19 λ where λ = the wavelength of the laser
light being used. This condition biases the reconstructing intensity field in the holographic system at the first zero of the
$J_0{}^2(x)$ (the zeroth order Bessel function squared) which is the function that describes the pattern of the normal time-average hologram.
Any displacement of the object now occurs about this point rather
than about the zero displacement point as is the usual case.

Frequency Modulated Reference Beam Holography (4): In this
technique the reference field is frequency modulated by doppler
shifting in an ultrasonic water cell; again at the same frequency
as the vibrating object, the $J_1{}^2(x)$ (the first order Bessel function squared) is selected and the dark field is implemented. This
technique is useful for frequencies above 1 megahertz. Figure 17
is a sketch of this optical system where the $J_1{}^2(x)$ term is used.

The sensitivity of both techniques is about the same.

The smallest vibrational amplitude that can be detected using
phase or frequency shifted reference beam holography remains to be
established both theoretically and experimentally; however, initial
experimental results indicate that surface vibrations less than
50 Å (50 x 10 $^{-8}$ cm) have been successfully recorded over an entire
surface at one time using these new holographic techniques.

It is well known that the propagation of an ultrasonic wave
through a material is significantly effected by the local material
property conditions in that structure. Such conditions as porosity,
voids, and cracks have a substantial affect on the radiation patterns
of the acoustical fields as they move through those areas.An attempt
to directly read out these signals by looking at the standing wave
patterns that develop on the structure's surface is presently being
conducted with the new modulated reference beam holography.

Some initial results are shown in the following examples.
Figure 18 is taken from a standard time-average hologram of a.120 cm
on .120 cm laminated structure being acoustically excited at
250 kHz. The surface displacements are substantially less than
approximately .32 x 10 $^{-4}$ cm and consequently, no vibrational
patterns are observable. However, by phase shifting the reference
field to increase the motion sensitivity of the holographic system,
Fig. 19 was obtained under the same acoustical excitation conditions.
A highly symmetric standing wave pattern is seen to emanate from
the transducer which is located on the opposite side of the panel.
Additionally, four 1.27 cm diameter disbonds located at the bond

Fig. 18.   .120 cm on 1.20 cm laminate; object vibrating at 250
           kHz; no apparent surface motion using standard time-
           average holography.

Fig. 19   Same structure as Fig. 18 under identical excitation
          conditions with phase shifted reference beam; surface
          displacement is approximately 50 Å.

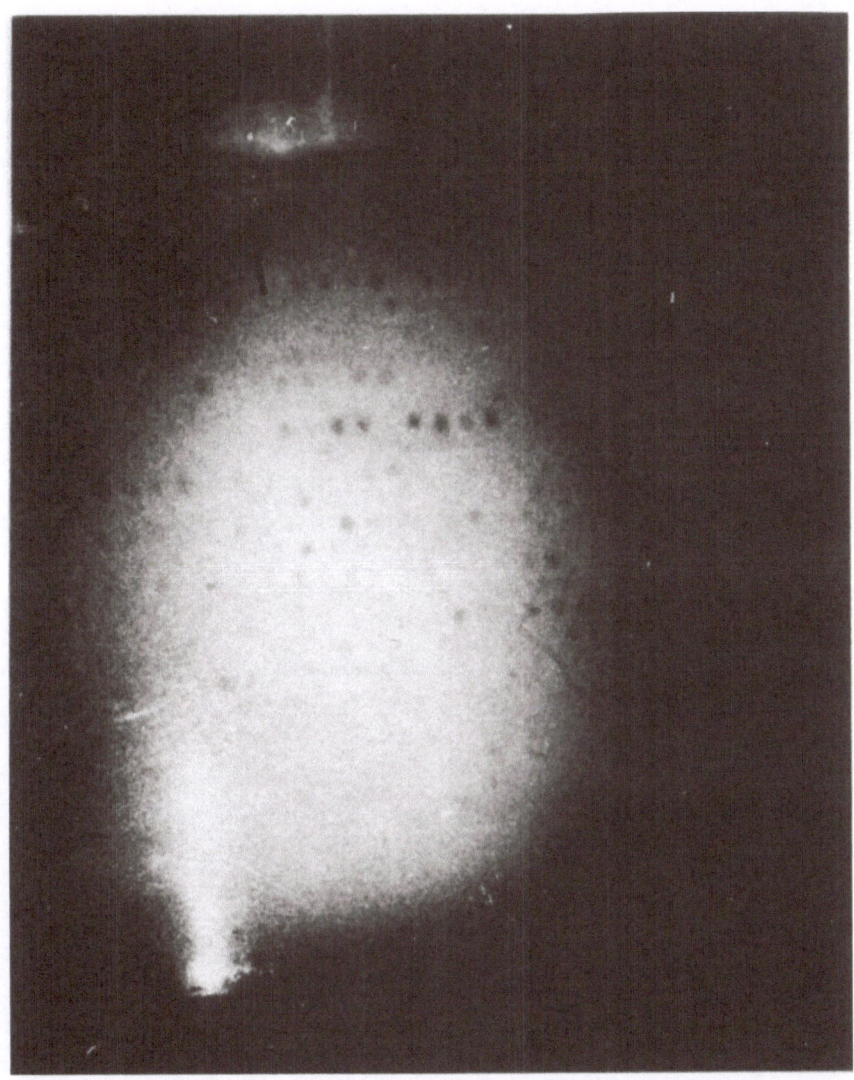

Fig. 20. Honeycomb-sandwich structure under acoustical excitation
at 237 kHz; no apparent motion with standard time-
average holography.

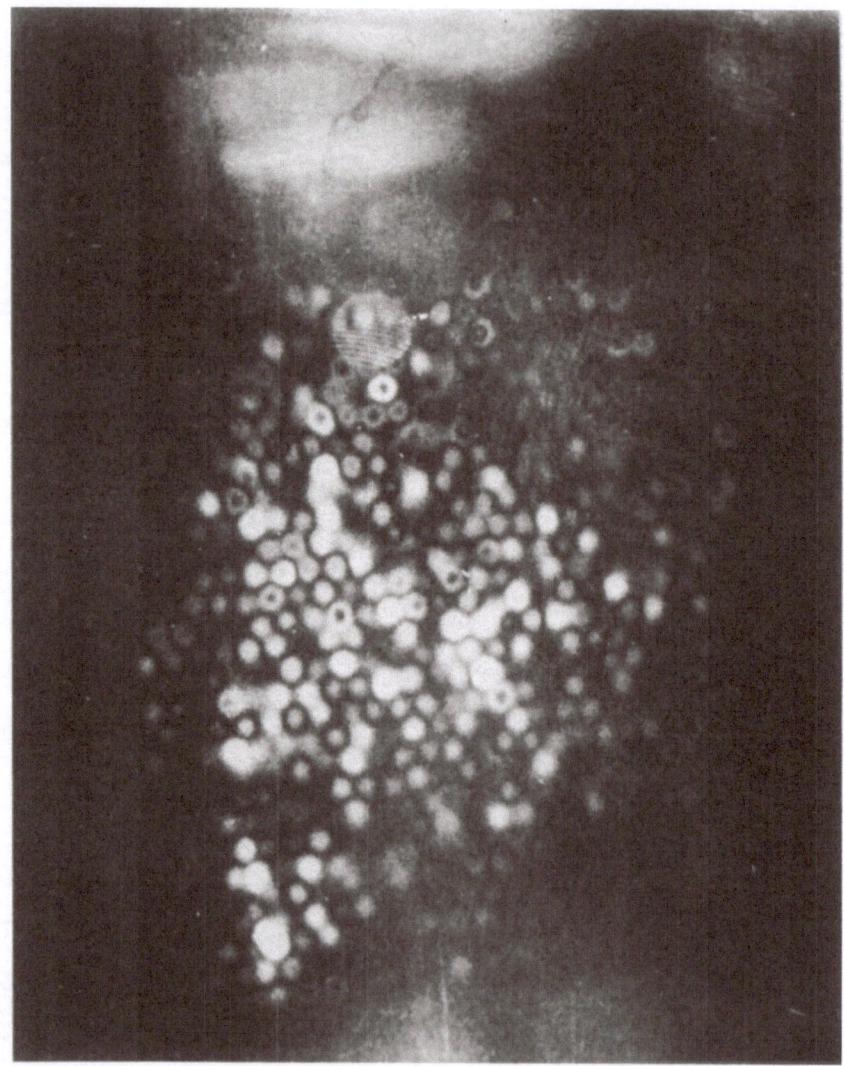

Fig. 21   Same as Fig. 20 except reference wave is phase shifted.

line interface are seen to significantly affect the resultant standing wave patterns in their respective areas. The surface displacements on the structure are estimated to be approximately 50 Å (50 x 10 $^{-8}$ cm).

A second example is shown in Figs. 20 and 21. Figure 20 is taken of a standard time-average hologram in which an adhesively bonded honeycomb structure with .154 cm face sheets and .635 cm cell structure is being excited at 237 kHz. Practically no motion is apparent. However, in Fig. 21, the reference beam has been phase modulated and a multitude of cell activity is present. The variation in the intensity over the field is due to geometric variations in the individual cells. These variations are attributed to lack or excess of adhesive and cell shape.

CONCLUSIONS

The high spatial frequency which results when an object is thermally stressed partially negates the ability of holography to detect flaws until the object has cooled down sufficiently and the fringe has reduced. Many of the flaws disappear when the object has cooled down and the detection capability is thus compromised. The addition of a fringe control system in the object illumination beam, however, allows the high spatial frequency fringe system to be reduced after stressing and the flaws are thus detectable.

Using a double-exposure, dual-reference holographic recording system and placing fringe control in one reference beam essentially alleviates the need for real-time holography. The advantages of this system are that it greatly increases the machine's speed, makes it easier to operate, and ultimately reduces the cost of operation.

In the areas of small amplitude vibration analysis, more experimental work is necessary to determine the ultimate detection sensitivity for these new types of holographic systems, and then to determine what applications for these new techniques are most feasible. However, it appears that a new and powerful NDT approach is evolving.

REFERENCES

(1) Wells, Donald R.,"NDT of Sandwich Structures By Holographic Interferometry, "Materials Evaluation, November 1969, Vol. XXVII, No. 11, p. 225-231.

(2) Champagne, Edwin and Kersch, Leonard, "Control of Holographic Interferometric Fringe Patterns, "Journal Optical Society of America, Vol. 59, No. 11, November 1969, p. 1535A.

(3) Metherell, A.F.; Spinak, S. and Pisa, E.J., "Subfringe Interferometric Holography for Linearly Recording Small Displace-

ments, "Journal Optical Society of America, Vol. 59, No.11, November 1969, p. 1534A.

(4) Aleksoff, Carl C., "Temporally Modulated Holography, "Institute of Science and Technology, University of Michigan (to be published in Journal of Applied Optics).

MICROWAVE HOLOGRAPHY

QUASI-HOLOGRAPHIC TECHNIQUES IN THE MICROWAVE REGION*

Emmett N. Leith

University of Michigan

Ann Arbor, Michigan (U.S.A.)

I. INTRODUCTION

Following the intensification of holographic activity in the 1960s. various researchers reported experiments that were direct microwave counterparts of the optical holography which preceded them (1) - (3). This work, which has recently become extensive, may be termed true microwave holography. As the title of our paper implies, we deal not with this rather restrictive field, but with a much broader one which embodies holographic-like techniques. With the broader license we gain access to a rather large body of material, of which we must discard all but a select portion.

There exists in the field of communication science a variety of techniques that resemble holography to various degrees, both in concept and in their mathematical formulation. Further when these processes are carried out with the aid of coherent optical systems, the resemblance to holography becomes striking indeed. Yet these processes have developed quite independently of holography and in no way depend upon principles originating in holography.

We select four such techniques for our discussion: the synthetic aperture, the chirp radar, the rotating-target-imaging system, and the beam-forming technique. Each in its own way resembles holography, and viewing them as holographic processes offers, in some cases, new insights that lead to new implementations. The first three constitute special cases of the range-Doppler radar system, so we begin by introducing this rather basic system concept.

*This paper was previously published under the same title in IEEE, Vol. 59, No. 9, September 1971, pp. 1305-1318).

II. THE RANGE-DOPPLER PRINCIPLE

In the range-Doppler radar system, shown in Fig. 1, the task is to measure the range, radial velocity, and relative reflectivi ty of a distribution of scatterers. A wave

$$f(t) = a(t) \; \text{expj} \left[\omega_o t + \phi(t) \right] \qquad 1.$$

is radiated: $f_o = \omega_o / 2\pi$ is the RF carrier frequency and a and ϕ are, respectively, the amplitude and phase modulations impressed on the wave. A point object at range r returns to the radar a signal,

$$g(t) = \sigma f \left(t - \frac{2r}{c} \right) = \sigma a \left(t - \frac{2r}{c} \right) \text{expj} \left[\omega_o \left(t - \frac{2r}{c} \right) + \phi \left(t - \frac{2r}{c} \right) \right] \quad 2.$$

which is just the radiated signal time delayed and Doppler shift- ed; σ is a complex constant containing such factors as the complex reflectivity of the object and the attenuation with distance which, in the radar case, because of the round trip factor, becomes the inverse fourth-power factor. The Doppler shift is implicit in the variable r, which is a time-varying function. Let the radial velo city be v_1, which for simplicity we suppose to be constant;

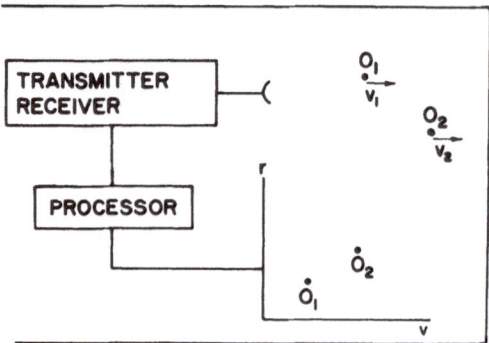

Fig. 1   Pulse-Doppler radar, showing two point objects imaged in a range-velocity display. Ranges are r_1 and r_2, radial velo cities are v_1 and v_2.

thus

$$r = r_1 + v_1 t \qquad 3.$$

where r_1 is the range at time $t = 0$. We insert (3) into (2) and make the customary narrow-band approximations[1].

$$a\left[t - \frac{2(r_1 + v_1 t)}{c}\right] \cong a\left(t - \frac{2r_1}{c}\right) \qquad\qquad 4a.$$

$$\phi\left[t - \frac{2(r_1 + v_1 t)}{c}\right] \cong \phi\left(t - \frac{2r_1}{c}\right). \qquad\qquad 4b.$$

The return signal $g(t)$ is then

$$g(t) = \sigma a\left(t - \frac{2r_1}{c}\right)\exp j\left[\omega_o\left(t - \frac{2r_1}{c} - \frac{2v_1 t}{c}\right) + \phi(t) - \frac{2r_1}{c}\right]$$

$$= \sigma a\left(t - \frac{2r_1}{c}\right)\exp j\left[\left(\omega_o - \frac{4\pi v_1}{\lambda_o}\right)t + \phi\left(t - \frac{2r_1}{c}\right) - \frac{4\pi}{\lambda_o}r_1\right]$$

$$= \sigma f\left(t - \frac{2r_1}{c}\right)\exp\left(-j\,\frac{4\pi v_1}{\lambda_o}\,t\right) \qquad\qquad 5.$$

where the bulk phase delay $\exp - j(4\pi r_1/\lambda_o)$ has been incorporated into σ. The round trip delay time $2r_1/c$ can be written as t_1, while the term $4\pi v_1/\lambda_o$ is a Doppler shift which we represent as ω_{d1}. Measurements are made of both the time delay and the Doppler shift, from which the range and the radial velocity are readily determined. One of the fundamental problems, then, is to determine optimum forms for $f(t)$ since our ability to make these measurements is strongly dependent on the form of this signal.

The range-Doppler measurement is performed by cross correlating the return signal with a reference function

$$r_c(t) = f(t)\exp\left(-j\omega_r t\right) \qquad\qquad 6.$$

---

(1) If a and ϕ are narrow-band functions, centered about zero frequency, then in a short time interval $(2v_1/c)^t$, these functions will have changed negligibly even though the RF carrier term $\exp(j\omega_o f)$ may have changed considerably.

which is a frequency-shifted replica of the radiated pulse, yielding

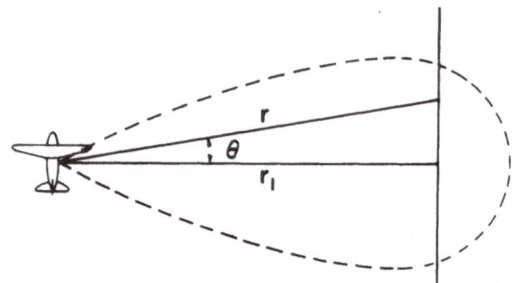

Fig. 2   Synthetic - aperture geometry.

$$u(t_r, \omega_r) = r_c \circledast g$$

$$= \sigma \exp(j\omega_r t_r) \int f(t - t_r) f^*(t - t_1)$$

$$\exp\left[-j(\omega_r - \omega_{d1})t\right] dt \qquad\qquad 7.$$

where $\circledast$ represents the cross-correlation operation. This equation is readily placed in the standard form

$$|u(\tau', w')|^2 = |\sigma|^2 \left| \int f(t' + \tau') f^*(t') \exp(-j\omega't)dt \right|^2 \qquad 8.$$

by the substitution $t' = t - t_1$, $\tau' = t_1 - t_r$, and $\omega' = \omega_r - \omega_{d1}$.
Equation (8) is known as the ambiguity function of f. The ambi-
guity function, somewhat analogous to an impulse response or a
point-spread function, describes how the system images the object
field into a two-dimensional range-Doppler space. The waveform
f(t) is designed to give an ambiguity function with a sharp peak
at $\tau' = \omega' = 0$ with low secondary responses. There are limitations to how well this can be achieved. Note that when there are
no Doppler shifts, the ambiguity function becomes just the magnitude squared of the autocorrelation function.

The similarity to holography is evident: simultaneously within the radar wave detected at the receiver may be many superimposed signals, originating from a range-velocity object space, and
the requirement is to unscramble the signals and thereby reconstruct the original object space. The correlation operation accomplishes this, displaying each object point as a response in a
$\tau_r \omega_r$ space, centered about the original location t_1, ω_{d1}. When the
data are stored in a manner such that they can be interrogated

with light, and when the range-Doppler analysis is carried out by
coherent optical-processing techniques, the resemblance to holography becomes strong. In some of the specific instances to be
discussed the holographic analogy is stronger yet.

III. THE SYNTHETIC ANTENNA

In the synthetic-aperture technique, a special case of range-
Doppler radar, the Doppler shifts are generated primarily by the
motion of the radar platform. The system can be described in a
variety of ways, with each offering its own special insight into
the process. Here we view the process in three rather different
ways: as a somewhat modified range-Doppler system; as a synthetic
antenna; and finally, as holography.

The radar may be carried by an airplane or, equally well, by
any other relatively fast-moving vehicle. In the usual embodi-
ment, shown in Fig. 2, the antenna is mounted in a sidelooking
configuration, although this is by no means always the case. The
radial velocity between an object point and the radar is proportional to the sine of the bearing angle θ of the point. Thus at
any instant there is a one-to-one correspondence between Doppler
shift and position in the beam. The Doppler analysis thus yields
resolution in the along-track dimension, and a range-Doppler display becomes an image of the object field over which the beam
passes.

In general, the Doppler analysis provedes azimuth resolution

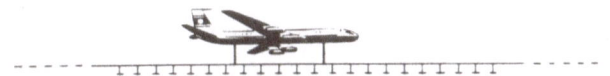

BEAMWIDTH (IN RADIANS) = $\frac{\lambda}{D}$

D = ARRAY LENGTH

Fig. 3 An aircraft utilizing a real antenna which achieves the
resolution of a synthetic antenna.

considerably better than that produced by the angular extent of
the beam so that, at any range r_1 a large number of object posi-

tions simultaneously in the beam can be resolved. If we compare
this situation with conventional radar,wherein azimuth resolution
is just the beamwidth, the resulting image seems to have been pro
duced by a radar system having a narrower beam, hence a longer
antenna, than is actually the case. This observation leads to

the synthetic-antenna concept.

A long receiving antenna yielding resolution comparable to
the Doppler processing technique could, indeed, be a very long antenna. It would have a length equal to the distance that our platform moves during the time that a signal is received from a scattering point. This distance would be the linear width of the radiated beam at the range of interest, or possibly less if we choose
not to utilize all the available data. In general, such an antenna
would have an unmanageable length, as Fig. 3 suggests.

The long antenna achieves a narrow receiving beam through the
coherent summation of the signals received on the individual elements. However, there is no fundamental requirement that the many
elements exist simultaneously; a single element carried on the air
craft could serve in sequence the function of each element. This
single element, through the forward motion of the aircraft, would
in turn occupy the position of each array element of the long antenna. The signal received at that position could be stored and,
when all the signals have been collected, they could be combined
just as the large array would have done. The result, in either ca
se, would be indistinguishable. The small antenna has then func-
tioned as a large array, or equivalently, has formed a synthetic
antenna.

A. Analysis

The analysis can proceed either from the Doppler analysis or
from the synthetic-aperture viewpoint. We choose the former, thus
utilizing the results already presented.

As previously, f(t) is the radiated wave and

$$g(t) = \sigma a\left(t - \frac{2r}{c}\right) \exp\left\{ j\left[\omega_o\left(t - \frac{2r}{c}\right) + \phi\left(t - \frac{2r}{c}\right)\right]\right\} \qquad 9.$$

is the signal received from a point object at range r, as in Fig.2.
Let the wave be unmodulated; thus, $a = a_o$ (a constant), and $\phi = 0$.
Resolution in range is, of course, not available with this waveform, but for the moment we consider only a single range. Later
we introduce pulsing for ranging purposes, and finally, we consider
an arbitrary waveform. The returned signal is thus

$$g(t) = \sigma a_o \exp\left[j\omega_o\left(t - \frac{2r}{c}\right)\right] = \sigma a_o \exp\left(j\omega_o t - \frac{4\pi}{\lambda_o} r\right). \qquad 10.$$

The Doppler shift is the time derivative of the phase delay,

$$\omega_d = -\frac{4}{\lambda_o}\frac{dr}{dt} \qquad\qquad 11.$$

which is readily found to be

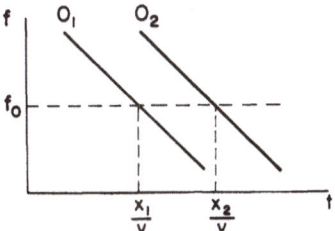

Fig. 4 Doppler frequency as function of time, for two point objects O_1 and O_2, located at positions x_1 and x_2.

$$\omega d = -\frac{4\pi}{\lambda} v \sin\theta \qquad\qquad 12.$$

where v is the aircraft velocity. If the observation time is sufficiently short that the object bearing, and hence the Doppler frequency remain constant, then the previously noted one-to-one correspondence between Doppler frequency and target-bearing angle applies, and a Doppler analysis yields the azimuthal resolution we seek.

In practice the integration times are much longer, perhaps over the entire duration that the point object is in the beam; consequently, the Doppler shift is not constant and this quite simple approach does not suffice. Instead, we make the small angle approximation,

$$r = \sqrt{r_1^2 + (x - x_1)^2} \cong r_1 + \frac{1}{2}\frac{(x - x_1)^2}{r_1} \qquad\qquad 13.$$

(satisfactory for the usual case of a beamwidth less than 10°) which yields

$$g(t) = \sigma a_o \; \exp j \left[\omega_o t - \frac{2\pi}{\lambda_o r_1}(x - x_1)^2 \right] \qquad\qquad 14.$$

where an inconsequential constant phase term, $-(4/\lambda_o) r_1$ has been, as before, incorporated into σ.

Taking the aircraft position to be $x = vt$, it is apparent that the received signal can be expressed purely as a function of either position x or of time:

$$g(t) = \sigma a_o \, \exp j \left[\omega_o t - \frac{2\pi}{\lambda_o r_1} (vt - x_1)^2 \right] \qquad \qquad \text{15a.}$$

$$g\left(\frac{x}{v}\right) = \sigma a_o \, \exp j \left[\omega_o \frac{x}{v} - \frac{2\pi}{\lambda_o r} (x - x_1)^2 \right]. \qquad \qquad \text{15b.}$$

We have use for both forms.

The Doppler-shifted signal has frequency

$$f = \frac{1}{2\pi} \frac{d}{dt} \left[\omega_o t - \frac{2\pi}{\lambda_o r_1} (vt - x_1)^2 \right] = f_o - f_d \qquad \qquad \text{16.}$$

where

$$f_d = \frac{2v}{\lambda_o r_1} (vt - x_1) \qquad \qquad \text{17.}$$

is the Doppler shift. Hence as depicted in Fig. 4, the signal reflected from object point O_1 undergoes a linear frequency shift as the beam sweeps past. The Doppler shift is maximum as the object enters the beam, since then the radial velocity is maximum. A beam of the aircraft the radial velocity component, and hence also the Doppler shift, is zero. In the trailing portion of the beam, the object is receding and the Doppler shift is thus negative. A second object O_2 which enters the beam later, produces a similar

Doppler history, except that each frequency occurs at a later time.

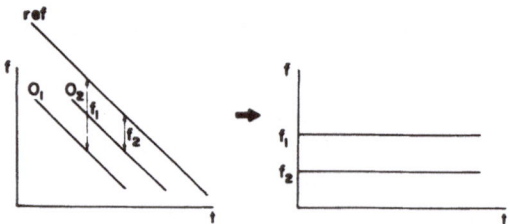

Fig. 5 Conversion of the signal histories to constant frequencies.

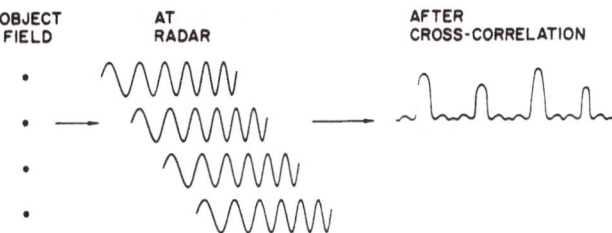

Fig. 6 The imaging process, showing an object field, the signal
received at the radar, and the signal after cross-correlation processing.

Since all object points generate the same Doppler spectrum as
they proceed across the beam, a Doppler analysis is no longer appropriate and the earlier analyses must be modified. One possible
solution is to mix the incoming signal with an internally generated signal swept in frequency at the same rate as each Doppler history, as illustrated in Fig. 5. Each signal history is converted
into a constant-frequency signal, resulting in a one-to-one correspondence between object position and frequency. The Doppler analysis now provides the desired resolution.

Rather than such mixing, let the received signal be cross cor
related with an internally generated signal r_c having the same
form as the signal from a point object:

$$u = \int_{-L/2}^{L/2} \sigma a_o \ \mathrm{expj}\left[\frac{\omega_o}{v} x - \frac{2}{\lambda_o r_1}(x - x_1)^2 \right]$$

$$\mathrm{exp} - j\left[\frac{\omega_o}{v}(x - x') - \frac{2\pi}{\lambda_o r_1}(x - x')^2 \right] dx = g \circledast r_c \qquad 18.$$

where L is the length of the data we process - perhaps one beamwidth, perhaps less. Note that the terms in x^2 cancel; thus the
correlation process in fact accomplishes the FM removal suggested
previously. Indeed, such an FM removal followed by the appropriate
Doppler filtering would constitute exactly the cross-correlation

operation of (18).   Integration of (18) yields

$$u = K \sigma \text{ sinc } \frac{2(x' - x_1)L}{\lambda_0 r_1} \qquad\qquad 19.$$

where K is a constant and sinc $\gamma = \sin(\pi\gamma)/\pi\gamma$.

The above cross correlation (which, of course, could alternatively be described as a matched filtering) completes the imaging process. The signal at the receiver is a superposition of linearly frequency modulated signals with relative displacements in accordance with the positions of the point objects producing them, as seen in Fig. 6. The correlation process compresses each elemental signal into a narrow spike; the signals, now resolved, constitute a fine-resolution image of the object field. The process, both in the broad concept and in the mathematical formulation, strongly resembles a conventional imaging process at optical wavelengths; the implementation, however, is remarkably different, although we note with anticipation that it need not be.

Having described the synthetic-aperture process as it occurs at a single range, we introduce pulse modulation to provide range resolution. The analysis previously given is not essentially altered. The pulse modulation constitutes a sampling process on the Doppler frequency histories, and if the sampling rate or pulse-repetition frequency, is adequate, the complete signals can readily be reconstituted from the samples. And, of course, range resolution is achieved in the usual manner, being determined by the pulse duration. The cross-correlation process is performed independently for each range element; the result is a two-dimensional radar image, in range and along-track coordinates, of the original object field.

B. Implementation: Optical Processing

In the early period of synthetic-aperture radar (1951-1955), many methods were devised for processing the stored data; processing techniques were developed around such devices as storage tubes, recirculating delay lines, and filter banks. Because the quantity of data produced by a synthetic antenna system is enormous, utilization of even a moderate fraction of the data results in difficult storage and processing problems. If integration times were to be sufficiently long as to require the linear FM approximation (14), instead of the constant-frequency approximation, and thereby produce a synthetic antenna focused at all ranges, then each range would have to be processed in a separate channel, since this FM rate is range dependent. Thus both the storage and the processing requirements are severe.

Photographic film, because of its high-density storage capabilities, was found to be an attractive storage medium; optical correlators, because of their multichannel capability stemming from their ability to image in two dimensions, appeared as attractive candidates for the processor. Additionally, these two solutions are quite compatible; photographically stored data are in an ideal form for the optical processor input.

The typical transducer for storing the electrical signal photographically consists of a cathode-ray tube and a camera, as illustrated in Fig. 7. Each range sweep is displayed as an intensity-modulated line and photographed. Successive line traces are recorded side-by-side, resulting in a two-dimensional raw data record, with range being distance across the film. Fig. 8, shows the object-space geometry, a range swath, a point object, and the resulting data record.

The stored data are a scaled record of the data collected along the flight path, where $x \to px$ and p is the scaling factor. The signal, recorded as photographic transmittance, is thus

$$g_o(x) = \sigma a_o \cos\left[\alpha x - \frac{2\pi p^2}{\lambda_o r_1}\left(x - \frac{x_1}{p}\right)^2\right] \qquad 20.$$

where $\alpha/2\pi$ is a spatial frequency related to the video-carrier frequency of the electrical signal at the recorder. Similarly, the reference function is

$$r_o(x) = \cos\left(\alpha x - \frac{2\pi p^2}{\lambda_o r_1} x^2\right). \qquad 21.$$

Although we neglect it in the equations, an amplitude scaling factor is needed since photographic transmittance cannot exceed unity. The cross-correlation process yields

$$g_o \circledast r_o = K\sigma \operatorname{sinc} \frac{2p\left(x' - \frac{x_1}{p}\right)L}{\lambda_o r_1} \qquad 22.$$

which is analogous to (19), except that, because of the scaling, distances are measured in a miniaturized image space.

The optical processing method is shown in Fig. 9. Observation of (18) shows the cross-correlation process to consist of three elementary suboperations: a multiplication; an integration; and a displacement of one function relative to the other. Each is readily performed optically.

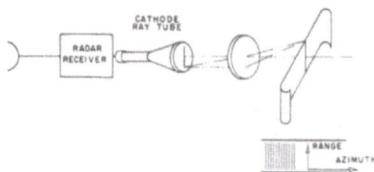

Fig. 7   Producing the data record.

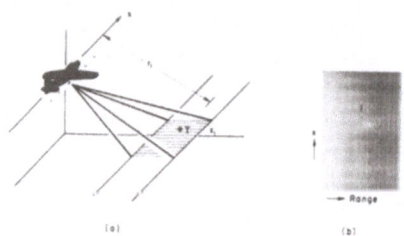

Fig. 8   The radar system geometry.(a) A point object.(b) The result
         ing data record.

Fig. 9   A coherent optical system for cross correlation.

Fig.10   A coherent optical system for multichannel processing.

Imaging of the transparency r_0 onto g_0 produces the multiplication. To produce the integration, we utilize the well-known
Fourier transforming properties of lenses. The signal g_0 is placed
in the front focal plane of L_3, and in the back focal plane the
light distribution (4)

$$q(x) = \int r_0 g_0 \exp\left(j \frac{\eta}{\lambda_1 F} x\xi \right) dx \qquad\qquad 23.$$

is formed, where λ_1 is the wavelength of the light and F the focal
length of the lens. If our observation is confined to the axial
point $\xi = 0$,(23) becomes a simple integration. Thus a pinhole is
placed on axis and the illumination in the pinhole is continuously

photographed as the signal record moves through the aperture. In this manner, the cross-correlation operation of (22) is formed. As the signal record moves, each elemental signal is brought into a-lignment with the reference function, and the output at that instant is a measure of the reflectivity σ of the image point that produced the signal. Thus the overlapping signals are unscrambled, producing a fine-resolution image from the raw data record.

But the optical implementation had its problems, which in hindsight may seem trivial, yet appeared quite challenging when they first arose. The first such difficulty now appears; we must write the signals r_o and g_o about a bias or grey level, since photographic film is restricted to positive values. The consequence is that we produce ($r_b + r_o$) ✷ ($g_b + g_o$) instead of r_o ✷ g_o, where r_b and g_b are the required bias terms. Four terms result, three of which are extraneous and have the potential for ruining our results.

We dispose of this threat by invoking again the Fourier transforming property of lenses. The spatial-frequency spectrum of $r_b + r_o$ is displayed at the back focal plane of lens L_1; r_b appears as a bright spot on axis, while r_o, having no zero-frequency component (a consequence of the spatial carrier α), is displayed as two symmetrical distributions of light about the axis. A stop on axis removes r_b, while r_o passes to the signal record. Two of the three offensive terms have at once been eliminated, including the potentially disastrous one r_b ✷ g_b. The surviving extraneous term r_o ✷ g_b is really innocuous since r_o and g_b have nonoverlapping spectra, and hence produce a cross-correlation value of zero.

Optical cross correlation is easily implemented, but so is e-lectronic cross correlation. What then is the motivation for performing the awkward and time-consuming recording procedure required for the optical method? The answer, already indicated, is that this correlation process must be performed separately for each resolution element in range, since the signal g is range dependent. For a reasonable range swath, thousands of range-resolution elements are utilized; hence, many correlators are needed. This consideration led to the optical approach, since the optical system can accommodate a two-dimensional input format; the yet unused dimension can thus create a multichannel processing capability. We may handle as many channels and perform as many independent correlations as there are resolvable elements across the input aperture. Fig.10 shows a pictorial view of the same optical system shown first in Fig. 9. The reference-transparency function now appears as a two-dimensional function $r_o(x, r)$ and when imaged onto the signal record, supplies the proper reference function of each range.

The optical system, in a sense, performs too much. The integrating lens L_3 performs an integration in two dimensions - and the integration in range would superimpose all range elements,thus destroying all resolution in range. The cylindrical lens corrects this difficulty. The cylindrical lens and spherical lens together image, in the range dimension, the raw data film onto the output plane. As before, the spherical lens L_3 integrates in the x dimension, displaying the result along a vertical line through the $\overline{\text{axis}}$. A recording film behind the slit records the image.

C. Further Evolution of the Optical Technique

At this stage, the processor is certainly a successful and useful device. The inventive process, however, did not stop here; improvements were indeed possible and were soon forthcoming. A shortcoming of the previously mentioned system is the large light attenuation loss; the reference-function transparency absorbs about half of the incident light; worse yet, most of the transmitted light remains in the zero order, only to be removed further downstream by the axial stop. Perhaps only 0.001 percent of the light incident on r_o reaches the output slit.

To correct this shortcoming, we start with the observation that the transparency r_o, being a real function, has redundancy in its spectrum in the form of conjugate sidebands. Thus we need not pass both sidebands and the axial stop in the focal plane of L_1 can be replaced with a half-plane stop which, in addition, removes one sideband. The resulting image is unaffected.

Except, of course, for the additional 50-percent light loss. Yet, pursuing this train of thought further eventually leads to considerable improvement of light utilization efficiency and to $\underline{\text{de}}$ velopments of even greater significance. Writing the reference function in its exponential form

$$K\left[r_b + \cos\left(\alpha x - \phi\right)\right] = K\left[r_b + \tfrac{1}{2}\exp\left[j\left(\alpha x - \phi\right)\right]\right.$$
$$\left. + \tfrac{1}{2}\exp\left[-j\left(\alpha x - \phi\right)\right]\right] \qquad\qquad 24.$$

where

$$\phi = \frac{2p^2\pi}{\lambda_o r}x^2$$

we can readily demonstrate that each sideband represents one of the two exponential terms; thus the single sideband mode is equiv$\underline{\text{a}}$

lent to operating with a complex reference function $\exp\left[\pm j(\alpha x - \phi)\right]$.
If the complex reference function could be generated directly instead of being derived from the cosine function, an enormous reduction of light attenuation would result, for the SSB-reference function considered as a transparency is a purephase function and transmits in principle all the light in its path. To produce the phase transparency is easy in principle; we have only to properly figure a glass blank, cutting its thickness in accordance with the argument of r_o.

But in practice the problem is formidable, for in general only a few simple shapes, such as spheres,cylinders, and some aspherics can be accurately produced; arbitrary thickness functions are beyond present-day art.

However, further inspection of the argument $\alpha x - \phi$ reveals some interesting possibilities. The term αx represents a linearly increasing phase shift and is supplied by a simple prism. Further, since prisms do nothing more than bend light rays, the prism term is clearly unneeded. This term is required in the cosine transparency for various reasons: for example, it makes possible the separation and subsequent removal of the bias term. In the complex transparency, it serves no purpose.

The second term of the argument, $\phi = 2\pi p^2 x^2 / (\lambda_o r)$, being quadratic in x, is realized as a cylindrical lens (4). Such lenses can be readily obtained; hence, the complex transparency can be synthesized. So it appears, until we note the term r in the denominator. The focal length is determined by the coefficient of x^2, and from basic principles (4) is found to be

$$F_x = \frac{1}{2p^2} \frac{\lambda_o}{\lambda_1} r \qquad\qquad 25.$$

where λ_1 is the wavelength of the light used. The "cylindrical" lens thus has a focal length which varies linearly along its length, and thus could better be described as a conical lens. One could plausibly expect that its fabrication is not feasible.

However, not only is this supposition incorrect, but indeed such lenses had already been produced for quite unrelated purposes. The conical lens can be derived from the axicon (5), a lens in the form of a rather flat cone (Fig. 11). A sector, or "pie" slice from an axicon of proper specification is an exceedingly accurate approximation to the conical lens. Obtaining the conical lens was then merely a matter of selecting the proper axicon from the dealer's stock. The feasibility and availability of the conical lens was a most fortuitous circumstance.

The resulting optical correlator, shown in Fig. 12, is improv
ed by a factor 20 in its light utilization efficiency. Additionally, the optical system is considerably simplied. Since there is no

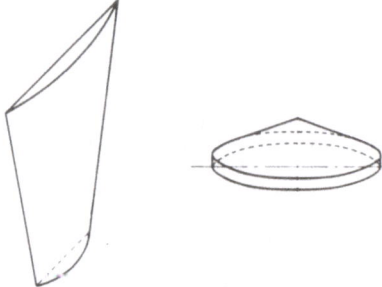

Fig.11   The conical lens and the axicon lens from which the former
was derived.

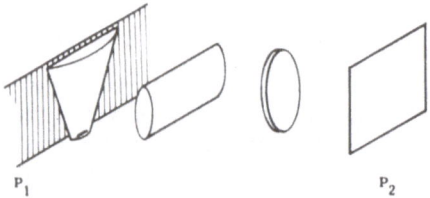

Fig.12   The optical cross correlator with conical lens.

bias term to remove, the conical lens may be placed in direct contact with the signal record, thereby eliminating two lenses and a
spatial filter. The system comprises a most unlikely combination
of lenses, but accomplishes a task that would otherwise be difficult indeed.

These developments were gratifying from an instrumentation
viewpoint, but they appeared to undermine the theory that had produced them. The correlator elements, including also now the reference function, are lenses. Lenses form images only from other
images or from original objects. The image we produce at the output slit must indeed exist quite independently of the lenses to
which theory had led.

Further considerations indicate that this supposition is true. For if the reference function is a lens, so must the signal from each object point also be a lens, since the reference function has the same functional form as such a signal. Each object-point signal is, of course, a Fresnel zone-plate lens, and illumination of such a structure with a collimated beam produces two primary foci - one real and one virtual (Fig. 13). The images formed by the Fresnel zone-plate signal can be shown to possess precisely the structure of the image produced by the correlation process previously described. Thus by reducing the optical system to only three lenses, we are, it appears, led to abolishing even these, as well as the correlation theory upon which all had been based.

At this stage a new viewpoint emerged which restored the rational basis of the process. In the ensuing months, beginning in late 1955, the theory of synthetic-aperture radar (in combination with optical processing) was recast in terms of holography. The various elements and functions of the system were now viewed in an altogether different way. The communication theory tools were laid aside, to be replaced with those of geometrical and physical optics. Terms such as linear filter, spectrum, and sideband gave way to optical ones like diffraction lens, focal length, and f number. Ray diagrams became the basic tool of analysis. As a result, the process became understood in a more pictorial way than before; problems that previously were solved using mathematical formulas could now be solved by optical diagrams and elementary ray tracing.

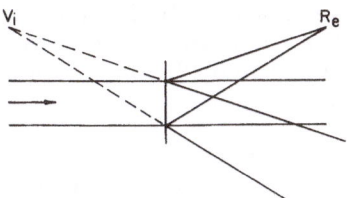

Fig.13 The focusing properties of a Fresnel zone plate. Collimated light impinges on the structure. Part of the emergent light forms a real image R_e, part forms a virtual image V_i, A portion emerges unchanged and continues as a collimated beam.

The signal record was no longer to be regarded as merely stored data, but as a scaled-down hologram of the wavefield impinging on the aircraft flight path. Illumination with coherent light produces, in minature, an optical replica of this wavefield. Images are produced, as in conventional holography, but these images have resolution corresponding to the data-collecting aperture rather than to the aperture of the real antenna. Hence the synthetic a-

perture is realized through conversion into an equivalent real a-
perture of a holographic optical system.

The image thus formed is defective in various respects. For
example, the signals have focal lengths proportional to the range
of their origin; thus the locus of focal positions is a tilted plane, as shown in Fig. 14. A similar tilted plane, corresponding to
the real image foci, forms on the other side of the signal record.
Since the signals are cylindrical zone-plate lenses, having focal
power in the x dimension only, the images are defocused in range.
Range resolution, formed by the pulse modulation, is present on the
raw data record, which is therefore the plane of range focus. The
radar system is thus anamorphic, and a compensating anamorphic property must be introduced into the optical system. The tilted plane
must be untilted and then brought into coincidence with the plane
of range focus.

As before, a conical lens, having a focal length equal but opposite to the negative focus of the signal record, is placed against
the signal record, thus reimaging the tilted plane at infinity and
erecting it in the process (Fig.15). Next, a cylindrical lens placed one focal length from the signal record images the signal record
at infinity, thereby bringing the two planes into coincidence. A
conventional spherical lens then forms a real image, free from the
original defects.

The optical system is identical to the previous one, but the
function of each element is described in a different way; the cross
correlator has become an anamorphic imaging system. Thus we find
that diverse theories from the two traditionally unrelated discipli
nes of communication theory and physical optics have produced an i-
dentical equipment synthesis!

The optical analysis can be carried much further (6). We may
consider, for example, the character of the image (or image data)
at various steps of the process. A square in object space, formed
by four point objects, is mapped onto the data record via the trans
formation $x \longrightarrow px$ and $r \longrightarrow qr$, and since in general $p \neq q$, the data
record is elongated in one dimension. Similarly, the image formed
from the hologram, under collimated illumination, will be elongated.

The lens system produces yet a further transformation, which
we can best explain by considering the anamorphic system as two dis
tinct imaging systems, one for each dimension, x and r (Fig.16). In
the r dimension, the cylindrical and spherical lenses (with focal
length F_{cy} and F, respectively) form a simple telescope with magnification

$$M_r = \frac{F}{F_{cy}} \qquad\qquad 26.$$

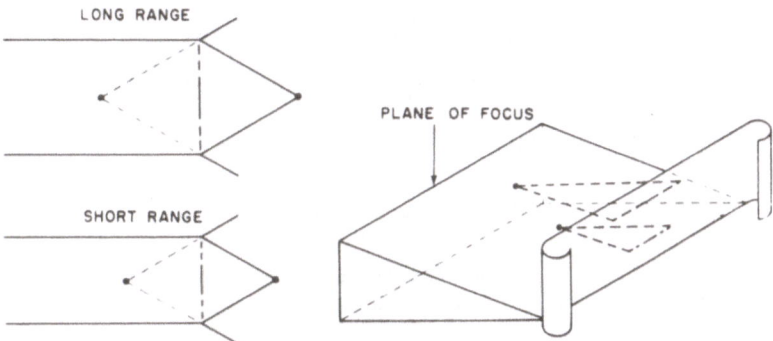

Fig.14 Focal properties of the signal record.

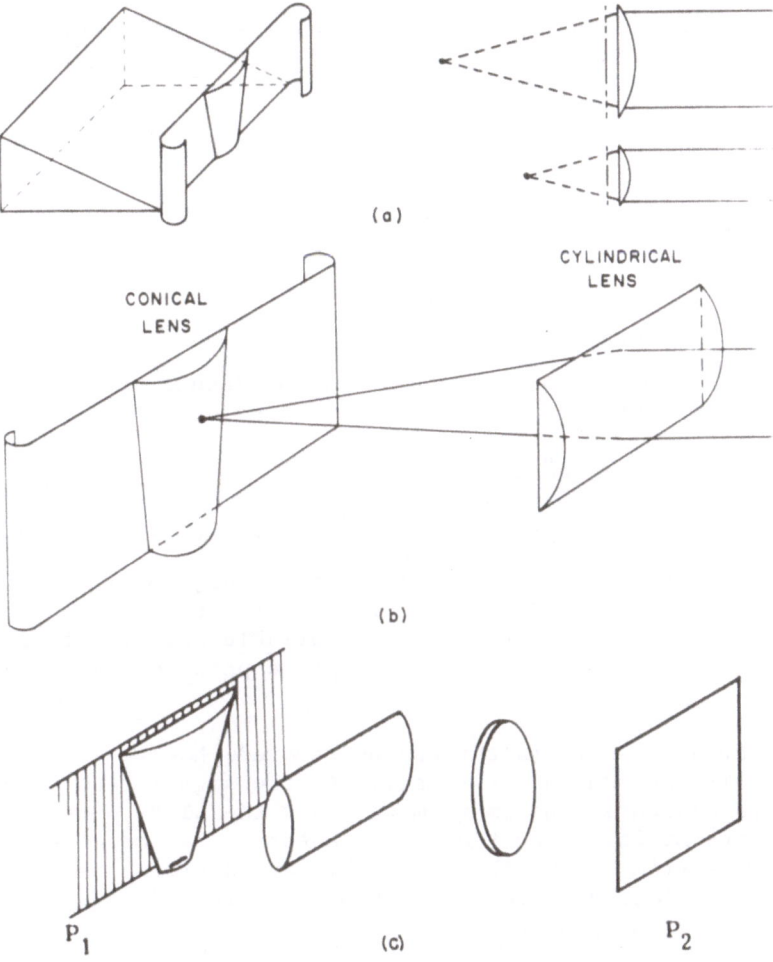

Fig.15 Development of an anamorphic lens system for compensation of the radar image.

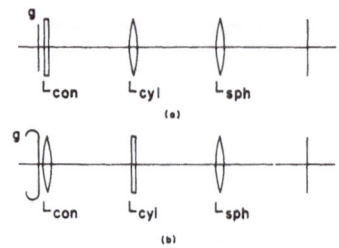

Fig.16   The anamorphic lens system as viewed in r and x meridians.
         (a) The r meridian. (b) The x meridian.

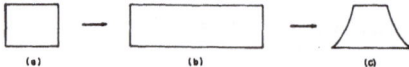

Fig.17   Transformations on the image. (a) The object. (b) The image
         formed by the data record. (c) The image formed by the ana-
         morphic lens system.

whereas, in the x dimension, the conical and spherical lenses form
a simple telescope with magnification

$$M_x = \frac{F}{F_x} = \frac{F}{\left(\frac{1}{2p^2} \frac{\lambda_o}{\lambda_1} r\right)}$$ 27.

where F_x is the focal length of the conical lens. The proper a-
spect ratio can be restored by choosing the magnification ratio so
as to compensate for the nonequality of p and q:

$$\frac{M_x}{M_r} = \frac{F_{cy}}{F_x} = \frac{p}{q} .$$ 28.

But since F_x is range dependent, the proper aspect ratio can
be attained at one range only. At other ranges, the image will be
distorted and the object-space square is ultimately mapped into the
peculiar shape shown in Fig. 17. The variable magnification is produced by the conical lens, and is the price paid for removing the
anamorphism.

The variable scale factor can be removed, however, using the
scanning slit introduced previously. As the signal record moves
through the aperture, the image moves past the slit, with the magnified parts moving proportionately faster. The image recorded
on the moving film behind the slit will have a proper aspect ratio
provided the velocity ratio of the two films is made equal to the
ratio p/q quite independently of the various lens focal lengths.

Alternatively, an equivalent conical lens, placed in a plane where the Fourier transform of the signal is formed, acts as a spatial matched filter and removes the anamorphism without introducing the variable magnification distortion. This assertion can be demonstrated either with geometrical optics, which is a rather tedious procedure, or with linear-filter theory, which yields the same result by a short verbal statement. For, contrary to the impression that may now have been created, the optical approach is not always the more succinct.[2]

The holographic viewpoint and the physical and geometrical optics analyses that thereby become available not only offer new insights into the synthetic-aperture process, they also give rise to implementations that are not readily suggested by other approaches. The holographic viewpoint has, therefore, been quite valuable.[3]

A radar image formed by optically processing data gathered by a synthetic-antenna radar is shown in Fig. 18. The radar picture, which is of a rural area in Northern Michigan, shows a variety of scenery, including rivers, wooded areas, and farmland. Highways and rural buildings are clearly visible, as well as shadows cast by wooded tracts. From the shadowing, it is obvious that the radar-carrying aircraft was to the north of the illuminated area, yet the picture appears to have been taken from directly above. This rendition of perspective is characteristic of the synthetic-antenna radar system.

---

(2) These demonstrations are left to the reader.

(3) The development related here obviously has a logic: each step seems to point to the next, as though the historical development may really have proceeded in this manner. In general it did, except that the conical lens was, in fact, suggested by the holographic viewpoint, and was the first of many innovations arising from this concept. Some of these, unlike the conical lens, are almost inexplicable except by holographic concepts.

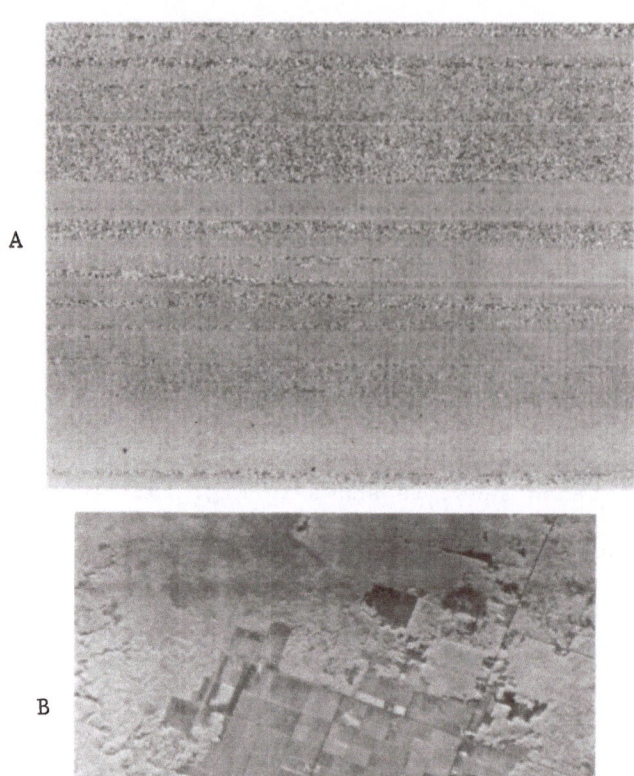

Fig.18 An example of synthetic-aperture radar imagery. (a) A raw
data record. (b) A processed image.

D. History of the Synthetic Antenna

The synthetic-aperture technique, developed under military
sponsorship, has a history stemming from 1951. Many hundreds of
persons contributed to its development. The most comprehensive ac
count of the early history is given in a paper by Sherwin, Ruina,
and Rawcliffe (5). The synthetic-antenna concept is generally cre
dited to Wiley, who noted in 1951 that stationary objects within
the beam of a moving radar could be resolved by Doppler filtering.
Similar conclusions were reached by Sherwin, Kovaly, Prothe, Newell,

Ruina, and Rawcliffe at the University of Illinois (5). The resolution improvement was explained by introducing the synthetic-antenna concept. In an extensive flight program, the group successfully demonstrated their ideas. Other significant contributions
include the "zero-beat" processing concept by Hausz, and the Redap,
or reentrant recirculating delay-line processor developed by Stein
berg and Sunstein (5).

In 1954 an extensive report describing many new and sophisticated ideas, including optical processing, was produced at the
University of Michigan, the primary contributors being Vivian and
Cutrona. The optical method, as originally described, utilized in
coherent illumination. Although cumbersome and never implemented,
it was the forerunner of the elegant version we have thus described.

Coherent optical configurations were examined in 1955 by Porcello and myself after annoying diffraction effects produced problems with the incoherent methods. The single sideband approach,
the interpretation of the reference function and signals as lenses,
and the subsequent fusion of holography and synthetic-aperture radar, are the author's work, performed during the period 1955-1956.
The Michigan system was first flown in 1957 and, after a few
flights, produced excellent results (6), (7).

The holographic viewpoint today has gained wide acceptance
and is the usual method of explaining the processing system (8).
Yet, when first introduced and for several years afterward, it
found little acceptance. This situation gradually changed when
theory was reduced to practice and experimental results became a-
vailable. By 1960 the holographic view became dominant, and has
so remained. This situation is not entirely satisfactory, for the
holographic view-point puts undue emphasis on the optical-processing aspect, which is always a minute part of the entire system.
Furthermore, the optical-processing method must be regarded as a
complement rather than a replacement for the electronic-processing
methods, for the latter have continued to flourish.

IV. PULSE COMPRESSION

Radar systems employing conventional pulsing are beset with
two conflicting requirements: to achieve fine resolution, the pulse should be narrow; to achieve a long-range capability, the avera
ge radiated power should be high. High power can be obtained by
a) increasing the pulse duration, which causes loss of resolution;
by b) increasing the pulse-repetition frequency, which results in
the ambiguity problem of pulses returning to the receiver simultaneously for two or more ranges; or by c) increasing the pulse intensity, or peak power, a process limited by insulation problems
in the transmitter.

The pulse-compression technique is a fourth possibility which avoids all of the previously mentioned difficulties. A coded pulse is radiated and, upon reception, is compressed by correlation tech niques to a fraction of its original duration. Thus high-average power is attained without the need for high-peak power and without reduction of range resolution.

The pulse-compression radar fits into the category of pulse-Doppler radar. Thus the holographic analogy noted earlier applies and indeed becomes quite striking when the coded waveform is of the linearly frequency modulated or "chirp" variety.

A. Analysis and Instrumentation

Let the radiated pulse be

$$f(t) = a \exp(j[\omega_0 t + \phi(t)]), \quad \text{for } \frac{T}{2} \le t \le \frac{T}{2}$$

$$= 0, \quad \text{for } |t| > \frac{T}{2}. \tag{29.}$$

The return signal $g(t) = \sigma f(t - t_1)$ is cross correlated with the reference function $r_c(t)$, which, since we consider only statio nary object points, is not frequency shifted and is, therefore, just a replica of the radiated pulse $f(t)$:

$$u = \sigma \int f(t - t_r) f^*(t - t_1)dt. \tag{30.}$$

This operation is usually performed passively using a matched filter. The result is the autocorrelation function of the transmitted pulse, centered about the time-delay value t_1. Evidently, we should choose a code $a \exp[j\phi(t)]$ having an autocorrelation function with a narrow main lobe and low sidelobes.

The width of the compressed pulse is approximately the reciprocal of the bandwidth W. The compression ratio is thus the time-bandwidth product TW of the pulse. A narrow uncoded pulse thus has a TW product of unity. The coding, then, has the purpose of increasing the TW product of the pulse, or equivalently, of increasing the bandwidth over that of an uncoded pulse of the same duration.

Two types of encoding have commonly been used: the multiphase pulse train and the chirp pulse. In the former type, the radiated pulse consists of a sequence of abutting subpulses, all of constant amplitude, but of varying phase (Fig. 19(a)). The phases,

Fig.19 Two forms of pulse coding. (a) A multiphase pulse train,
with phases ϕ_1, ... ϕ_n. (b) A chirp pulse.

ϕ_1, ... , ϕ_n, are chosen so as to yield a desirable autocorrela-

tion function. Sophisticated techniques exist for finding such co
des.

     The well-known shift register codes are often used. Here the
ϕ's assume two values, 0 and π: the code then consists, to within
a multiplicative constant, of a sequence of + 1's and - 1's arrang
ed in what can be described as a pseudorandom ordering. The autocorrelation function (actually the sequence correlated with its pe
riodic extension) is n when the sequences match and - 1 otherwise,
where n is the number of pulses in the sequence.

     The chirp or linearly frequency-modulated pulse (Fig. 19(b))
is the more commonly used waveform for pulse compression. Ease of
generation and ease of processing make it attractive. Furthermore,
the autocorrelation function of a linearly FM pulse is, in general,
just as satisfactory as that of the more complex codes. Either ty
pe of pulse may,of course, be compressed by optical means.

     Optical correlators of various types are available to compress
the pulse. A basic system, shown in Fig. 20(a), is applicable to
any type of waveform. The signal from the radar receiver is converted into an ultrasonic wave, which passes through an ultrasonic
delay line and is absorbed at the other end. Illumination of the
delay line with a coherent light beam results in spatial modulation of the light beam in a manner analogous to the behavior of
the photographic signal record previously described. The signal
may then be correlated with a reference mask exactly as was done
with the synthetic-aperture data. An alternative technique, the
frequency-domain synthesis, is shown in Fig. 20(b). The Fourier
transform $F(\omega)$ of the wavefront is displayed as a spatial-frequency distribution at P_2. A transparency bearing $F^*(\omega)$ is introduced

at P_2, and the compressed pulse appears at the output plane P_3.

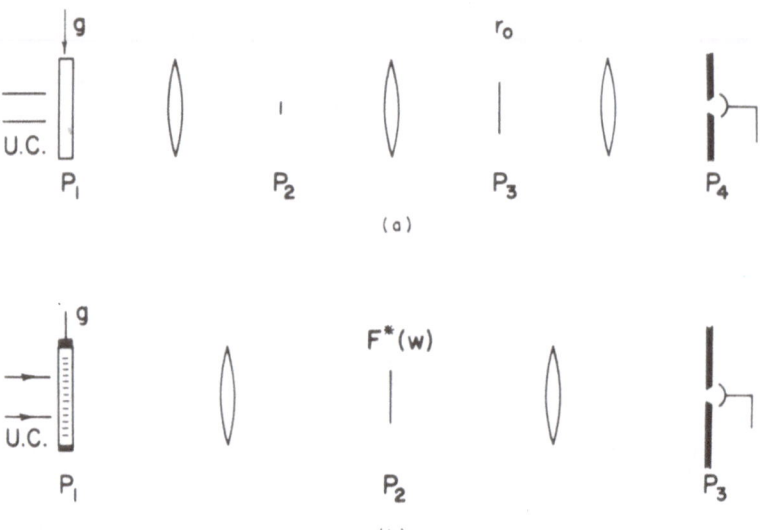

(a)

(b)

Fig. 20   Two forms of optical processor. (a) The spatial-domain synthesis. (b) The spatial-domain synthesis. in (a), g (in an
ultrasonic cell U.C.) is imaged, after spatial filtering
at P_2 onto r_o. The correlated output is sensed by a photo
detector at P_4. In (b), the signal g is imaged through a
spatial-matched filter at P_2 and the filtered signal is detected at P_3.

A major problem, that of implementing the complex filter
$F^*(\omega)$ made the frequency-domain synthesis only a conceptual solution until 1961 when Kozma and Kelly, using a technique which today is known as computer-generated holography, synthesized on photographic film a transparency bearing the spectrum of a shift register code (9). Results are shown in Fig. 21.

The systems of Fig. 20 are general, applying to any pulse
form. However, we stress here the chirp pulse because then the
resemblance to holography is strongest. The return from an object
field is then a superposition of superimposed but displaced chirp
signals that, on recording, become the familiar zone plates of holography.

For the chirp pulse, the autocorrelation function may be readily calculated. Indeeed we did exactly this in spatial coordinates
(18) when discussing the synthetic antenna, since that process also gives rise to linearly frequency modulated signals.

The optical system is considerably simplified for the chirp

Fig.21   Spatial-matched filtering for a shift register code. (a) A
         shift register sequence, written on a spatial carrier.
         (b) The spatial-matched filter for the code. (c) The result
         of imaging the structure (a) through the filter (b).
         The compressed pulse is shown on the left. The structure
         on the right is the zero-order term produced by the filter
         and is of no significance.
         (Courtesy of A. Kozma and D. Kelly.)

pulse, as then we may advantageously use the zone-plate or self-compressing properties of the signals. These properties arise whe ther the signal is stored in a delay line, on photographic film, or on any medium that can be interrogated with coherent light. The recorded signal from an object field is then indistinguishable from a hologram (a one-dimensional one, of course), and when coherently illuminated forms the usual true and conjugate images of the holographic process corresponding to the two primary foci of the Fresnel zone plate. As with the synthetic-aperture case, these images are identical to the result obtained using cross correlation or matched filtering. This conclusion, while intuitively expected, can be readily demonstrated by regarding free space as a matched filter for the chirp pulse.

The chirp signal from a point object,

$$g(t) = \sigma a_o \exp j\left[\omega_o\left(t - \frac{2r_1}{c}\right) + \frac{1}{2}k\left(t - \frac{2r_1}{c}\right)^2\right]$$

$$= \sigma a_o \exp j\left[\frac{4\pi}{\lambda_o}\frac{c}{2}\left(t - r_1\right) + \frac{2k}{c^2}\left(\frac{c}{2}t - r_1\right)^2\right]$$

31.

which may be expressed in spatial coordinates by using the relation $r = (c/2)t$, can be recorded using a scaling process $r \longrightarrow qr$, producing

$$\sigma a_o \cos\left[\alpha\left(r - \frac{r_1}{q}\right) + \frac{2kq^2}{c^2}\left(r - \frac{r_1}{q}\right)^2\right].$$

32.

Using the Fourier transform relation

$$\mathscr{F}\left\{\exp\left(j\gamma x^2\right)\right\} = \frac{\pi}{\gamma}\exp\left(j\frac{\omega^2}{4\gamma}\right)$$

the Fourier transform of the signal is readily shown to be

$$K\left[\exp j\beta(\omega_r - \alpha)^2 \exp - j\omega_r\frac{r_1}{q} + \text{complex conjugate}\right]$$

33.

where K is a constant and $\beta = c^2/(8q^2k)$.

Free space can be regarded as a spatial filter, dispersive in nature and having the transfer function (4),

$$H(\omega_x, \omega_y) = \exp(jkz) \exp\left[-j\frac{\lambda z}{4\pi}(\omega_x^2 + \omega_y^2)\right] \qquad 34.$$

a result readily obtained by Fourier transforming the usual Kirch-hoff expression for Fresnel diffraction (4). Let the pulse be re-corded along the x dimension; then $\omega_r \sim \omega_x$. At a position in the diffraction field a distance z from the signal transparency, the spectrum (33) is multiplied by the transfer funtion of the inter-vening space. The first term of (33) has its quadratic phase fac-tor perfectly compensated when

$$z = \frac{\pi c^2}{2kq^2\lambda_o} \qquad 35.$$

We expect this value of z to be just the focal length of the Fresnel zone plate, a hypothesis readily confirmed by inspection of (32), and by noting that a lens of focal length F introduces a quadratic phase factor $\exp j[(\pi/\lambda F)(x^2 + y^2)]$.

The demonstration may be completed by Fourier transforming the product of signal spectrum and transfer function, and convert-ing back to time coordinates, whereupon the result is identical to that obtained by direct matched filtering of the temporal signal g(t).

There are several significant differences between the free-space filter and some of the electronic chirp filters. The chirp filters (at least in the 1950s) were expensive, bulky, heavy, high-ly attenuating, and not always linear; sometimes they became some-what detuned. Free space, on the other hand, has none of these shortcomings.

In converse to the present discussion, the holographic pro-cess may be modeled as a pulse-compression process, an approach u-sed by Upatnieks and me in our earliest holography paper (11).

B. Combined Beam Sharpening and Pulse Compression

Obviously, pulse-compression and synthetic-antenna techniques can be combined. A chirp or other coded waveform can be radiated and then compressed upon reception, whereupon the beam-sharpening operation can be carried out just as if the radiated pulse had not been coded.

More interesting, however, is the case where the coded pulse is stored and later processed by optical means. Under this circum-

stance the beam-sharpening and pulse-compression operations can be
treated as orthogonal aspects of a single two-dimensional operation. An optical system, because of its inherent capability for
treating two-dimensional signals, can readily perform the combined
operation (12).

As previously, we stress the chirp method. The recorded signal from a point object has the form

$$\sigma a \, \cos \left[\alpha x - \frac{2\pi p^2}{\lambda_o r_1} \left(x - \frac{x_1}{p} \right)^2 + \frac{2kq^2}{c^2} \left(r - \frac{r_1}{q} \right)^2 \right] \qquad 36.$$

where the first quadratic phase factor is due to the Doppler shift
and the second arises from the chirp waveform. The signal is a two
dimensional zone plate with a different focal length in each dimension (12). Inspection shows the focal length to be

$$F_x = \pm \frac{1}{2p^2} \frac{\lambda_o}{\lambda_1} r_1 \qquad 37.$$

$$F_r = \pm \frac{\pi c^2}{2kq^2 \lambda_1} \qquad 38.$$

for the x and r dimensions, respectively.

The zone plate is generally anamorphic since the factors determining the focal length are different for each dimension. Proper selection of the FM rate and other parameters can make the focal lengths equal, but for one range element only.

For an extended target field, the plane of azimuth focus is
the tilted plane described previously. The plane of range focus,
however, is no longer the signal record itself, but is a plane parallel to the signal record and located one focal length F_r in

front of or behind the record, depending on whether we utilize the
real or virtual image term.

The one-dimensional multichannel optical processor of Fig.12
is converted into a two-dimensional multichannel processor merely
by moving the cylindrical lens a distance F_r along the optic axis

to that the range focal plane lies in the front focal plane of the
cylindrical lens. This quite simple expedient has, at once, given
us a two-dimensional processor. The pulse-compression operation

has indeed been obtained without cost.

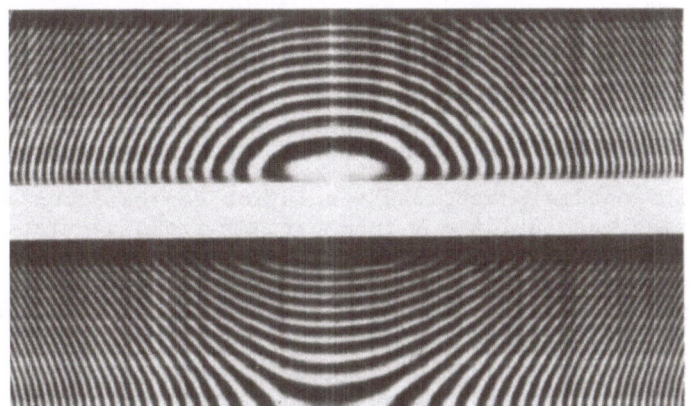

Fig.22   Elliptical and hyperbolic zone plates produced by a pulse-
Doppler radar system.

The quadratic coefficients of $(x - x_1/p)^2$ and $(r - r_1/q)^2$ may
have like or unlike signs, depending on whether the frequency sweep
rate k is positive or negative. If the coefficients have like
signs, the constant-phase contours are ellipses; if the signs are
unlike, the contours are hyperbolas. Signals of each type are
shown in Fig. 22. The elliptical type if conceptually simpler. One
order has a positive focal length in each meridian; that is, F_x and
F_r although not equal, are each positive for that order. The other
order has a negative focal length in each meridian.

The hyperbolic type has F_x and F_r with unlike signs. One order has a focal length that is positive in one meridian and negative in the other, while the other order has the signs reversed.

Either type of zone plate can be handled, but the elliptical
type is preferable since the anamorphic properties of the correcting optical system can be less severe.

C. Terminologies: Electrical and Optical

There is a well-established mathematical formalism derived
from linear-filter theory for describing the action of the chirp
filter. Likewise, the focusing action of a zone plate is well described by the concepts of physical optics. Since the two phenomena are essentially identical, we can compare them and draw some ana
logies between the two sets of terminologies. We might, for example, describe the pulse compression in the usual way and then apply
the concepts of physical optics to a zone plate formed by recording

the pulse.

Comparison of the two terminologies reveals interesting relations. The focal-length expression F_r (38) shows focal length to
correspond in a proportional way to the reciprocal of the FM rate
k. Similarly, bandwidth and f number are inversely related. The
electrical signal has bandwidth W_t = kT, where T is the pulse dura
tion. For the optical case, the resolution can be found by the u-
sual expression resolution = $\lambda \cdot$ f-number, or from calculation of the
bandwidth. The results, as to be expected, will agree.

The time-bandwidth product (TW), important in communication
theory, has no established counterpart in optics, although the
term space-bandwidth production is gaining acceptance. The TW pro
duct is an important characteristic of a signal, partly because it
defines the number of degrees of freedom of a signal, and partly
because of its invariance to many transformations. The reader can
verify from a few simple calculations that the TW product has remained invariant under the recording transformation.

D. Historical Summary

The pulse-compression method was invented independently by
several researchers during the 1940s (13). This work continued in
to the 1950s at a number of establishments, including Sperry and
Bell Laboratories. The earliest detailed description in the published literature is by Klauder, Price, Darlington, and Albersheim (10). Investigation of optical methods for compression of
the pulse was initiated by the author in 1957, the salient facets
of the effort being the utilization of the zone-plate properties
of the chirp pulse, and the combining of pulse compression and syn
thetic aperture into a single two-dimensional operation, performed
by coherent optical methods (14), (15), (10). This two-dimensional optical-processing method was independently suggested by Curtis
and Smith at RCA (16). Optical-processing methods, along different
lines, were carried out by Lambert and his group at Columbia University (17), using ultrasonic delay lines to provide, in real time, a range-Doppler analysis of the incoming signal. Dickey of Ge
neral Electric reported on optical techniques for pulse compression
at a symposium in 1959 (18).

V. RANGE-DOPPLER RADAR FOR ROTATING BODIES

The range-Doppler principle, when applied to rotating bodies,
can yield excellent resolution, as evidenced by the dramatic radar
imagery of planets that has been thus obtained. The technique is
akin to the synthetic-aperture principle, but with some additional
aspects of considerable significance.

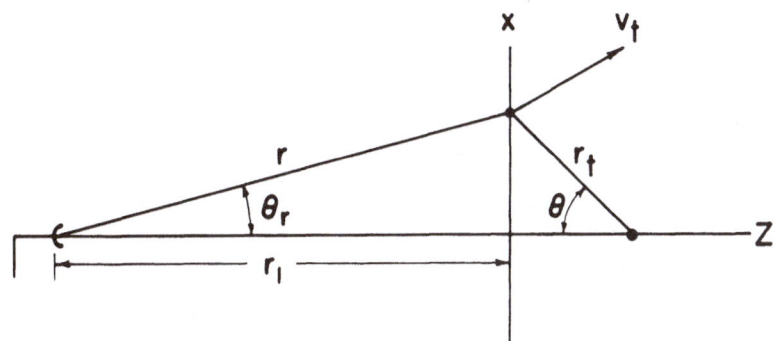

Fig.23 Geometry of the rotating body range-Doppler radar system.

A system of particles, shown in Fig. 23, rotates with constant angular velocity about a common axis. The particles could represent point scatterers on a single rigid body. A pulse-Doppler radar a distance r from the object point O_1 radiates a signal

$$f(t) = a \exp\left[j(\omega_0 t + \phi) \right]$$ 39.

and receives from O_1 a signal

$$g(t) = \sigma a\left(t - \frac{2r}{c} \right) \exp j\left[\omega_0\left(t - \frac{2r}{c} \right) + \phi\left(t - \frac{2r}{c} \right) \right].$$ 40.

If $r \gg r_t$, the approximations

$$a\left(t - \frac{2r}{c} \right) = a\left(t - \frac{2r_1}{c} \right)$$

$$\phi\left(t - \frac{2r}{c} \right) = \phi\left(t - \frac{2r_1}{c} \right)$$ 41.

are valid. If the observation time is sufficiently short, then r_1 and the radial velocity v_r (which we take as being approximately v_z) can be considered constant, and the range $r(t)$ can be written

$$r(t) = r(t = 0) + v_z t$$ 42.

where $v_z = \omega_t r_t \sin \theta_0$, and ω_t is the angular velocity. The signal $g(t)$ may then be written

$$g(t) = \sigma f \left(t - \frac{2r_1}{c} \right) \exp - j \left[\frac{4\pi}{\lambda_o} r(t = 0) + \omega_d t \right] \qquad 43.$$

where

$$f_d = \frac{\omega_d}{2\pi} = \frac{2}{\lambda_o} \omega_t r_t \sin \theta_o \qquad 44.$$

is the Doppler frequency shift. Contours of constant Doppler shift are found by setting

$$\omega_d = \frac{4\pi}{\lambda_o} \omega_t x_t = C \qquad 45.$$

where C is a constant and $x_t = r_t \sin \theta_o$. The Doppler contours, given by

$$x_t = \frac{C \lambda_o}{4\pi \omega_t} \qquad 46.$$

are thus planes parallel to the plane defined by the axis of rotation and the line between radar and the rotating system axis. Iso-range controus are planes normal to this line and thus normal to iso-Doppler planes, provided the radar is sufficiently distant. Thus a range-Doppler analysis produces an x, r image of the rotating-particle system.

Implicit in the foregoing analysis is an axis of rotation normal to the radius vector r_o. A minor extension of the preceding analysis shows an essentially similar result when this condition is not met.

The Doppler resolution $\Delta \omega_d$ is determined by the time of observation

$$\Delta \omega_d = \frac{2\pi}{T} \qquad 47.$$

from which the x resolution is found to be

$$\Delta x = \frac{\lambda_o}{2\omega_t T} . \qquad 48.$$

The resolution in range is, as usual, roughly the reciprocal of the bandwidth of f (t)

$$\Delta r \; \cong \; \frac{1}{\Delta W} \; . \qquad\qquad 49.$$

Both range and cross-range resolution are independent of distance; therefore the system has potential for excellent resolution at large distances.

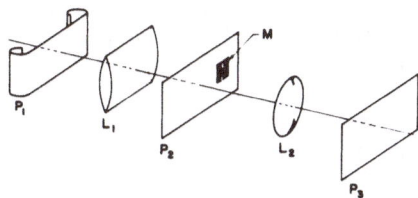

Fig.24   Optical processing system for range-Doppler analyses of rotating body data. M is a spatial-matched filter, which operates on one sideband of the signal.

Rotating-target returns have been range-Doppler processed by various methods, including digital, electronic analog, and optical. The optical-processing procedure is similar to that for combined pulse compression and synthetic-antenna beam sharpening. A pulsed waveform f(t) is radiated repetitively and successive pulse returns are recorded adjacently, thus forming a two-dimensional signal record. The optical-processing system is shown in Fig. 24. A cylindrical lens produces a Fourier transformation on the range variable, thus forming the pulse spectrum at P_2. The pulse is compressed by means of a spatial-matched filter at P_2 and the processed images is displayed at P_3 after a transformation by lens $L2$. The same lens forms at P_3, through its Fourier transforming properties, a Doppler analysis along the x direction.

For the x dimension, the data record is indistinguishable from a Fourier transform hologram. Under coherent illumination, the record forms, at infinity, an image which is displayed as a real image in the back focal plane of a lens.

The velocity tangential to the line between radar and object is

$$v_x = v_t \cos \theta_o = \omega_t r_t \cos \theta_o \qquad\qquad 50.$$

but since $r_t \sin \theta_o = z_t$,

$$v_x = \omega_t x_t \qquad\qquad 51.$$

showing that all particles at constant range r from the radar have the same cross-range velocity (although, for different ranges, v_x will be different). If we view the particles in a coordinate system that moves with the same tangential velocity v_x, the radar resembles a synthetic-aperture system, and the signal received at the antenna can be regarded as a sampling of wavefronts impinging on the antenna flight path. The holographic viewpoint thus applies equally well to both types of systems.

The resolution, however, is considerably better than with the conventional synthetic antenna. The latter has resolution

$$\triangle x = \frac{\lambda_o r}{2A} \qquad\qquad 52.$$

where A is the aperture generated. The rotating-body system has resolution given by (48) and since the travel distance, or synthetic-antenna length, is $A' = r_t \omega_t T$, the resolution can be expressed as

$$\triangle x = \frac{\lambda_o r_t}{2A'} . \qquad\qquad 53.$$

Combining (52) and (53) yields

$$A = A' \frac{r}{r_t} . \qquad\qquad 54.$$

The resolution is better than for an equal synthetic aperture by the ratio of the object distance to the length of the rotation axis. For distant objects, such as a planet, this resolution-improvement factor can be several thousand. The resolution is essentially as good as if the synthetic aperture were at the axis of rotation.

An object point at angular position θ_r, with respect to the antenna, has a Doppler shift

$$\omega_d = \frac{4\pi}{\lambda_o} v \sin \theta_o \qquad\qquad 55.$$

thus appearing to arrive at an angle θ_o. Hence the reconstruction
process converts an object field angle θ_r into an image field angle θ_o, a viewpoint consistent with regarding the synthetic antenna as being effectively at the axis of rotation.

One of the most fascinating examples of rotating-target imagery is that carried out on the planet Venus at the Jet Propulsion
Laboratory (JPL) of the California Institute of Technology (19).
The JPL radar transmitted, at a wavelength of 12.5 cm, a binary-
coded pulse of duration 51 100 μs; the pulse was divided into 511
subpulses of phase 0 or π. The code was repeated without interrup-
tion for 5 min corresponding to the round trip transit time between Earth and Venus. At the end of this time, the transmisssion
ceased and the reflected signal from Venus was picked up by the
receiver and the results stored on magnetic tape. The Doppler
bandwidth of the received signal was approximately $\pm$ 15 Hz.

The stored data were range-Doppler analyzed by means of a
digital computer which compressed the pulse and performed the
Doppler analysis.

The data were also processed by optical means, which relates
the process more to holography (20). The data, stored on tape,
were recorded photographically through the display of a line trace
on a cathode-ray tube, in the manner described in connection with
the synthetic-aperture technique. The photographic record was
then processed optically, yielding an image, shown in Fig. 25,
with resolution of about 15 km in range and about 180 km in the
cross-range direction.

Note that the surface of a sphere has two points with the
same range and velocity, one in the upper hemisphere and the other
in the lower. Unless special means are used to suppress one set of
images, the two hemispheres will not be imaged separately, a defi-
ciency present in the image of Fig. 25.

The signal-to-noise ratio of the resulting image, always a problem because of the large signal attenuation resulting from the
enormous distances involved, can be significantly increased by the
use of post-detection integration. This is accomplished simply by
continuing the exposure while moving the data film through the a-
perture until all of the data have been utilized. All of the data
in the aperture at one time are coherently integrated, thus producing an image in accordance with the theory previously given. The
length of data that can be usefully integrated coherently is limited by various phase errors to about 2 s; beyond this, there is no
improvement in Doppler resolution. A continuum of 2-s sections of
data were summed incoherently (i.e., on an intensity basis).

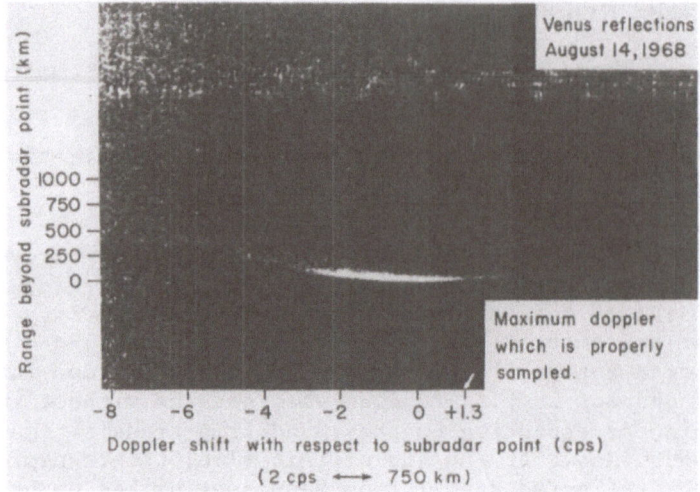

Fig.25 Image of a portion of the planet Venus obtained from a pulse-
        Doppler radar system at the Jet Propulsion Laboratory of
        the California Institute of Technology, with optical pro-
        cessing at the University of Michigan. (Courtesy of L.
        Porcello, C. Heerema, and N. Massey.)

Thus each 5-min radiation period results in the incoherent summation of 150 images, thus producing considerable S/N improvement
over a single exposure.

VI. PHASED-ARRAY BEAMFORMING

Another significant holographic-like technique, beam formation on phase-array systems, unlike the other previously described,
does not fall into the category of range-Doppler radar; indeed,
phased-array systems need not be radar devices, and they may be
passive as well as active.

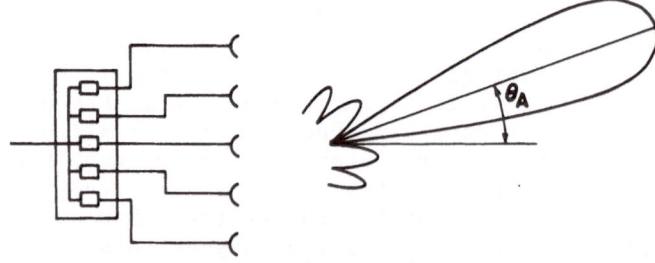

Fig.26 Beamforming and steering on a phase-array system. The beam
        is aimed in the direction of ϕ_A.

The phased-array antenna system we describe here is an array of receiving elements in which the received signals, instead of being summed at the antenna, are separately fed into a processing system, either analog or digital, which can process the signals in a variety of different ways. In particular, the signals may be summed with progressive linear-phase delays; e.g., the signal on each array element may be delayed 10° with respect to that on the adjacent element on the left. Such a pattern of summation gives the array a somewhat skewed receiving pattern (Fig. 26). Summing with no phase delay produces a beam pattern centered about the normal to the array.

Generation of a beam pattern in this manner is called beam-forming. By controlling the delay pattern, the beam pattern can be steered in any desired direction - a process called beam steering. The signals can each be divided into many portions, each sent into a separate beamforming device which forms a beam in a different direction. Thus unlike the conventional array, the phased-array system can form a narrow beam simultaneously in many directions.

Let $g(x_k, t)$ be the signal received in the kth element. The beam-forming process then produces

$$u = \sum_k w(x_k) g(x_k, t) \qquad 56.$$

where $w(x_k)$ is a complex weighting function whose phase component represents the phase delay and whose modulus represents an apodization factor. If the number of elements is large, the summation may be approximated by the integral

$$\int_{-\infty}^{\infty} w(x) g(x, t) dx. \qquad 57.$$

For a uniform weighting fuction and a linear-phase delay,

$$w = \exp (j\omega x) \left(\frac{-L}{2} \le x \le \frac{L}{2} \right) = 0, \qquad |x| > \frac{L}{2} \qquad 58.$$

and for a plane wave incident at angle θ

$$g = \exp j \left(\frac{2\pi}{\lambda_o} \sin \theta \right) \qquad 59.$$

Equation (57) becomes

$$u = \int_{-L/2}^{L/2} \exp\, j \left(\alpha + \frac{2\pi}{\lambda_o} \sin\theta \right) x$$

$$= K\, \mathrm{sinc} \left(\frac{\alpha}{\pi} + \frac{2}{\lambda_o} \sin\theta \right) \frac{L}{2} \,. \qquad\qquad 60.$$

The array thus has a receiving pattern with a maximum in the direction θ_A such that

$$\alpha + \frac{2\pi}{\lambda_o} \sin\theta_A = 0. \qquad\qquad 61.$$

It is evident that quadratic-phase factors may be introduced so as to cause the array to be focused at a point short of infinity.

The beamforming process, as given in (57) with weighting function exp $(j\alpha x)$, is a Fourier transformation. This result is expected in view of the well-known Fourier transform relationship between the antenna illumination function and the far-field antenna receiving pattern. Each point of the Fourier transform represents one of the beams; hence, a device, analog or digital, that Fourier transforms the incident signal $g(x)$ thus forms all the beams.

The beamforming process is obviously akin to the holographic process; the wave-field incident on the array is recreated as an electrical signal, and the second (the electrical) step of the process inverts the process, thereby forming an image of the source distribution.

This simularity suggests carrying out the second step by recording the wave field as a hologram, with subsequent reconstruction at optical frequencies. Alternatively, from the viewpoint of optical data processing, the Fourier-transform relation of (57) also suggests the optical approach, since lenses readily produce Fourier transformations.

The data received on each array element are recorded on photographic film or other device that can serve as the input to an optical system. The recording process, in general, should preserve the geometry so that the reconstruction process duplicates the propagation process that formed the data. The signal thus recorded is coherently illuminated and Fourier transformed by means of a lens. In the Fourier-transform plane, each point represents a beam; or stated equivalently, the Fourier-transform plane is an image of the far-field source distribution sensed by the array. The latter view implies that the entire space sensed by the array is mapped

into an optical space, with mapping equations similar to those go-
verning synthetic-aperture systems, or in general, holographic systems.

The process of transferring the data to film may be performed
in many ways. For example, the signal on each element may be sampled, the sampled function displayed as an intensity-modulated tra
ce on a cathode-ray tube, and the data transferred to film in the
manner described for the synthetic aperture.

Numerous variants of the optical method are possible. The
two-dimensional processing capability of the optical system permits two-dimensional arrays to be handled. The signal recorded on
each spot on the data film can represent the signal at a single
instant, or, in the case of a highly monochromatic signal, the
point can represent an integration over time, as in the case of op
tical holography. In the case where the array is one-dimensional,
and we desire to display, as a function of time, the signal $g(x,t)$
on each array element, we may let one dimension of the recording
film, say the width dimension, represent the dimension x, while
letting distance along the length of the film represent time. This
leads to the multichannel system wherein the beamforming process
is performed in one dimension, while each signal (a time function)
can be filtered, cross correlated with another signal, or otherwise processed. Cylindrical optics can be useful in such situations. An example of the flexibility of the optical approach is
given by Beste and Leith (21), wherein a single optical system
performs simultaneously on two separate arrays and correlates the
signal on one array with that on the other.

Arrays other than linear or planar may be treated. Circular
arrays provide an interesting example. The recording process,
through its considerable flexibility, permits transformations of
the geometry; thus a circular array may be transformed upon record
ing into a linear array.

The phased-array system may be either active or passive. Coherence between the source and the equivalent "reference function"
in the receiver is not necessary, nor need there be coherence between any of the radiators in the object field. Thus a flexibility
exists that is generally absent from optical holography.

VII. OTHER HOLOGRAPHIC-LIKE SYSTEMS

The quasi-holography systems already discussed are probably
the major ones, but the list is by no means exhausted.

The so-called "hologram radar", suggested by Brown in 1964
as a direct microwave counterpart of optical holography, is, as
Brown later noted, better regarded as a union of the synthetic-aper

ture and the phased-array techniques (22). An airborne antenna
aimed directly below is used in the synthetic-aperture mode. This
antenna is an array of dipole elements oriented normal to the
flight path, and resolution in the cross-track dimension is produc
ed by the beamforming techniques of Section VI. The radar either
is not pulsed or is pulsed only so that the transmitter will be
off during the receiving time. Resolution in range is achieved
by the three-dimensional imaging effects well known in holography.

A modification of this technique, using the phased-array method in combination with the rotating-target method, was reported
by Larson, Johansen, and Zelenka (23).

Kock has recently reported on a variety of microwave-processes
that fall into the category of quasi-holography, including end-fire
synthetic arrays and moving-target-indication radars (24) - (31).

Some early work by Rogers (32) is worth noting, although it
was not done at microwave frequencies and was proposed as a direct
extension of Gabor's original work and, therefore, qualifies as
true rather than quasi-holography. Rogers described the formation
of holograms from radio waves reflected from inhomogeneities in
the ionosphere. As the inhomogeneity moves, its diffraction pattern on the ground moves across a receiver. The detected signal
is then recorded on film as a hologram, from which an image of the
inhomogeneity can be formed. Images of both atmospheric inhomogeneities and an aircraft in flight were thus obtained.

VIII. CONCLUDING COMMENTS

Viewing these various processes as holographic often is helpful in increasing one's understanding of them, particularly through
being able to visualize them in a very physical and pictorial way.
In general, the techniques of geometrical and physical optics are
at once brought to bear. Occasionally, techniques are suggested
that would have had little likelihood of arising from a communication theory view.

ACKNOWLEDGMENT

The author wishes to thank L.J. Cutrona, who perhaps more
than any other individual contributed to the success of synthetic-
aperture radar; A. Ingalls, who in the years 1956-1959 helped enlarge on the holographic viewpoint; A. Kozma, who most successfully
used the holographic viewpoint to develop new optical-processing
concepts for synthetic-aperture radar; and C.J. Palermo, whose con
tributions to coherent optics include, among many others, the sug
gestion of optical beamforming.

REFERENCES

(1) R.P. Dooley, "X-band holography", Proc. IEEE(Corresp.)., vol. 53,Nov.1965, pp. 1733-1735.

(2) D.E. Duffy, J. Opt. Soc. Amer., vol.56, 1966, p.832.

(3) G. Tricoles and E.L. Rope, J. Opt. Soc. Amer., vol.56, 1966, p.542; also vol.57, 1967, p.97.

(4) J.W. Goodman, Introduction to Fourier Optics. New York: Mc Graw-Hill, 1968, see for example p.60.

(5) C.W. Sherwin, J.P. Ruina, and R.D.Rawcliffe, "Some early developments in synthetic aperture radar systems, "IRE Trans. Mil. Electron., col. MIL-6, Apr. 1962, pp.111-115.

(6) L.J. Cutrona, W.E. Vivian, E.N. Leith, and G.O. Hall, "A high-resolution radar combat-surveillance system, "IRE Trans.Mil. Electron., vol. Mil-5, Apr. 1961, pp. 127-131.

(7) L.J. Cutrona, E.N. Leith, L.J. Porcello, and W.E. Vivian, "On the application of coherent optical processing techniques to synthetic-aperture radar", Proc. IEEE, vol. 54, Aug. 1966, pp. 1026-1032.

(8) E.N. Leith and A.L. Ingalls, Appl. Opt., vol. 7, 1968, p.539.

(9) A. Lozma and D.L. Kelly, Appl. Opt., vol.5, 1965, p.387.

(10) J.R. Klauder, A.L. Price, S. Darlington, and W.J. Albersheim, Bell Syst. Tech. J., vol. 34, 1960, p.745.

(11) E.N. Leith and J. Upatnieks, J. Opt. Soc. Amer., vol. 52, 1962, p. 1123.

(12) E.N. Leith, "Optical processing techniques for simultaneous pulse compression and beam sharpening", IEEE Trans. Aerosp. Electron. Syst., vol. AES-4, Nov. 1968, pp.879-885.

(13) Modern Radar, R. Berkowitz, Ed. New York: Wiley, 1965, p.212.

(14) E.N. Leith, in Proc. 5th Annu. Radar Symp. Ann. Arbor, Mich.: University of Michigan, Feb. 1959.

(15) ———, in Proc. 2nd Conf. Pulse Compression. Rome, N.Y.: Rome Air Development Center, Aug. 1959.

(16) W.C. Curtis, private communication, 1959.

(17) L. Lambert, "Optical correlation", see for example, ch.3 of (13).

(18) F. Dickey, in Proc. 2nd Conf. Pulse Compression. Rome, N.Y.: Rome Air Development Center, Aug. 1959.

(19) R.M. Goldstein and R.L. Carpenter, Science, vol. 139, 1963, p. 910.

(20) L.J. Porcello, C.E. Heerema, and N.G. Massey, J. Geophys.Res., vol. 74, 1969, p.27.

(21) D.C. Beste and E.N. Leith, "An optical technique for simultaneous beamforming and cross-correlation", IEEE Trans.Aerosp. Electron. Syst., vol. AES-2, July 1966, pp. 376-384.

(22) W.M. Brown, private communication, 1964.

(23) R.W. Larson, E.L. Johansen, and S.J. Zelenka, "Microwave holography", Proc. IEEE (Lett.), vol. 57, Dec. 1969, pp. 2162-2164.

(24) W.E. Kock, "Stationary coherent (hologram) radar and sonar",

Proc. IEE (Lett), vol. 56, Dec. 1968, pp.2180-2181.

(25) ——, "A hologram form of bistatic radar or sonar", Proc. IEEE (Lett.), vol. 57, Jan. 1969, p. 100.

(26) F. Tuttle and W.E. Kock, "A holographic pulse compression technique employing amplitude modulation", Proc. IEEE (Special Issue on Computers in Industrial Process Control) (Lett.) vol. 58, Jan. 1970, pp. 153-154.

(27) W.E. Kock, "Passive (cooperative) hologram radar", Proc. IEEE (Lett.), vol. 58, Aug. 1970, p. 1297.

(28) ——, "Pulse compression with periodic gratings and zone plate gratings", Proc. IEEE (Lett.), vol. 58, Sept. 1970, pp. 1395-1396.

(29) ——, "Holographic amplitude pulse compression for synthetic aperture radar", Proc. IEEE (Special Issue on Optical Communication) (Lett.), vol. 58, Oct. 1970, pp. 1773-1774.

(30) ——, "Synthetic end-fire hologram radar", Proc. IEEE (Lett.), vol. 58, Nov. 1970, pp. 1858-1859.

(31) ——, "Holographic techniques in continuous wave bistatic radars", Proc. IEEE (Lett.), vol. 58, Nov. 1970, pp.1863-1864.

(32) G.L. Rogers, Nature 177 613 (1956).

ACOUSTICAL HOLOGRAPHY

Byron B. Brenden

Holosonics, Inc., Richland, Washington (U.S.A.)

CHAPTER I :   PROPAGATION OF ACOUSTICAL WAVES

1.1   Introduction

This chapter is devoted to a review of the basic characteristics of the propagation of acoustical waves in liquids and solids. It provides no new information but seeks instead to review in the reader's mind those elements of acoustical wave propagation which will be useful in the discussion of liquid surface and scanned acoustical holography.

A few words regarding terminology may be useful. The terms "acoustic," "sound," and "sonic" are considered to be unrestricted in frequency range. Whereas, the term "audio" defines a frequency range from approximately 15 Hz to 20kHz over which the human ear is responsive, and the term "ultrasonic" applies to frequencies above the audio range, no such restrictions are placed on the terms "acoustic," "sound," or "sonic."

1.2   Description of the Wave

For the purposes of the subject matter to be presented in these chapters, a simple description of an acoustical wave given by

$$w = W \cos\left(\Omega t - K z\right) \qquad\qquad 1.1$$

will suffice. In this description w represents the instantaneous value of the displacement of an elemental volume of liquid
(a particle) from its normal or undisturbed position when no wave
is present, W represents the maximum displacement (amplitude), t
represents the time measured from some arbitrary beginning time,
and z is the distance along the z-axis of and x, y, z coordinate
system.

The quantities Ω and K are defined by

$$\Omega = 2\pi/\Lambda \qquad\qquad\qquad 1.2.$$

and

$$K = 2\pi c/\Lambda . \qquad\qquad\qquad 1.3.$$

where Λ is the wavelength and c is the velocity of the wave.

Most of the discussion to follow will deal with compressional
(longitudinal) waves, however, Equation 1.1 also adequately descri
bes the scalar features of transverse (shear) waves.

Equation 1.1 provides a description of the wave in terms of
particle displacement. It is also useful to have a description in
terms of pressure and to establish the relationship between the
particle displacement amplitude W and the pressure amplitude.

1.3 Wave Description in Terms of Pressure

A description of an acoustical wave in terms of pressure can
be derived with reference to Figure 1.1. Because of the presence
of the wave, the pressure throughout any small volume, $V = \Delta x \Delta_y \Delta z$
is not constant. The result is that the z-dimension Δz is changed
from its initial value to a new value, $\Delta z(1 + \frac{\partial w}{\partial z})$. The resulting

change in volume is

$$\Delta V = V \left(1 + \frac{\partial w}{\partial z}\right) - V = V \frac{\partial w}{\partial z}$$

i.e.,

$$\frac{\Delta V}{V} = \frac{\partial w}{\partial z} \qquad\qquad\qquad 1.4.$$

The change in volume per unit volume ($\Delta V/V$) produced by a change in pressure ΔP is known as the compressibility β. There are two types of compressibility. Isothermal compressibility is measured while keeping the system at a constant temperature. In order for the system to remain at constant temperature, heat must flow out of the volume V while it is being compressed and flow in while it is being expanded. If no heat flow occurs during the compression or expansion of the volume, the resulting measured value

$$\beta = - \frac{\Delta V}{V \Delta P} \qquad\qquad 1.5.$$

is known as the adiabatic compressibility. Heat transfer is normally very small during the rapid oscillations of an acoustical wave so that it is the adiabatic conditions that prevail.

Substituting from Equation 1.4 into Equation 1.5 and replacing ΔP by p, we obtain

$$p = -\frac{1}{\beta} \frac{\partial w}{\partial z} . \qquad\qquad 1.6.$$

Upon differentiation of Equation 1.1, we find that

$$p = -\frac{K}{\beta} W \left(\sin \Omega t - Kz \right) \qquad\qquad 1.7.$$

and setting

$$p_o = -\frac{K}{\beta} W \qquad\qquad 1.8.$$

we have

$$p = p_o \sin\left(\Omega t - Kz \right) \qquad\qquad 1.9.$$

Equation 1.8 provides us with the desired relationship between the particle displacement amplitude W and the pressure amplitude p_o.

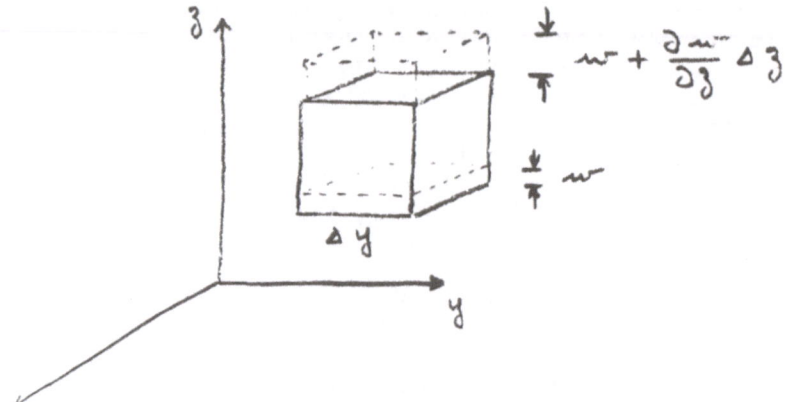

Fig. 1.1 - Motion of an Elemental Volume of Fluid

A longitudinal acoustical wave travelling in the z-direction moves the elemental volume V = ΔxΔyΔz a distance w along the z-axis. It also elongates or compresses the z-dimension of the volume by an amount $\frac{\partial w}{\partial z} \Delta z$.

1.4 Velocity of Propagation in Liquids

Referring to Figure 1.1, we see that the force Fz on the bottom of the elemental volume is

$$F_z = P \ \Delta x \Delta y .\qquad\qquad 1.10.$$

and that on the top is

$$F_z + \frac{\partial F_z}{\partial z} \Delta z = \left(P + \frac{\partial P}{\partial z} \ \Delta z\right) \Delta x \Delta y .\qquad 1.11.$$

Thus, the net force in the z-direction is

$$\Delta F_z = - \frac{\partial F}{\partial z}\, z = - V \frac{\partial P}{\partial z}\,. \qquad\qquad 1.12.$$

Similarly

$$\Delta F_x = -V \frac{\partial P}{\partial x} \qquad\qquad 1.13.$$

and

$$\Delta F_y = -V \frac{\partial P}{\partial y}\,. \qquad\qquad 1.14.$$

The net force per unit volume is given by

$$\overline{F} = (\overline{i}\Delta F_x + \overline{j}\Delta F_y + \overline{k}\Delta F_z)/V \qquad\qquad 1.15.$$

so it follows that

$$\overline{F} = - \nabla P \qquad\qquad 1.16.$$

where

$$\nabla = \overline{i}\,\frac{\partial}{\partial x} + \overline{j}\,\frac{\partial}{\partial y} + \overline{k}\,\frac{\partial}{\partial z} \qquad\qquad 1.17.$$

and $\overline{i}$, $\overline{j}$, and $\overline{k}$ are unit vectors in the x, y, and z directions respectively.

The pressure P was considered to be the total pressure, but if
we consider P to be made up of a constant static pressure Po and a
varying component p which may be either positive or negative then,

$$P = Po + p \qquad\qquad 1.18.$$

and

$$\nabla P = \nabla p \qquad\qquad 1.19.$$

so the force per unit volume may also be written

$$\overline{F} = -\nabla p. \qquad\qquad 1.20.$$

Since we are considering a plane wave propagating in the z-direction for which $\frac{\partial p}{\partial x} = \frac{\partial p}{\partial y} = 0,$ we may write

$$\overline{F} = -\frac{\partial p}{\partial z}. \qquad\qquad 1.21.$$

According to Newton's Third Law of Motion, the force represented by Equation 1.21 produces an acceleration $\frac{\partial^2 w}{\partial z^2}$ such that

$$-\frac{\partial p}{\partial z} = \rho \frac{\partial^2 w}{\partial z^2} \qquad\qquad 1.22.$$

where ρ is the density of the liquid.

The expressions for p and w are given by Equations 1.1 and 1.9. When these expressions are substituted into Equation 1.22, we find that

$$p_o K = -\rho \Omega^2 W \qquad\qquad 1.23.$$

or

$$\qquad\qquad\qquad\qquad\qquad\qquad 1.24.$$
$$\frac{K^2 W}{\beta} = \rho \Omega^2 W.$$

From this we see that

$$\frac{\Omega^2}{K^2} = \frac{1}{\rho \beta} \qquad\qquad 1.25.$$

but

$$\frac{\Omega^2}{K^2} = c^2 \qquad\qquad 1.26.$$

so the velocity of the wave is given by

$$c = \sqrt{\frac{1}{\rho\,\beta}}\,. \qquad\qquad 1.27.$$

1.5 Velocity of Propagation in Solids

The propagation of sound in solids may be analyzed on the basis of Hooke's Law and Newton's Third Law of Motion. Hooke's Law will be discussed by reference to Figure 1.2. A strain w in the rod of diameter d produces a stress Sz opposite in direction to the force Fz. According to Hooke's Law, the magnitudes of the strain and the stress are related through the equation

$$S_z = -\frac{4Fz}{\pi d^2} = -y_o\,\frac{w}{z} \qquad\qquad 1.28.$$

where y_o is Young's modulus.

Hooke's Law states that the elongation per unit length w/z is linearly proportional to the force per unit area $4Fz/\pi d^2$ producing it. In the case of an infinitesimal volume of length ∂z and strain $\partial w/\partial z$ Hooke's Law becomes

$$S_z = -y_o\,\frac{dw}{dz}\,. \qquad\qquad 1.29.$$

Young's modulus is measured using rods for which z d. The circumference of the rod is unrestrained. Thus, a strain $\frac{\partial w}{\partial z}$ in the z-direction also produces a change in diameter and the ratio

$$\sigma = \frac{dr}{dw} \qquad\qquad 1.30.$$

is known as Poisson's ratio.

In bulk material, the sides of each elemental volume are con strained. This affects the value of the constant of proportionali ty so that instead of Equation 1.29 we must write

$$Sz = -y_B \frac{dw}{dz}$$ 1.31.

where the bulk modulus y_B replaces Young's modulus as a constant
of proportionality. Young's modulus and the bulk modulus y_B are
related by the expression

$$y_o/y_B = 1 - b$$ 1.32.

where

$$b = \frac{2\sigma^2}{1 - \sigma}.$$ 1.33.

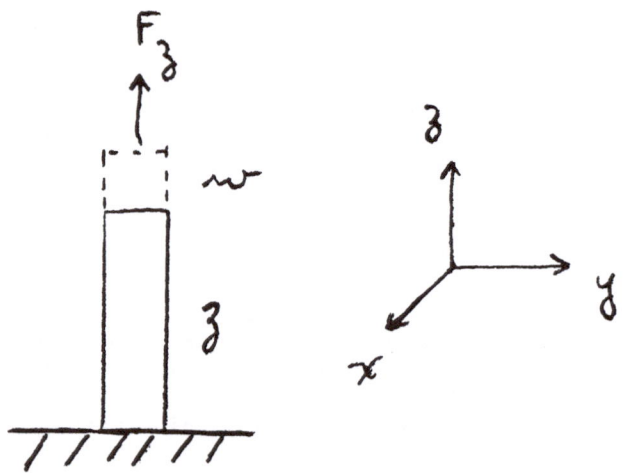

Fig. 1.2 - Hooke's Law

Hooke's Law states that the elongation per unit length w/z is
linearly proportional to the force per unit area $4/Fz/\pi d^2$ producing
it.

A typical value of Poisson's ration for hard solids is

$$\sigma = 0.3 \qquad\qquad 1.34.$$

so that

$$b \simeq 0.26. \qquad\qquad 1.35.$$

The velocity of propagation of an acoustical wave through a solid may be determined from the application of Hooke's Law and Newton's Third Law of Motion. Consider the elemental volume of length dz in Figure 1.3. The net force tending to increase the length of this volume is

$$F_2 - F_1 = \frac{dF_z}{dz}\, dz. \qquad\qquad 1.36.$$

The corresponding strain is dw^2/dz where

$$d\left(\frac{dw}{dz}\right) = \frac{d^2w}{dz^2}\, dz. \qquad\qquad 1.37.$$

Applying Hooke's Law we have

$$\frac{dFz}{dz}\, dz = \pi r^2 y_0 \frac{d^2w}{dz^2}\, dz. \qquad\qquad 1.38.$$

The force $\frac{dFz}{dz}\, dz$ produces an acceleration on the mass $\rho\pi r^2 dz$ such that

$$y_0 \frac{d^2w}{dz^2} = \rho\, \frac{d^2w}{d\tau^2}\,. \qquad\qquad 1.39$$

Equation 1.39 is the familiar wave equation in which the velocity of propagation is given by

$$c_r = \frac{y_0}{\rho} \qquad\qquad 1.40.$$

i.e., c_r is the velocity of propagation of a compressional wave

in a rod for which the wavelength is much greater than the diameter
of the rod. In bulk material where the diameter is much greater
than the wavelength

$$c_B = \frac{y_B}{\rho} = c_r \left[\frac{1 - \sigma}{(1 + \sigma)(1 - 2\sigma)} \right]^{1/2} \qquad 1.41.$$

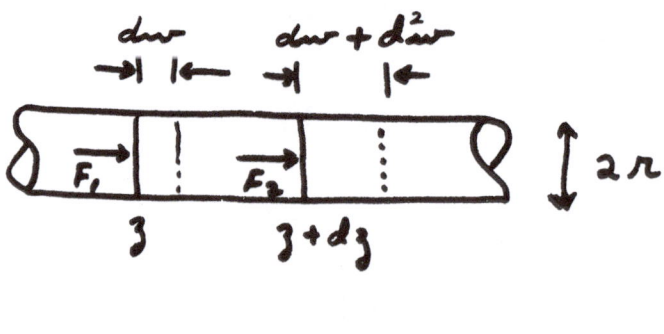

$$F_1 = F_3 \qquad\qquad F_2 = F_3 + \frac{\partial F_3}{\partial 3} d3$$

Fig. 1.3 - Motion of an Elemental Volume in a Rod

An elemental volume of cross-sectional area πr^2 in a rod of
solid material of diameter d moves a distance w under the action of forces F_1 and F_2. Since $F_2 \neq F_1$ in general, the surface
at z + dz moves a greater (or lesser) distance than the surface
at z. Under the action of an acoustical wave inertial forces
also come into play.

Solids will support shear waves as well as compressional waves. The velocity of propagation of the shear wave is given by

$$c_s = c_r \sqrt{\frac{1}{2(1 + \sigma)}}. \qquad 1.42.$$

Acoustical holography can be performed using either compres-

sional or shear waves, but the fact that both types of wave exist
sometimes makes the interpretation of images difficult. In some
instances, one may use the difference in velocities of propagation
to discriminate against one wave or the other. Note that the ratio of velocities is

$$\frac{c_B}{c_s} = \sqrt{2}\left(\sqrt{\frac{1-\sigma}{1-2\sigma}}\right),\qquad\qquad 1.43.$$

a quantity which is always greater than 1.4. For "hard" solids
for which $\sigma = 0.3$, the ratio is 1.87 showing that the shear velocity is always appreciably less than the velocity of the compressional wave.

   1.6 Refraction and Reflection at Liquid-Solid Interfaces

   An acoustical wave incident on an interface between two media
of differing acoustical properties is in general split into three
waves. The manner in which the energy is partitioned among the
three waves requires a more extensive discussion than can be given
here but has been studied in detail by Mayer (W.G. Mayer, "Energy
Partition of Ultrasonic Waves at Flat Boundaries," Ultrasonics,
April-June, 1965, pp. 62-68) and Brekovskikh (L.M. Brekovskikh,
Waves in Layered Media, Academic Press, New York, 1960, Chapter I).
Figure 1.4 illustrates some features of energy partition for the
particular case in which the acoustical wave is incident from the
liquid side of a liquid-solid interface and the velocities of propa
gation are such that $c_B \gg c_s \gg c$, where c_B is the velocity of the
transmitted longitudinal wave, c_s is the velocity of the transmitted shear wave and c is the velocity of the incident longitudinal wave. There are a total of nine cases similar to that illustrat
ed in Figure 1.4. These nine cases correspond to three different
conditions on velocity, namely

$$c_B > c_s > c$$

$$c_B > c > c_s$$

$$c > c_B > c_s$$

which apply to each of the three types of incident wave, namely

Incident Longitudinal in the Liquid,
Incident Longitudinal in the Solid,
Incident Shear in the Solid.

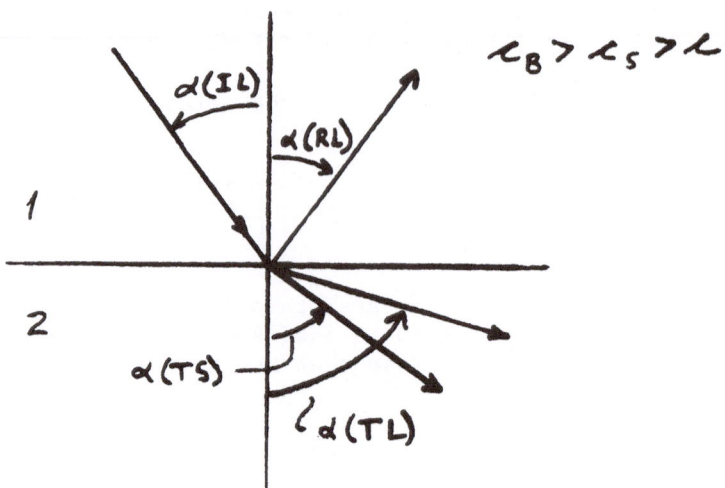

Fig. 1.4 - Energy Partition at Interfaces

A longitudinal wave incident from the liquid side (Side 1) of a liquid-solid interface is partially transmitted. The transmitted energy is further partitioned between a shear wave and a longitudi‍nal wave.

For the case illustrated in Figure 1.4, the angle of inciden‍ce, reflection and refraction are related by the equations

$$\frac{c}{\sin \alpha (IL)} = \frac{c_B}{\sin \alpha (TL)} = \frac{c_s}{\sin \alpha (TS)} = -\frac{c}{\sin \alpha (RL)} \qquad 1.44.$$

where the identifying letters IL, TL, TS, and RL stand for incident longitudinal, transmitted longitudinal, transmitted shear, and re‍flected longitudinal respectively. The expressions describing the relative amplitudes of these waves are in general quite complex. They simplify for longitudinal waves at normal incidence to

$$\frac{RL}{IL} = \frac{\rho_2 c_B - \rho_1 c}{\rho_2 c_B - \rho_1 c}. \qquad 1.45.$$

The product ρc is known as the characteristic impedance and if often designated by the letter Z. Thus, $\rho_2 c_B$ would be Z_2 and $\rho_1 c$ would be Z_1.

1.7 Energy and Intensity

Referring again to Figure 1.1, we note that the infinitesimal volume $\Delta x \Delta y \Delta z$ is moving with a velocity

$$U = \frac{dw}{dt} \qquad\qquad 1.46.$$

where the instantaneous particle displacement w is given by Equation 1.1. The kenetic energy per unit volume is therefore given by

$$KE = \tfrac{1}{2}\rho U^2 = \tfrac{1}{2}\rho \Omega^2 W^2 \sin^2(\Omega t - Kz). \qquad 1.47.$$

The total energy in the wave is simply the maximum value of the kenetic energy, thus

$$E = \tfrac{1}{2}\rho \Omega^2 W^2. \qquad\qquad 1.48.$$

By intensity we mean the rate of flow of energy through a unit area. Since the wave is propagating with a velocity c, the rate of flow of energy through a unit area is given by

$$I = \tfrac{1}{2}\rho c \Omega^2 W^2. \qquad\qquad 1.49.$$

In liquids

$$W^2 = \left(\frac{\beta \rho_o}{K}\right)^2 \qquad\qquad 1.50.$$

so

$$E = \tfrac{1}{2}\rho \left(\frac{\Omega}{K}\right)^2 \beta^2 p_o^2. \qquad\qquad 1.51.$$

Substituting from Equations 1.25 and 1.27, we see that

$$E = \frac{P_o{}^2}{2\rho c^2} \qquad\qquad 1.52.$$

and

$$I = \frac{P_o{}^2}{2\rho c}. \qquad\qquad 1.53.$$

Note the Equation 1.7 may be written

$$p = -\frac{K}{\Omega}\frac{W}{\beta}\,\Omega\,\sin(\Omega t - Kz) \qquad\qquad 1.54.$$

and since

$$\frac{K}{\Omega} = \frac{1}{c}, \qquad\qquad 1.55.$$

$$\beta = \frac{1}{\rho c^2}, \qquad\qquad 1.56.$$

and

$$\frac{dw}{dt} = -W\Omega\sin(\Omega t - Kz) \qquad\qquad 1.57.$$

we have

$$p = \rho c U \qquad\qquad 1.58.$$

where U represents the particle velocity $\frac{dw}{dt}$. To introduce second order effects, we use the expression

$$\rho = \rho_o + \Delta\rho \qquad\qquad 1.60.$$

where $\Delta\rho$ represents the increase (decrease) in density when the original volume V_o is decreased (increased) by ΔV.

1.8 Radiation Pressure

An acoustical wave carries with it a certain momentum. When this momentum is altered by reflection or absorption, a pressure is exerted on the reflector or absorber. Thus, the wave exerts a radiation pressure Π on all reflectors and absorbers. We can describe this pressure in terms of the parameters of the acoustical wave and of the medium in which it is propagating, if we consider second order effects. In Equation 1.27, we did not, for example, consider the fact that the density ρ is not constant throughout the volume that the wave occupies, but varies with the instantaneous pressure p. We will now consider the effect of this varying density.

If the mass of material contained in a volume V_o is M, then

$$\Delta \rho = \frac{M}{V_o + \Delta V} - \frac{M}{V_o} = - \frac{M \Delta V}{V_o^2} \qquad 1.61.$$

i.e.,

$$\Delta \rho = -\rho_o \frac{\Delta V}{V_o}. \qquad 1.62.$$

Therefore, we can write instead of Equation 1.5

$$\beta = \frac{\Delta \rho}{\rho_o p} \qquad 1.63.$$

or

$$\Delta \rho = \rho_o \beta p. \qquad 1.64.$$

Thus for Equation 1.58, we write

$$p = \rho_o (1 + \beta p) \, cU. \qquad 1.65.$$

We now approximate the term βp by

$$\beta p \simeq (\rho_o c^2)^{-1} (\rho_o cU) = \frac{U}{c} \qquad 1.66.$$

so that

$$p = -\rho_o cW\Omega \sin(\Omega t - Kz) + \rho_o W^2 \Omega^2 \sin^2(\Omega t - Kz). \qquad 1.67.$$

We may now identify the radiation pressure as the average value of the excess pressure p, i.e.

$$\Pi = \frac{1}{T} \int_0^T pdt \qquad 1.68.$$

where

$$T = \frac{2\pi}{\Omega}. \qquad 1.69.$$

When the integration is carried out, we find that

$$\Pi = \tfrac{1}{2}\rho_o \Omega^2 W^2 \qquad 1.70.$$

and since

$$P_o = -\mathbf{\rho}_o c W \qquad 1.71.$$

we may also write

$$\Pi = \frac{P_o^2}{2\rho c^2}. \qquad 1.72.$$

Equation 1.72 represents the pressure on a perfect absorber, since in that case, the energy and momentum are reduced to zero at a rate such that the pressure is produced. At normal incidence, a perfect reflector reverses the direction of the momentum producing twice the rate of change of momentum. Hence, for a perfect reflector with the wave at normal incidence,

$$\Pi r = \frac{P_o^2}{\rho c^2}. \qquad 1.73.$$

1.9   Interaction of an Acoustical Wave
with a Free Liquid Surface

Consider an acoustical wave incident upon a free liquid surface as shown in Figure 1.5. The wave is considered to be at normal incidence. Such a surface is a perfect reflector so that in
accordance with Equations 1.73 and 1.53, we note that the radiation pressure upon the surface is

$$\Pi_r = \frac{2I}{c}. \qquad\qquad 1.74.$$

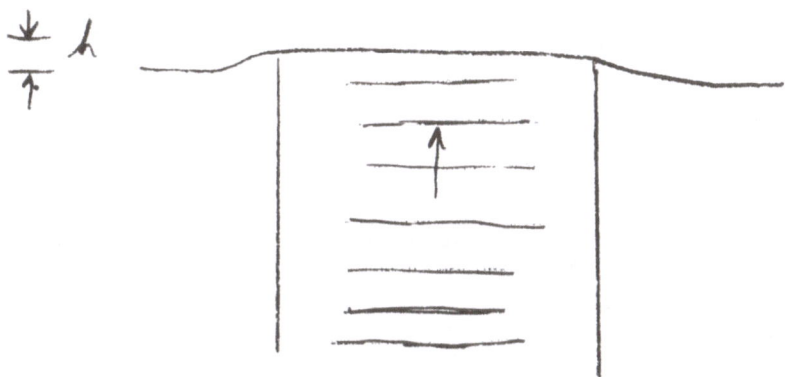

Fig. 1.5 – Elevation of a Free Liquid Surface Produced
by Acoustical Radiation Pressure

An acoustical wave is totally reflected at a liquid-gas inter
face. The resulting reversal of the direction of the wave produces a rate of change of momentum and a corresponding radiation
pressure which elevates the liquid surface a distance h.

If the diameter of the incident beam is greater than a centimeter, the effects of surface tension will be negligible except at
the edge. The radiation pressure is opposed chiefly by gravitation
al pressure which is given by

$$\Pi_g = \rho g h. \qquad\qquad 1.75.$$

Equating these two pressures and solving for h yields the

expression

$$h = \frac{2I}{\rho g c}. \qquad\qquad 1.76.$$

It will be instructive to compare the magnitude of the surface displacement as given by Equation 1.76 with the amplitude of particle displacement W which according to Equation 1.49 is given by

$$W = \frac{1}{2\rho u} \sqrt{\frac{2I}{\rho c}}. \qquad\qquad 1.77.$$

In this case, we have written for the angular frequency Ω, the expression $2\pi u$, where u is the frequency of the acoustical wave.

Using,

$$I = 0.1 \text{ watt/cm}^2 = 10^6 \text{ erg/sec/cm}^2 \quad \gamma = 10\text{MHz}$$

$$\rho = 1 \text{ gm/cm}^3 \text{ (water)} \quad c = 1.5 \times 10^5 \text{ cm/sec (water)}$$

and

$$g = 980 \text{ cm/sec}^2$$

we find that

$$h = 1.36 \times 10^{-2} \text{ cm}$$

and

$$W = 5.8 \times 10^{-8} \text{ cm}$$

Thus, for the conditions stated above, the surface displacement caused by radiation pressure is more than five orders of magnitude greater than the particle displacement amplitude in the same wave from which the radiation pressure is derived.

CHAPTER II : ACOUSTICAL HOLOGRAPHY, SCANNING TECHNIQUES

2.1 Introduction

One major impediment to the development of optical holography
from the time of its inception by Gabor (1) to the time of renew-
ed interest generated by Leith and Upatnieks (2) was the lack of
sufficiently coherent and intense sources of light. No such limitation impeded the development of acoustical holography. Coherent
transmitters of acoustical waves at all frequencies have been available for many years. It is necessary, however, to devise a substitute for the square law area detector used in optical holography, since photographic plates are of only marginal use in recording acoustical holograms (3). A variety of substitutes have been
demonstrated to be effective (4-8).

One of these employs a microphone or piezoelectric receiver
having an effective diameter less than or approximately equal to
the fringe spacing in the interference pattern generated by the reference and object beams in the plane of the hologram. Such a receiver responds to the instantaneous amplitude of the acoustical
field and not just to the intensity as is the case with photographic plates and optical holography. Consequently, it is possible
to substitute an electronic reference signal for the acoustical reference beam which would be used in a system analogous to an optical holography system.

The acoustical hologram must be built up by point by point
scanning of the hologram plane. As illustrated in Figure 2.1, the
signal derived from the scanning transducer is mixed with the reference signal in such a way that the phase and amplitude information
characteristic of the object is transmitted to a small lamp. The
brightness of the lamp thus varies in accordance with the phase and
amplitude information transmitted to it, and as it travels throught
the raster pattern with the receiver, the brightness of the lamp as
a function of position is recorded on photographic film. When the
film is developed, the resulting transparency serves as a hologram.

Optical images of the object as viewed with sound can be produced by illuminating the hologram with coherent light as shown in
Figure 2.2. A typical viewing system employs a 7 cm diameter, 60 cm
focal length lens adjusted to focus the point source of light at a
distance D of about 7 meters. The hologram is held in a liquid
gate to reduce the effects of the uneven surface of the film.

In acoustical holography as in optical holography, one can
arrange the system so that the intensity of the lamp is determined
by

Ref. 1) D. Gabor, Proc. Roy. Soc. London A 197:545 (1949); 2) E. N.
Leith and J. Upatnicks, J. Opt. Soc. Am. 52:1123-1130 (1962); 53:1377
(1963); 3) P. Greguss, Research Film 5:330 (1965); 4) R. B. Smith

$$I = \left| V_1 + V_2 \right|^2 = \left| V_1 \right|^2 + \left| V_2 \right|^2 + V_1 V_2^* + V_1^* V_2 \qquad 2.1.$$

where V_1 represents the wave scattered from the object and V_2 represents the wave scattered from the image. We can, in fact, arrange the electronics so that the biasing terms $(V_1)^2$ and $(V_2)^2$ are filtered out so that

$$I = V_1 V_2^* + V_1^* V_2 \qquad 2.2.$$

The film which is exposed to light from the lamp is assumed to have a transmissivity which is linearly related to the intensity of the lamp. When the film is properly illuminated with coherent light characterized by the complex amplitude V_a, the transmitted light is characterized by

$$H = V_a I. \qquad 2.3.$$

That portion of the light contributing to the true image is characterized by

$$H_T = V_a \cdot V_1 \cdot V_2^* \qquad 2.4.$$

and that contributing to the conjugate image is given by

$$H_c = V_a \cdot V_1^* \cdot V_2. \qquad 2.5.$$

We shall, for the purposes of the analysis to follow, consider that

$$V = A \exp\{i(\Omega t - K \cdot r)\} = A \exp\{i(\Omega t - \phi)\} = U \exp\{i\Omega t\} \qquad 2.6.$$

and B. B. Brenden, Ultrasonics 7:125-126 (1969); 5) J. D. Young and J. E. Wolfe, Appl. Phys. Letters 11:294-296 (1967); 6) A. Korpel, Appl. Phys. Letters 9:425-427 (1966); 7) R. K. Mueller and N. K. Sheridan, Appl. Phys. Letters 9:328-329 (1966); 8) A. F. Metherell, H. M. A. El-Sum, and L. Larmore, Acoustical Holography, Vol. 1, Plenum Press, New York (1969).

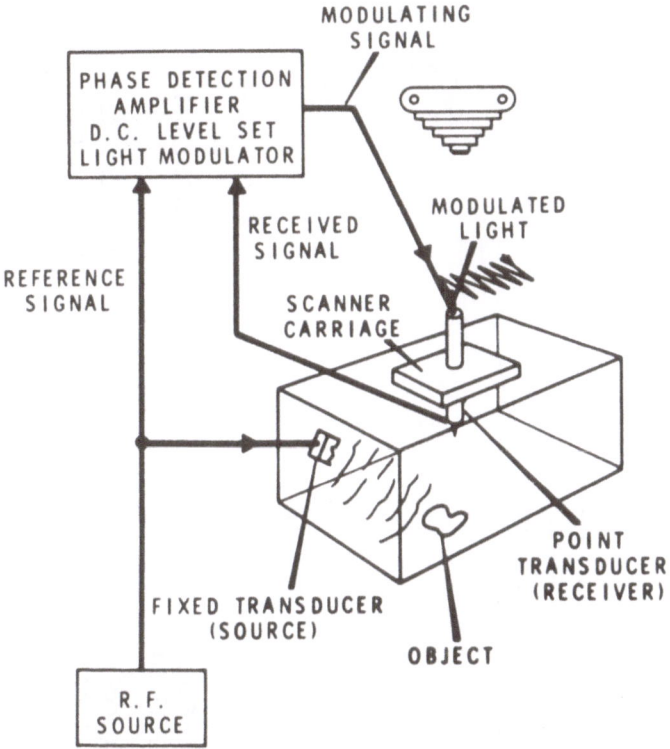

MODULATING
SIGNAL

PHASE DETECTION
AMPLIFIER
D.C. LEVEL SET
LIGHT MODULATOR

MODULATED
LIGHT

RECEIVED
SIGNAL

REFERENCE
SIGNAL

SCANNER
CARRIAGE

POINT
TRANSDUCER
(RECEIVER)

FIXED TRANSDUCER
(SOURCE)

OBJECT

R.F.
SOURCE

Fig. 2.1 - Acoustical Holographic System

One of the features of scanned holography systems is that an acoustical reference beam is not needed. It may be replaced by an electronic reference signal derived from the same source as used to drive the source transducer. In the system diagramed in this figure, a small piezoelectric receiver is used to scan the hologram plane. It is directly coupled to a point source of light which is modulated in intensity by the amplified and bias adjusted signal obtained by multiplying the reference signal with the receiver signal.

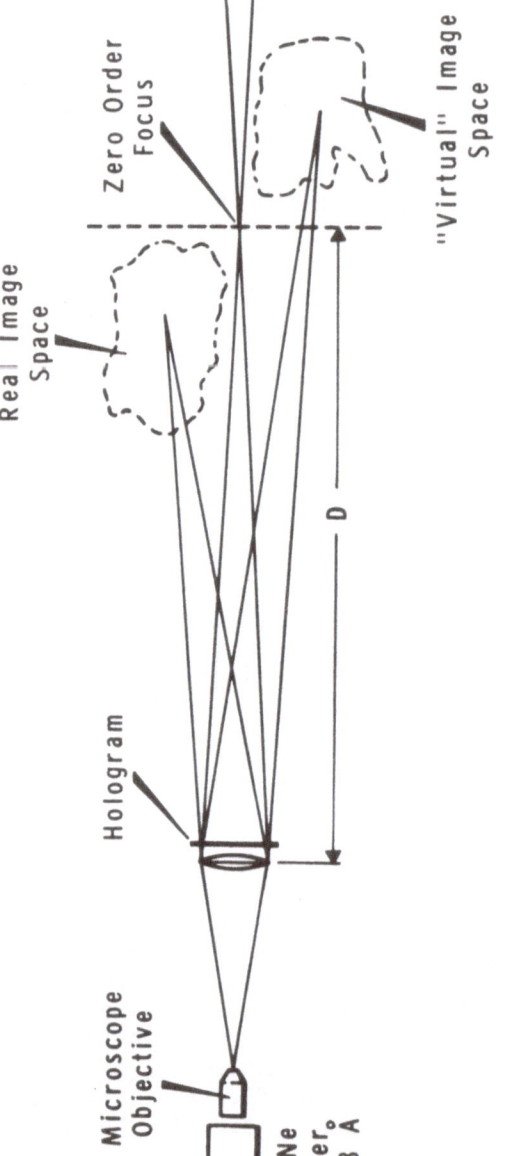

Fig. 2.2 – SYSTEM FOR PRODUCING IMAGES FROM AN ACOUSTICAL
HOLOGRAM

Systems for producing images from acoustical holograms differ considerably from those used for producing images from optical
holograms. The differences are necessary because the angles of
diffraction are so small and the hologram to image distances are
so great except for focused image holograms. A lens is used near
the hologram to bring the undiffracted light to a focus at a distance of a few meters from the hologram. This brings the true
and conjugate images closer to the hologram and facilitates the
separation of the orders of diffracted light.

2.2   Source-Receiver Reciprocity

Recording the hologram by using a scanning receiver is a logi cal adaptation from optical holography. However, holographic recordings may also be obtained by scanning the source (9-10). Source-receiver reciprocity may be deduced from Figure 2.3 wherein a source S, and object O, and a receiver R are shown at distances r_S, r_O, and r_R respectively from the origin of a coordinate system. The receiver generates a signal V_O which is mixed with a coherent reference signal V_R. The reference signal V_R may be direct radiation from the source S or it may be derived from an electronic reference signal, which we shall characterize by the equation

$$V_R = A \cos(\Omega t + \phi_R).$$

2.7.

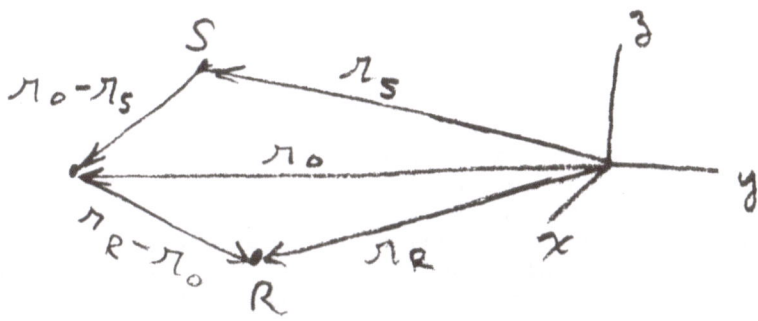

Fig. 2.3 - Source-receiver Reciprocity

The object point O at a distance r_O from the origin of a coordinate system scatters energy from the   source S to a receiver R. If the receiver is scanned over a two dimensional planar area, a ho logram can be formed which, when properly illuminated by coherent light, will produce an image of the object O. This is to be expect ed by analogy with optical holography. Source-receiver reciprocity arguments show that if the source and receiver positions are interchanged and the source scanned over the planar area, the resulting hologram will be identical to the receiver scanned hologram.

Ref. 9) A. F. Metherell and S. Spinak, Appl. Phys. Letters, 13:22 (1968); 10) V. I. Neeley, Phys. Letters, 28A(7):475-476 (1968).

The signal V_O is made up of the sum of all the wave amplitudes reflected from different parts of the object, i.e.

$$V_O = \int_{object} A(r_O) \cos\left\{ \Omega t + k_{SO} \cdot (r_O - r_S) + \right.$$
$$\left. + k_{OR} \cdot (r_R - r_O) + \phi_O \right\} dr_O \qquad 2.8.$$

where k_{SO} and k_{OR} are wave propagation vectors having a magnitude $2\pi/\Lambda$, and $A(r_O)$ is the amplitude of the radiation scattered from the object point at r_O to the receiver.

If the source and receiver positions are exchanged so that the source is at r_R and the receiver at r_S, the new receiver signal V'_O is

$$V'_O = \int_{object} A(r_O) \cos\left\{ \Omega t + k_{RO}' (r_O - r_R) + \right.$$
$$\left. + k_{OS} \cdot (r_S - r_O) + \phi_O \right\} dr_O. \qquad 2.9.$$

However $k_{RO} = -k_{OR}$ and $k_{OS} = -k_{SO}$, so

$$V'_O \equiv V_O \qquad 2.10.$$

and since the reference signal V_R is not affected by this inter-change, the resulting hologram is invariant to an exchange of the positions of the source and receiver, i.e., holograms may be ob-tained by scanning the source instead of the receiver.

It has also been shown (11) that the source and receiver may be scanned simultaneously and at different velocities providing on ly that the velocities are linearly related. Thus, a great amount of flexibility exists in scanning arrangements.

2.3 The Geometry of Imaging

Among the first questions to be answered with regard to acous tical holography is, "At what distance from the hologram do the images lie?" Given an object distance r_1, a reference source di-stance r_2, and reconstruction source distance r_a, at what distance r_b may I expect to find the image? The basic analytical study of this question together with a treatment of third order aberrations

Ref. 11) B. P. Hildebrand and K. A. Haines, J. Opt. Soc. Am. 59: (1969).

has been given by Meier (12) and Champaigne (13). These analysis
treat the object as if it were a single point. A more general
treatment in which the object is considered to be complex structure
is given by Hildebrand and Brenden (14). For the purposes of this
paper, the use of a single point object will suffice.

Expressions of the same form as Equation 2.6 will be used to
characterize the wave originating at point P in Figure 2.4.

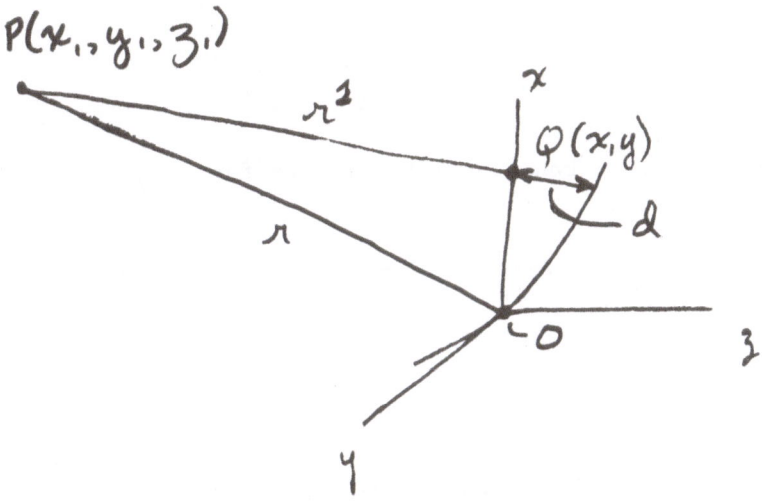

Fig. 2.4 - Geometrical Description

The distance r^1 from point $P(x_1, y_1, z_1)$ to the scanning elemen
location $Q(x, y)$ differs from the distance r by an amount d.

The time dependent term $\exp\{i\Omega t\}$ cancels out of all equations and, therefore, is omitted in the discussion. The wave is
characterized by the expression

$$U = A \exp(-i\phi) = A \exp\left\{-iK(r^1 - r)\right\} \qquad 2.11.$$

The expression equivalent to Equation 2.4 thus becomes

$$H_T = A_a A_1 A_2 \exp\left\{-i(\phi_1 - \phi_2 + \phi_a)\right\} \qquad 2.12.$$

Ref. 12) R. W. Meier, J. Opt. Soc. Am. 55(8): 987 (1965); 13) E. B.
Champagne, J. Opt. Soc. Am. 57(1): 51 (1967); 14) B. P. Hildebrand
and B. B. Brenden, An Introduction to Acoustical Holography, Plenum
Press, New York (1972).

The amplitude terms A_a, A_1, and A_2 are of no interest for the present. All the geometrical relationships are derived from the phase term. Our first problem is to express the phase in terms of the coordinates x, y, and z. To do this, we note that

$$r^1 = \left\{ (x - x_1)^2 + (y - y_1)^2 + z_1^2 \right\}^{\frac{1}{2}} \qquad 2.13.$$

and

$$r = \left\{ x_1^2 + y_1^2 + z_1^2 \right\}^{\frac{1}{2}}. \qquad 2.14.$$

Equation 2.13 can be written in the form

$$r^1 = \left\{ x^2 - 2xx_1 + x_1^2 + y^2 - 2yy_1 + y_1^2 + z_1^2 \right\}^{\frac{1}{2}} \qquad 2.15.$$

$$= \left\{ r^2 + x(x - 2x_1) + y(y - 2y_1) \right\}^{\frac{1}{2}}$$

Using the binominal expansion (or Taylor's series expansion) we recall that

$$(a + b)^n = a^n + na^{n-1}b + \frac{n(n-1)}{2} a^{n-2}b^2 + \ldots . \qquad 2.16.$$

If we set $n = 1/2$, $a = r^2$ and $b = x(x - 2x_1) + y(y - 2y_1)$, we find that

$$r^1 - r = \frac{x^2 + y^2}{2r} - \frac{xx_1}{r} - \frac{yy_1}{r} + 0(\frac{x^4}{r^3}) \qquad 2.17.$$

where $0(\frac{x^4}{r^3})$ indicates terms of the magnitude x^4/r^3 or less. Whenever we have a phase term in the form $K(r^1 - r)$, we may therefore use Equation 2.17 to replace $r^1 - r$.

We shall now consider a more general case which is quite useful in studying the various scanning arrangements characteristic of acoustical holography. Figure 2.5 illustrates the nomenclature to be used and is drawn to include the reconstruction process as well as the hologram formation process. A double coordinate system is used. There is an x, y, z coordinate system for receiver space and an ξ, η, ζ system for source space, but these are coupled by the fact that the origin of the ξ, η, ζ system is located at the point x_o, y_o, z_o. Note that

$$r_o^1 = \left\{ (\xi + x_o - x_1)^2 + (\eta + y - y_1)^2 + (\zeta + z_o - z_1)^2 \right\}^{\frac{1}{2}} \qquad 2.18.$$

and

$$r_o = \left\{ (x_o - x_1)^2 + (y_o - y_1)^2 + (z_o - z_1)^2 \right\}^{\frac{1}{2}}$$

so

$$r_o^1 - r_o = \frac{\xi^2 + \eta^2}{2r_o} - \frac{\xi(x_1 - x_o)}{r_o} - \frac{\eta(y_o - y)}{r_o} + 0(\frac{x^4}{r_o^3}). \qquad 2.19.$$

We are now in a position to describe scanning arrangements in which both the receiver and the transmitter scan simultaneously. To do this, we must choose functions which describe ξ, η and ζ in terms of x, y and z. These functions might be quite complex, but among the most useful are

$$\xi = px, \quad \eta = py \text{ and } \zeta = o \qquad 2.20.$$

in which p is a constant. If we substitute Equation 2.20 into Equation 2.19, we see that

$$r_o^1 - r_o = \frac{p^2(x^2 + y^2)}{2r_o} - \frac{px(x_1 - x_o)}{r_o} - \frac{py(y_1 - y_o)}{r_o}. \qquad 2.21.$$

Referring again to Figure 2.5, we see that the phase of the reconstructed waves H_T or H_c is given by

$$\phi_{\pm} = \pm K(r_o^1 - r_o + r_1^1 - r_1 - r_2^1 + r_2) + k(r_a^1 - r_a) \qquad 2.22.$$

where

$$K = \frac{2\pi}{\Lambda} \ , \ k = \frac{2\pi}{\lambda} \qquad\qquad 2.23.$$

and Λ is the wavelength of the acoustical wave and λ the wavelength of the light used for image formation.

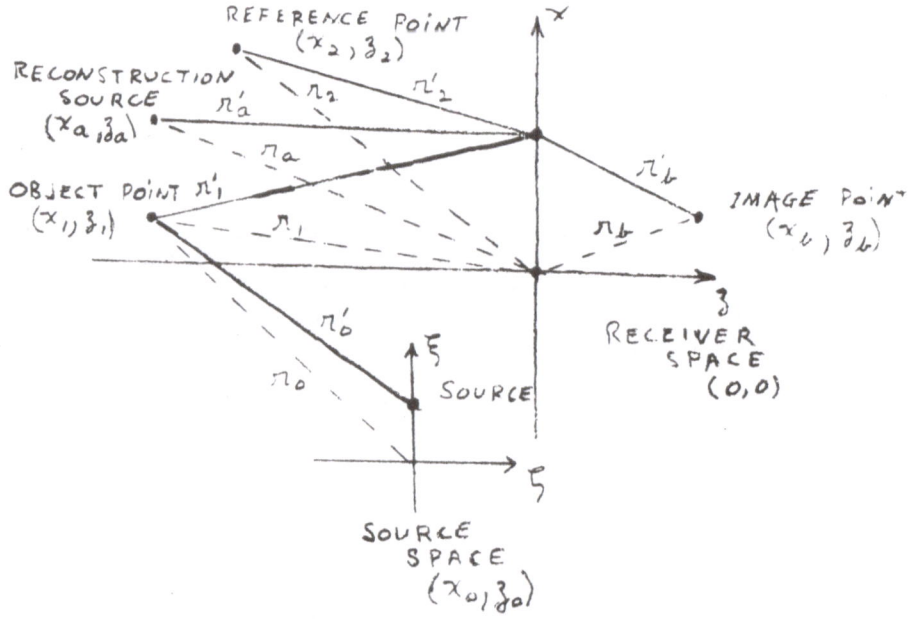

Fig. 2.5 - Generalized Scanning Schematic

The generalized schematic diagram which describes scanned holography includes a source space through which the source may move, and a receiver space through which the receiver may move. The schematic permits an analysis which accounts for the effects of relative motion between the source and receiver and which permits calculation of image distances, resolution and aberrations.

We note with reference to Figure 2.1 that a modulated light must travel over the scan path in synchronism with the receiver.

We note further that in accordance with the discussions of Section 2.2, the scan may be done with the source in which case the lamp must be coupled to the source. Actually the lamp need not be coupled directly to either the source or receiver, but its motion must be related to the motions of the source and/or receiver. The lamp could, for instance, move in an ξ' η' ζ' coordinate system such that

$$\xi^1 = mx, \quad \eta^1 = my \text{ and } \zeta^1 = 0 \qquad\qquad 2.24.$$

where m is the effective magnification of the final hologram due to linkage between the scanning element and any optical reduction or magnification that is introduced by the camera.

The coordinates x and y have been used to indicate the location of the scanning element. Now it is more convenient to identify them with the coordinate system in the final hologram and to exchange the roles of ξ^1, η^1, ζ^1 and x, y, z. When this is done, Equation 2.22 becomes

$$
\begin{aligned}
\phi_{\pm} = {}^{+}_{-}\frac{K}{2}\Bigg\{ & \frac{p^2(x^2 + y^2)}{m^2 r_o} - \frac{2px(x_1 - x_o)}{mr_o} - \frac{2py(y_1 - y_o)}{mr_o} \\
& + \frac{x^2 + y^2}{m^2 r_1} - \frac{2xx_1}{mr_1} - \frac{2yy_1}{mr_1} - \frac{x^2 + y^2}{m^2 r_2} + \frac{2xx_2}{mr_2} + \frac{2yy_2}{mr_2} \Bigg\} \qquad 2.25. \\
& + k\Bigg\{ \frac{x^2 + y^2}{r_a} - \frac{2xx_a}{r_a} - \frac{2yy_a}{r_a} \Bigg\} + O(\frac{x^4}{r^3}) \cdot
\end{aligned}
$$

A perfect imaging system or "perfect" hologram would diffract light from the reconstruction source in such a way that light converging on the image point would have a perfectly spherical wavefront describable in terms of phase by

$$2.26.$$

$$\phi g = k(r_b^{\ 1} - r_b).$$

or

$$\phi g = \frac{k}{2}\Bigg\{ \frac{x^2 + y^2}{r_b} - \frac{2xx_b}{r_b} - \frac{2yy_b}{r_b} \Bigg\} + O(\frac{x^4}{r^3}). \qquad 2.27.$$

The actual wave front is given by Equation 2.25. Imaging occurs under the condition that

$$\phi_{\pm} \simeq \phi_g. \qquad\qquad 2.28.$$

Any difference $\phi_g - \phi_{\pm}$ represents an imaging defect or aberration. Comparison of Equations 2.25 and 2.27 shows that the conditions for imaging are ($\mu = \dfrac{K}{k} = \dfrac{\lambda}{\Lambda}$)

$$\frac{1}{r_b} = \frac{1}{ra} + \frac{\mu}{m^2}\left(\frac{p^2}{r_o} + \frac{1}{r_1} - \frac{1}{r_2}\right) \qquad\qquad 2.29.$$

$$\frac{x_b}{r_b} = \frac{x_a}{r_a} + \frac{\mu}{m}\left(\frac{p(x_1 - x_o)}{r_o} + \frac{x_1}{r_1} - \frac{x_2}{r_2}\right) \qquad\qquad 2.30.$$

$$\frac{y_b}{r_b} = \frac{y_a}{r_a} + \frac{\mu}{m}\left(\frac{p(y_1 - y_o)}{r_o} + \frac{y_1}{r_1} - \frac{y_2}{r_2}\right). \qquad\qquad 2.31.$$

2.4 Magnification

The preceding equations allow us to calculate radial and lateral magnifications. For the radial magnification, we have

$$M_R = \frac{dr_b}{dr_1} = \pm \frac{\mu}{m^2}\left(\frac{r_b}{r_1}\right)^2 \left[1 + p^2\left(\frac{r_1}{r_o}\right)^2\left(\frac{dr_o}{dr_1}\right)\right] \qquad\qquad 2.32.$$

The value of $\dfrac{dr_o}{dr_1}$ may be deduced by reference to Figure 2.6 where we see that

$$\frac{dr_o}{dr_1} = \cos\theta = \frac{r_1^2 + r_o^2 - d^2}{2r_1 r_o} . \qquad\qquad 2.33.$$

A particular case of interest is one in which the source and the receiver are very close together in which case $d \simeq o$, $\dfrac{dr_o}{dr_1} = 1$, $r_o = r_1$ and $p = 1$. These conditions describe a system in which the source and receiver move together at the same rate. In this case, the radial magnification is twice that for a single element scan system for which p = o.

When z >>> x,y, the lateral magnification is given by

$$M_L = \frac{dx_b}{dx_1} = \pm \frac{\mu}{m}\left(\frac{r_b}{r_1}\right)\left[1 + p\left(\frac{r_1}{r_o}\right)\right] \qquad 2.34.$$

Note that for single element scan, i.e., when p = o, the ratio of radial to lateral magnification is given by

$$\frac{M_R}{M_L} = \frac{1}{\mu} M_L \qquad 2.35.$$

and since the ratio of wavelengths, $\mu = \frac{\lambda}{\Lambda}$, is, in general, very small, the image is greatly stretched in radial direction with respect to the lateral directions.

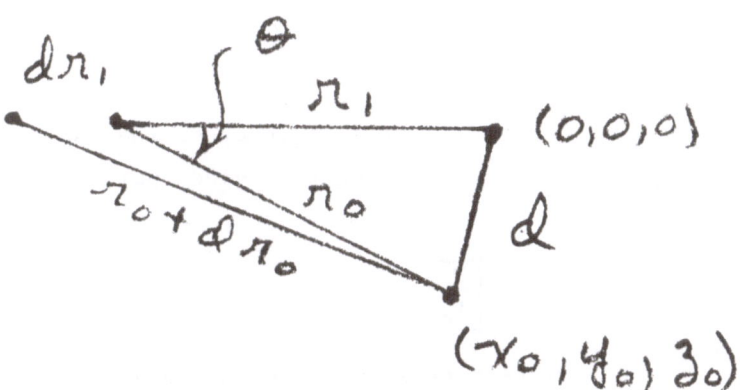

Fig. 2.6 - Source-receiver Separation

Separation of the source and the receiver introduces correcti̲ve terms depending upon dr_o/dr_1 which can be deduced by applying the law of cosines.

2.5 Resolution

The resolution of an imaging system is defined as the least separation Δx_1 between two points in the object which can be detected in the image. In the case of holograms, we can define resolution as the distance Δx_1 which an object point may be moved before the phase of the finest fringe in the hologram is reversed. The phase at the hologram recording plane is given by

$$\phi_H = \pm K(r_o^1 - r_o + r_1^1 - r_1 - r_2^1 + r_2).$$

$$
= \pm \frac{K}{2}\left[\frac{p^2(x^2 + y^2)}{r_o} - \frac{2px(x_1 - x_o)}{r_o} - \frac{2yy_1}{r_o} \right.
$$

$$
+ \frac{x^2 + y^2}{r_1} - \frac{2xx_1}{r_1} - \frac{2yy_1}{r_1} - \frac{x^2 + y^2}{r_2}
$$

$$
\left. + \frac{2pxx_1}{r_2} + \frac{2pyy_1}{r_2} \right] + O(\frac{x^4}{r^3})
$$

2.36.

The incremental change in phase caused by an incremental change in the position of the object point is

$$\Delta\phi_H = \frac{\delta\phi_H}{\delta x_1} \Delta x_1 \simeq Kx\left[\frac{p}{r_o} + \frac{1}{r_1}\right] \Delta x_1$$

2.37.

The phase of the finest fringe is reversed when $\Delta\phi_H = \pi$. Thus the least detectable movement Δx_1 of the object point is

$$\Delta x_1 \simeq \frac{\Lambda}{2x_{max}\left[\dfrac{p}{r_o} + \dfrac{1}{r_1}\right]}$$

2.38.

where x_{max} is the maximum dimension of the hologram in the x-direction. In taking the derivative, it was assumed that x, $y \ll r_o$, r_1. Equation 2.38 shown that the resolution is twice as good for simultaneous source-receiver scanning (p = 1) as for single element scanning (p = o).

2.6 Conclusions

Scanned acoustical holography may be carried out by moving either the source, the receiver or the object or any two of these elements. Simultaneous source-receiver scanning can increase the resolution by as much as a factor of 2.

CHAPTER III :   ACOUSTICAL HOLOGRAPHY, LIQUID SURFACE TECHNIQUES

3.1 Introduction

While scanned holography methods have the advantage of excellent sensitivity, they do have the disadvantage of having a low information handling rate and greater electronic complexity. If a single scan element is used, the scanning times for a single hologram range from 10 seconds to 30 minutes. If an array of scanning elements is used, considerable electronic circuitry is needed to sample each of the 100,000 or more elements either sequentially or in parallel. Essentially all the practical scanning systems reported on to date are two step systems in which a hologram is formed on photographic film and subsequently illuminated with coherent light to form an image.

On the other hand, a liquid surface has a dynamic response to acoustical energy which permits real-time imaging. We begin by treating the situation as though continous acoustical waves were used whereas in fact, both the acoustical waves and the light used to read out the liquid surface hologram are pulsed. Typically the acoustical pulses have a duration of about 80 μ s and are repeated 60 times per second while the pulse of light is 10μs to 20μs duration and timed with respect to the acoustical pulse to read out the hologram at an optimum time.

Figure 3.1 is a schematic diagram of the basic liquid-surface system. As discussed in the previous chapter, the radial magnification of the image formed by an acoustical hologram is greater than the lateral magnification by the ratio of the wavelength of sound Λ to the wavelength of light λ . This ration Λ/λ is normally very large and the practical result is that one is forced to view two dimensional slices of the image. Furthermore, the best imaging is achieved when the system of Figure 3.1 is modified by using an acoustic lens to image the object into the hologram plane. This is the equivalent of the focused image hologram of optical holography. This arrangement also produces an image of approximately the same size as the object without the need for the long optical path lengths normally used in reconstructing acoustical holograms.

The liquid-surface hologram is a phase grating and affects only the phase of the reflected light. Nevertheless, there is amplitude imaging because the amount of light diffracted into the first order varies as the amplitude of the pattern on the liquid surface which in turn varies as the amplitude of the wave in the object beam.

Fig. 3.1 - LIQUID-SURFACE HOLOGRAPHY SYSTEM

A liquid-surface acoustical holography system closely paral-
lels optical holography systems, but replaces film with a liquid-
surface which is read out instantaneously producing a real-time
image of the acoustical field in the object.

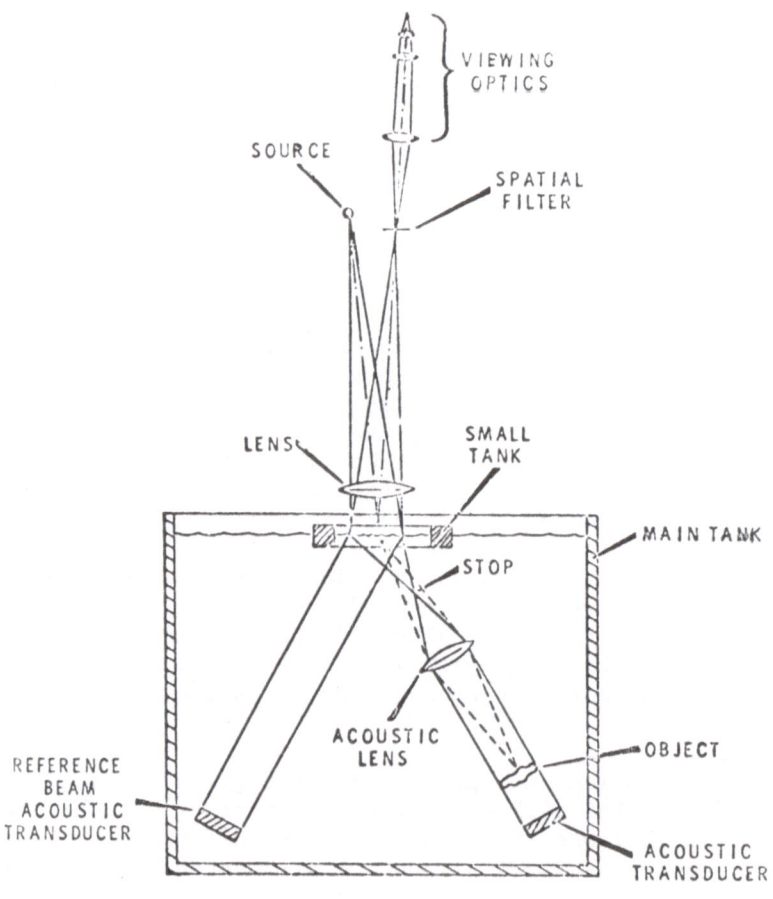

3.2 Interaction of the Acoustical Wave
with the Liquid Surface

Our first task will be to describe the interaction of the a-
coustical wave with the liquid surface. We consider two beams incident upon the surface at equal but opposite angles θ . The coor
dinate system to be used for the analysis has the surface $z = o$ co
incident with the undisturbed liquid surface. All holography systems have a reference beam and an object beam. In this case, the
reference beam at the surface ($z = o$) may be characterized by

$$U_r = P_r \exp(i\eta y) \qquad\qquad 3.1.$$

where

$$\eta = \frac{2\pi}{\Lambda} \sin\theta .$$

Similarly, the object beam is characterized by

$$U_o = P_o(x, y) \exp\left\{-i \left[\eta y + \phi(x, y)\right]\right\} \qquad\qquad 3.2.$$

These two waves add at the surface to produce an intensity I
given by

$$I(x,y) = \left| U_o + U_r \right|^2 / 2\rho c . \qquad\qquad 3.3.$$

The radiation pressure corresponding to this intensity is given by

$$\Pi(x, y) = \frac{2I}{c} \qquad\qquad 3.4.$$

Gravity and surface tension pressures act to oppose the radia
tion pressure. These are given respectively by

$$\Pi_g = \rho g z \qquad\qquad 3.5.$$

and

$$\Pi_t = -\gamma\left(\frac{\partial^2 z}{\partial y^2} + \frac{\partial^2 z}{\partial x^2}\right) \qquad\qquad 3.6.$$

so that the force balance equation is

$$\Pi(x, y) = \rho g z - \gamma\left(\frac{\partial^2 z}{\partial y^2} + \frac{\partial^2 z}{\partial x^2}\right) \qquad\qquad 3.7.$$

The radiation pressure may be expressed in terms of the wave pressures by

$$\Pi(x, y) = \left(\frac{2P_r}{\rho c^2}\right) P_o(x, y) \cos\left[2\eta y + \phi(x, y)\right]$$

$$+ \frac{\left\{P_r^2 + P_o(x, y)^2\right\}}{\rho c^2} . \qquad\qquad 3.8.$$

If we assume a stationary solution for Equation 3.7 of the form

$$z(x, y) = 2A \cos\left[2\eta y + \phi(x, y)\right] + B \qquad\qquad 3.9.$$

we find that

$$A = \frac{P_r P_o(x, y)}{\rho c^2\left[\rho g + 4\gamma\eta^2\right]} \qquad\qquad 3.10.$$

and

$$B = \frac{P_r^2 + P_o^2}{\rho^2 c^2 g} . \qquad\qquad 3.11.$$

This solution is valid only if

$$\frac{\partial^2 A}{\partial y^2}, \frac{\partial^2 A}{\partial x^2} \ll A \text{ (or B)} \qquad\qquad 3.12.$$

$$\frac{\partial^2 B}{\partial y^2}, \frac{\partial^2 B}{\partial x^2} \ll B \text{ (or A)} \qquad\qquad 3.13.$$

and

$$\frac{\partial^2 \phi}{\partial y^2} + \frac{\partial^2 \phi}{\partial x^2} \ll 4\gamma\eta^2 \qquad 3.14.$$

but these conditions are not too restrictive. Furthermore, in many practical cases

$$\rho g \ll 4\gamma\eta^2 \qquad 3.15.$$

so

$$A \simeq \left(\frac{P_r}{4\gamma\rho c^2\eta^2}\right) P_o(x, y) \qquad 3.16.$$

The previous discussion indicates a physical picture of the liquid surface in which there is a bulge of height B in the area over which the sound energy is incident. The interference pattern is impressed upon this bulge in ripples of amplitude 2A. At the maxima of these ripples, the surface is oscillating at the frequency of the sound wave and the particle displacement in that area has an amplitude W.

In order to compare the relative magnitudes of B, A, and W, consider the case in which both the reference and object beams consist of point sources at infinity. Let the pressure amplitudes P_r and P_o be equal so that we may write

$$2\rho c I_a = P_o^2 = P_r^2 = P_o P_r. \qquad 3.17.$$

We note that

$$\rho g = 980 \ \text{gm cm}^{-2} \ \text{sec}^{-2}$$

and that

$$4\gamma\eta^2 = 4\gamma\left(\frac{2\pi}{\Lambda} \sin\theta\right)^2. \qquad 3.18$$

Let

$$\theta = 30°$$

$$\gamma = 73 \text{ dyne cm}^{-1}$$

$$c = 1.5 \times 10^5 \text{ cm sec}^{-2}$$

$$\Lambda = 0.15 \text{ cm(} = 1MHz)$$

then

$$4\gamma\eta^2 = 1.2 \times 10^5 \text{ gm cm}^{-2} \text{ sec2}.$$

Note that the inequality of Equation 3.15 is satisfied. With the help of Equations 3.11, 3.16, and 3.17 we obtain

$$2A = \frac{4I_a}{4\gamma c\eta 2} \qquad\qquad 3.19.$$

and

$$B = \frac{4I_a}{\rho gc} \qquad\qquad 3.20.$$

so that

$$\frac{B}{2A} = \frac{4\gamma\eta^2}{\rho g} \approx 120 \qquad\qquad 3.21.$$

Taking I_a to be 0.1 watt cm^{-2}

$$A = 1.1 \times 10^{-4} \text{ cm} \qquad\qquad B = 2.5 \times 10^{-2} \text{ cm}$$

and, using Equation 1.77

$$W = 5.8 \times 10^{-7} \text{ cm}.$$

3.3 The Dynamics of Liquid-Surface Motion

It is a characteristic of simple harmonic motion that the frequency (f), mass (M), amplitude (y) and total energy E are related through the equation

$$f = \frac{1}{2\pi y} \sqrt{\frac{2E}{M}}.$$ 3.22.

Although the liquid-surface oscillations are not truly simple harmonic in nature, Equation 3.22 is useful in providing a reasonable picture of the dynamic response of a liquid-surface to pulses of ultrasound. We will use Equation 3.22 to determine the frequencies of oscillation of the bulge and of the ripple pattern.

For the bulge

$$E = \tfrac{1}{2}MgB$$ 3.23

and

$$y = B \simeq 10^{-2} \text{ cm}$$ 3.24

so

$$f_b = \frac{1}{2\pi} \sqrt{\frac{g}{B}} \simeq 50 \text{ Hz}$$ 3.25.

The magnitude of displacement used here corresponds to the maximum value that is obtained when the sound is applied continously under the conditions listed for the previous example. When short bursts or pulses of sound are applied the magnitude of B would decrease and the characteristic frequency would increase. If, as is usually true, the pulse duration is Δ ($\simeq 80 \times 10^{-6}$ sec) and the pulse repetition rate is ff ($\simeq 50$ Hz) the effective value for the bulge height would be

$$B_e = B.\Delta.ff \simeq 10^{-4} \text{ cm}$$ 3.26.

which is about equal to the value of the ripple height A. The corresponding value for the bulge frequency f_b is 500 Hz.

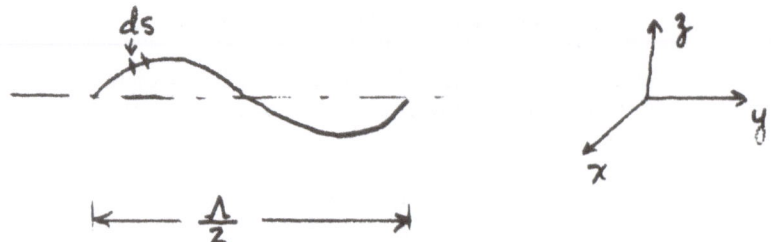

Fig. 3.2 - Energy in the Ripple Pattern

Deformation of the liquid surface by the sound wave increases the area of the surface. Considering a unit depth in the x-direction, the work done in distorting the surface is

$$W = (s - \frac{\Lambda}{2}) \gamma.$$

The natural frequency for the ripple pattern is determined in the same manner. Consider that one cycle of the ripple pattern is described by

$$z = 2A \cos 2\eta y. \qquad\qquad 3.27.$$

The total energy stored in the ripple is equal to the work done (Fig. 3.2) in distorting the flat surface into the form described by Equation 3.27. The total energy is equal to the work done in increasing the area of the surface from $\Lambda/2$ cm^2 per cm. in the x-direction to $\Lambda/2 + \Delta s$. An elemental path length along the ripple surface is given by

$$ds = \sqrt{(dz)^2 + (dy)^2} \qquad\qquad 3.28.$$

i.e.,

$$\frac{ds}{dy} = \sqrt{1 + (\frac{dz}{dy})^2} \simeq 1 + \frac{1}{2}(\frac{dz}{dy})^2. \qquad 3.29.$$

Upon integrating, we find that

$$s = \frac{\Lambda}{2} + \frac{8\pi^2}{\Lambda} A^2 \sin^2\theta. \qquad 3.30.$$

so that the work done per unit depth in the x-direction to deform this surface is

$$E = \frac{8\pi^2}{\Lambda} \gamma A^2 \sin^2\theta \qquad 3.31.$$

The effective mass per unit length in the x-direction we will take to be

$$M = \rho(\frac{\Lambda}{2})^2, \qquad 3.32.$$

i.e., the volume affected is $\frac{\Lambda}{2}$ deep in the z-direction. Substitut ing for Λ, we may also write

$$M = \frac{\rho c^2}{4\nu^2}. \qquad 3.33.$$

The amplitude of displacement is taken to be

$$y = 2A \qquad 3.34.$$

so that

$$f_r = \frac{2\pi}{c} \sqrt{\frac{\gamma \nu^3}{\rho c}} \sin\theta. \qquad 3.35.$$

Table 3.1 shows the variation of the ripple relaxation frequen cy f_r as a function of the frequency of the acoustical wave. The se figures agree with the observed behavior of the liquid surface. Experience has shown that the optimum pulse length Δ at $\nu = 3$ MHz

RIPPLE RELAXATION FREQUENCY
AS A
FUNCTION OF THE ACOUSTICAL FREQUENCY

Assume: $\gamma = 72$ dyne cm^{-1}

$\rho = 1$ gm cm^{-3}

$c = 1.5 \times 10^5$ cm sec^{-1}

$\theta = 30^o$

$$f_r = 4.38 \times 10^{-7} \sqrt{\nu^3}$$

ν	ν^3	f_r
3 MHz	5.2×10^9	2280 Hz
5 MHz	11.2	4900
7 MHz	18.5	8100
9 MHz	27.0	11800 Hz

is about 100 μs. If it is assumed that this is about a quarter of
the period of the ripple, the corresponding ripple frequency would
be 2500 Hz in good agreement with the calculated value of 2280 Hz.

Finally one should take note of the implications of Equation
3.26 with respect to the ratio B/2A which seemed so detrimental to
holography in the continous wave case as indicated by Equation
3.21. The ratio

$$\frac{B_e}{2A} = 0.48$$ 3.36.

is much more favorable and indicates why liquid-surface holography
is possible.

3.4 Interaction of Light with the Liquid Surface

The preceding analysis has provided a description of the liquid surface hologram. Now it is necessary to show how light inter
acts with this hologram to form an image. Image a beam of coherent
light of plane wavefront and amplitude D incident upon the liquid
surface at an angle θ_1. This beam of light will be characterized
by the equation

$$V(y, z) = D \exp i\ (\eta_1 y - \zeta_1 z)$$ 3.37.

where

$$\eta_1 = \frac{2\pi}{\lambda} \sin \theta_1$$ 3.38.

and

$$\zeta_1 = \frac{2\pi}{\lambda} \cos \theta_1.$$ 3.39.

After reflection from the liquid surface hologram the amplitu
de of the wave is reduced to R and the phase is modified by
$2z(x,y)$ where $z(x\ y)$ is the description of the liquid-surface given in Equation 3.9. Thus, after reflection

$$V(x, y) = R \exp (2i\zeta_1 B)$$

$$x \exp\left\{4i\zeta_1 A \cos \left[2\eta y + \phi(x, y)\right]\right\}$$

$$x \exp\left\{i(\eta_1 y + \zeta_1 z)\right\} \qquad\qquad 3.40.$$

We now make use of the identity

$$\exp(i\sigma \cos a) \equiv \sum_{n = -\infty}^{\infty} i^n\, J_n (\sigma)\, \exp (-ina) \qquad 3.41.$$

in which J_n is the nth order Bessel Function and set

$$\sigma = 4\eta_1 A \qquad\qquad 3.42.$$

and

$$a = 2\eta y + \phi(x, y). \qquad\qquad 3.43.$$

Then

$$V(x, y) = \exp i (\zeta_1 z + 2B)$$

$$x \sum_{n = -\infty}^{\infty} i^n\, J_n\, 4\zeta_1 A \qquad\qquad 3.44.$$

$$x \exp i \left[(\eta_1 - 2n\eta)y - n\phi(x, y)\right].$$

Each value of n represents a diffracted order of light. When $4\zeta_1 A$ is sufficiently small J_o approaches 1.0, J_{+1} approaches $2\zeta_1 A$, J_{-1} approaches $2\zeta_1 A$ and J_n for n not equal to -1, 0 or 1 approaches 0.0. Thus for $4\zeta_1 A$ sufficiently small

$$V(x, y) = R \exp i\zeta_1(z + 2B) \left\{ \exp i\eta_1 y \right.$$
$$+ 2i\zeta_1 A \exp i \left[(\eta_1 - 2\eta) y - \phi \right]$$
$$\left. + 2i\zeta_1 A \exp i \left[(\eta_1 + 2\eta) y + \phi \right] \right\}$$

3.45.

or

$$V(x, y) = Vo(x, y) + iV_{+1} + iV_{-1}.$$

3.46.

If we compare the expression for V_{+1} with the expression for Vo in Equation 3.2, we see that the phase component of the object information is duplicated in the light wave and since A is linearly proportional to Po(x, y) according to Equation 3.10, the amplitude information is also preserved and we conclude that the liquid surface serves as a true real-time hologram capable of transforming acoustical fields into equivalent optical fields.

CHAPTER IV : COMMENTS ON APPLICATIONS

4.1 Introduction

Acoustical holography (1, 2, 3) is a young science, having been first conceived in 1964. It was immediately thought of as offering great potential for nondestructive testing (4, 5) of industrial components, metal and plastic parts, for medical diagnostics (5, 6, 7) and research, for oceanography (8, 9), the location of objects on the ocean floor and for geophysics (10, 11) in the mapping of subterranean structures, the location of oil fields and mineral deposits. Non-holographic sonic techniques have been used in all these areas. Holographic techniques offer the possibility of better synthesis of the information and the advantage of forming an image.

4.2 Geophysical Applications

The earth and earth materials such as rock and overburden absorb sonic vibrations rather severely so that distances at which reflections can be detected vary from approximately one wavelength in overburden to 100 wavelengths (12, 13) in rock having very low absorption. A useful operating range for general purposes would normally range from 10 to 40 wavelengths. A hologram 30 wavelengths square is quite small. It can store only 900 bits of information. Normally one would like holograms to be at least 100 wavelengths square. This means that in geophysical applications, the object lies rather close to the hologram. This is an advanta

ge, since as was shown in Equation 2.38, the resolution improves
as the object to hologram distance decreases.

It is almost inconceivable that liquid surface methods could
be applied to geophysical holography. Scanning methods, on the o-
ther hand, seem especially suited. The hologram sampling plane
may be several kilometers square, and if a full two dimensional re
ceiving array is used, as many as 100,000 or more seismometers
would be required. According to Farr (10) such large sized arrays
are not prohibitive. There are, however, other alternatives such
as the crossed linear arrays suggested by Wells (11). Furthermore,
the principles of source, simultaneous source-receiver and receiver
scanning may be used to reduce the number of deployed seismometers.

The fact that many interfaces of interest in geophysical application of holography act as specular reflectors makes interpretation of images more difficult. The situation would be much improved if diffuse illumination could be employed. Simultaneous
source-receiver scanning provides a form of diffuse illumination
and is probably the preferred form of scanning for geophysical
work.

The frequencies used in geophysical holography are, of course,
much lower than for other applications. They fall in the range
from 10 Hz to at most 50 kHz. Frequencies over 500 Hz are, in
fact, not commonly used.

4.3 Applications in Oceanography

Applications of holographic concepts to imaging in the ocean
can employ frequencies from 500 Hz to 250 kHz. Attenuation is
much less than in rock, so ranges up to 15,000 wavelengths are pos
sible while ranges as low as 2 meters are of interest (8). At fre
quencies towards the 500 Hz end of the spectrum, the arrays requir
ed are large and unwieldy. Scanning by towing linear arrays or
single source-receiver transponders is possible, but there is a
problem in maintaining the transponder position within a fixed pla
ne. The scanning velocity is also limited due to the finite velocity of sound.

At the high frequency end of the spectrum, arrays on the order of 1 meter square containing more than 100,000 elements are
possible and would be useful.

4.4 Medical Applications

Both scanned and liquid surface holographic techniques are ex
pected to be applied to medical research and diagnostics. Non-holographic imaging techniques have been extensively studied and applied with some success(14). Holograms have been (15) made of ob-

jects within the cranial cavity of a pig using scanned holography. Experiments (6) with liquid surface holography systems have been shown that growing tumors in live rats are readily imaged. Some of the results of these experiments are shown in Figure 4.1 while Figures 4.2 through 4.4 demonstrate other acoustical images obtained using the liquid surface system. These images have also been recorded on video tape to capture the real-time aspect of the imaging.

Acoustical imaging provides good soft tissue differentiation as evidenced by the fact that tendons, muscle structure, ligaments blood vessels and tumors are clearly distinguished from other tissues. Imaging of this type is carried out using power levels and experimental conditions which provide a large margin of safety.

4.5 Applications in Nondestructive Testing

The liquid surface system of Figure 4.5 is being used extensively for nondestructive testing (4, 16, 17). Typical applications are the inspection of laminar structures including honeycomb panels for nonbonds, inspection of welds for voids and inclusions and inspection of nuclear fuel elements for cladding-to-fuel bonding.

Scanned acoustical holography systems have been proposed (4) for use in the inspection of large castings, welds and walls of large pressure vessels and viewing inside liquid sodium reactors (Fig.4.6 through 4.8).

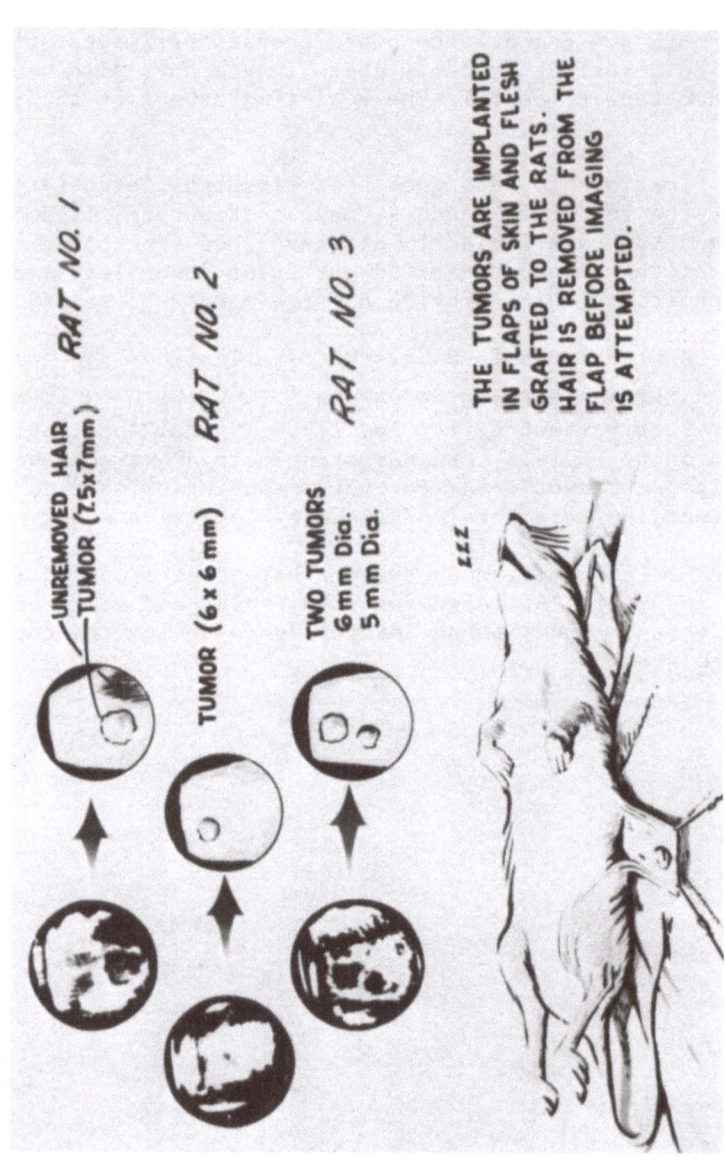

Fig. 4.1 — Images of Tumors in Rats

Experiments performed at Roswell Park Memorial Institute in Buffalo, New York demonstrated that tumors as small as 2mm in diameter could be imaged by liquid surface acoustical holography methods. Rates of growth were determined by imaging daily and measuring the change in size as shown by the acoustical image. The results shown about were obtained in preliminary experiments run at Battelle Northwest in Richland, Washington by Dr. Holyoke and B. B. Brenden.

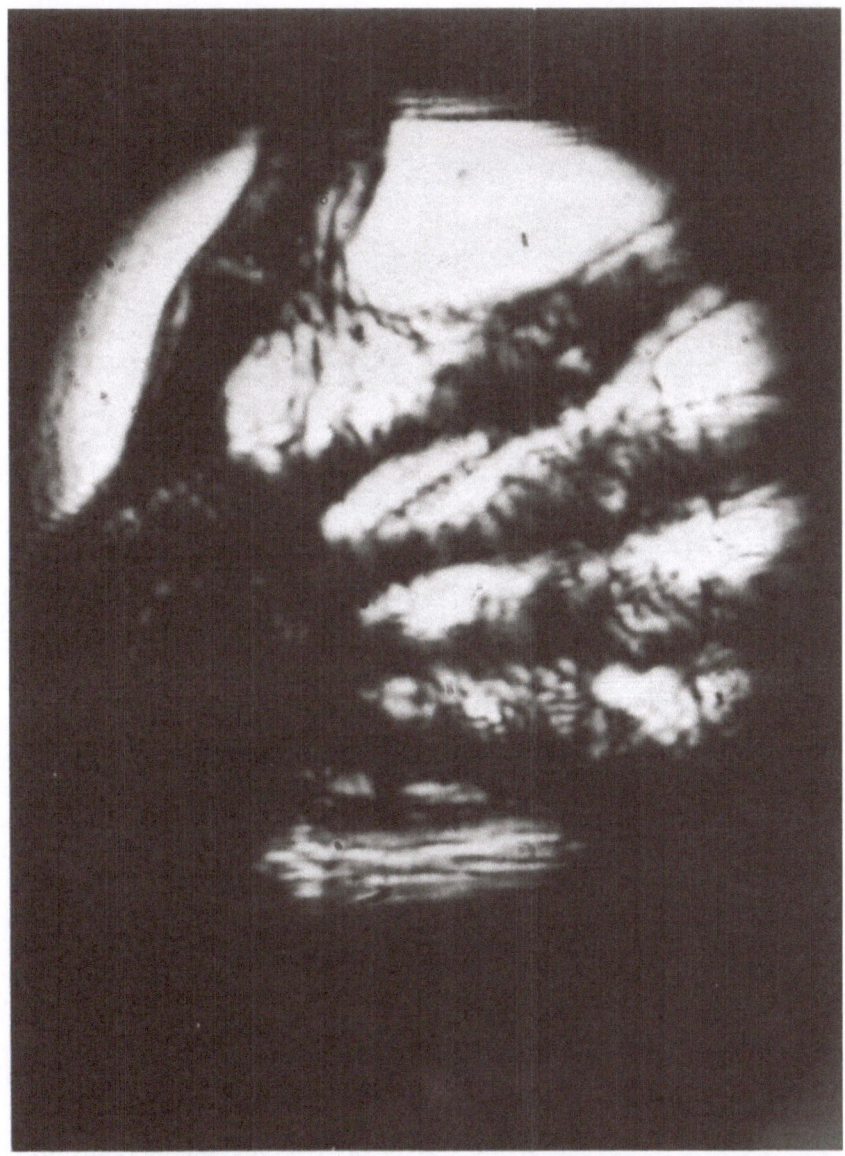

Fig. 4.2 - Acoustical Image of the Hand

This picture and the others in this series were made using the instrument shown in Figure 4.5 operating at 3 MHz ultrasonic frequency. It shows the palm of the hand. The metacarpals are imaged dark. Connective tissue at the joints, flexor tendons between the thumb and the wrist and between the finger and the wrist also show clearly. Picture is courtesy of Holosonics, Inc., Richland, Washington, U.S.A.

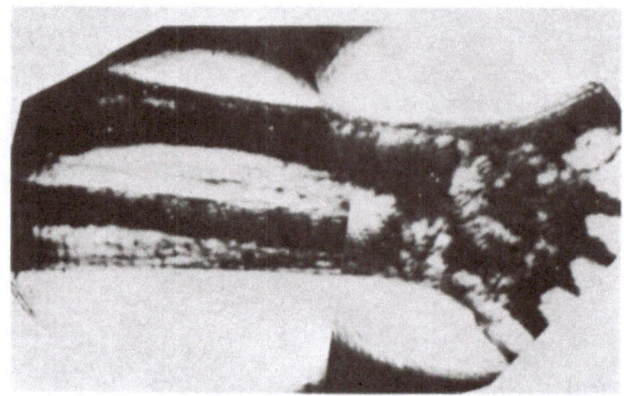

Fig. 4.3 - Composite Showing Forearm and Wrist

In addition to showing the radius, the ulna and the carpal
bones, two blood vessels are imaged, one between the radius and
ulna and one below the ulna. Images at 3 MHz of blood vessels are
characterized by the fact that the two walls show as dark lines with
the center of the vessel remaining transparent.

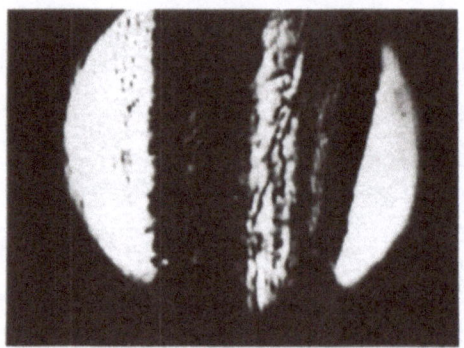

Fig. 4.4 - Radial Artery in the Forearm

This picture at MHz again demonstrates the capability of the
liquid surface holography system for imaging blood wessels. The
arm is rotated 90° from the orientation used in Figure 4.3 and the
radius and ulna lie one behind the other on the left side of the
pcture. The dark region to the right and below the artery is muscle
tissue. Note that the artery branches and goes out of focus. Picture is courtesy of Holosonics, Inc., Richland, Washington, U.S.A.

Fig. 4.5 - Acoustical Imager

The images shown in this picture was used to produce Figures 2-4 and Figures 6-8. The hand, arm or other object is inserted into the tank at the left and the image may be viewed instantaneously on the ground glass screen or on a video monitor not shown. Picture is courtesy of Holosonics, Inc., Richland, Washington, U.S.A.

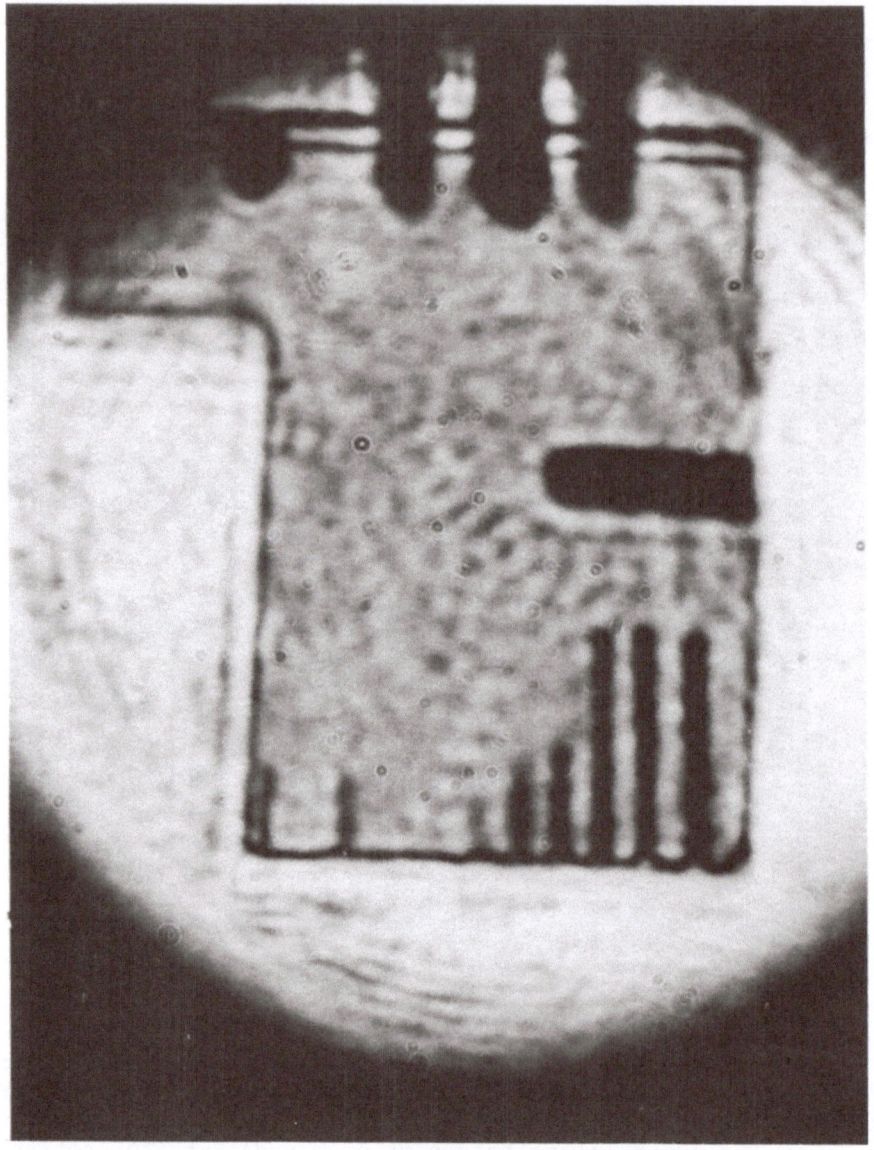

Fig. 4.6 – Acoustical Image of a Plastic Block

The acoustical image of an optically opaque, 1 cm thick plastic block clearly shows the position and depth of eight holes drilled in from the bottom edge and one large 6 mm diameter hole drilled in from the side edge. Thelargest of the holes drilled in from the bottom is 2.5 mm diameter, while the smallest is 0.5 mm diameter. The acoustical frequency is 5 MHz. Picture is courtesy of Holosonics, Inc., Richland, Washington, U.S.A.

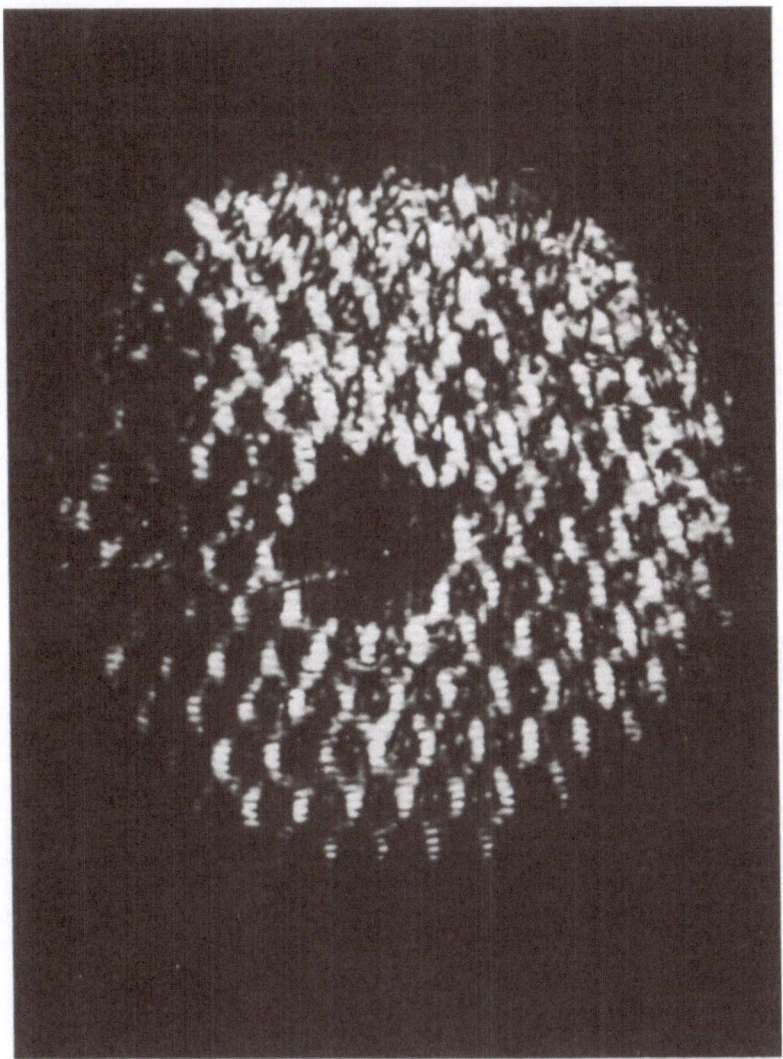

Fig. 4.7 - Acoustical Image of Honeycomb Panel

  Honeycomb sandwich panels consist of a core of aluminum
structure resembling a honeycomb which has been sliced perpendicular to the channels. Sheets of aluminum are bonded to each side
of this core. An unbonded area is imaged in the center of this
picture.

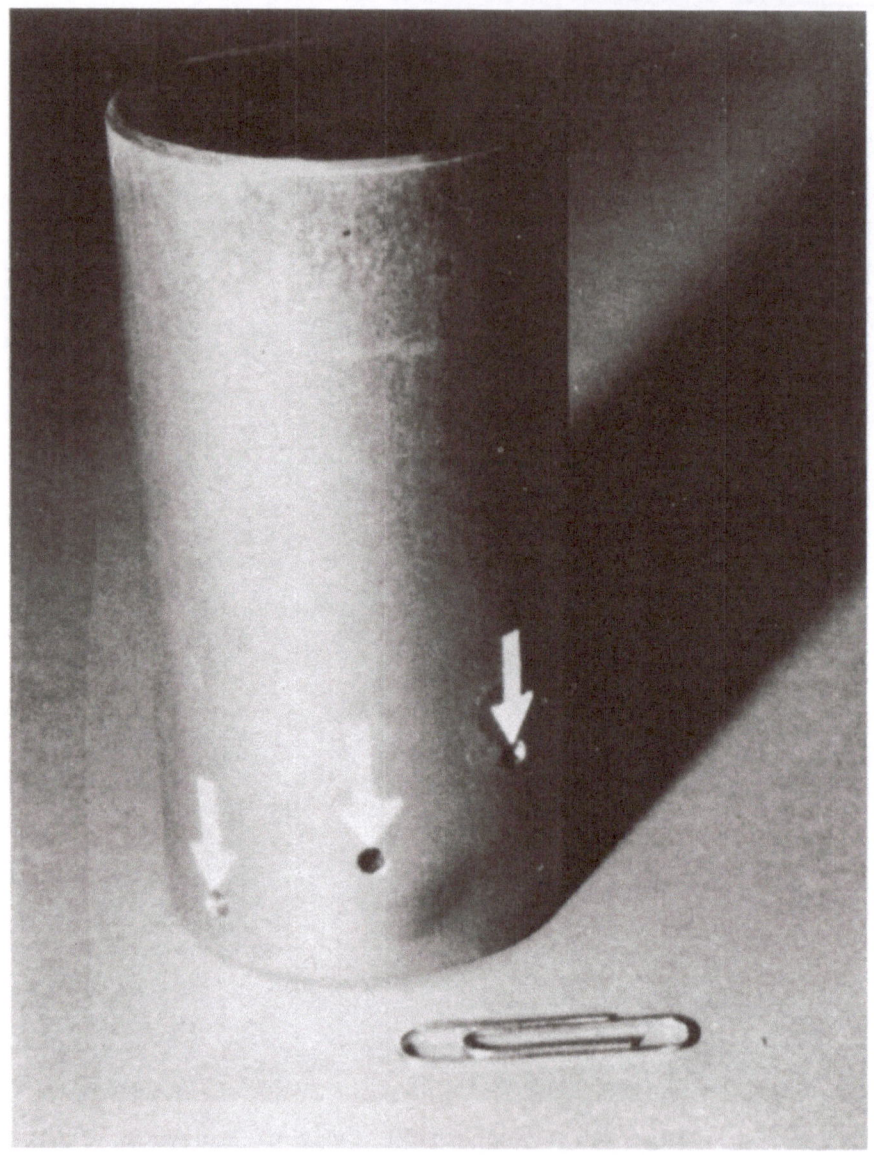

Fig. 4.8 - Steel Cylinder Optical.

The optical and acoustical views of a steel cylinder, 10 cm long and 6 cm in diameter, demonstrate the ability of the liquid surface acoustical holography system to penetrate metals and focus on features such as the 3 mm diameter hole drilled along a radius toward the center of the cylinder. Other radially drilled holes are out of focus.

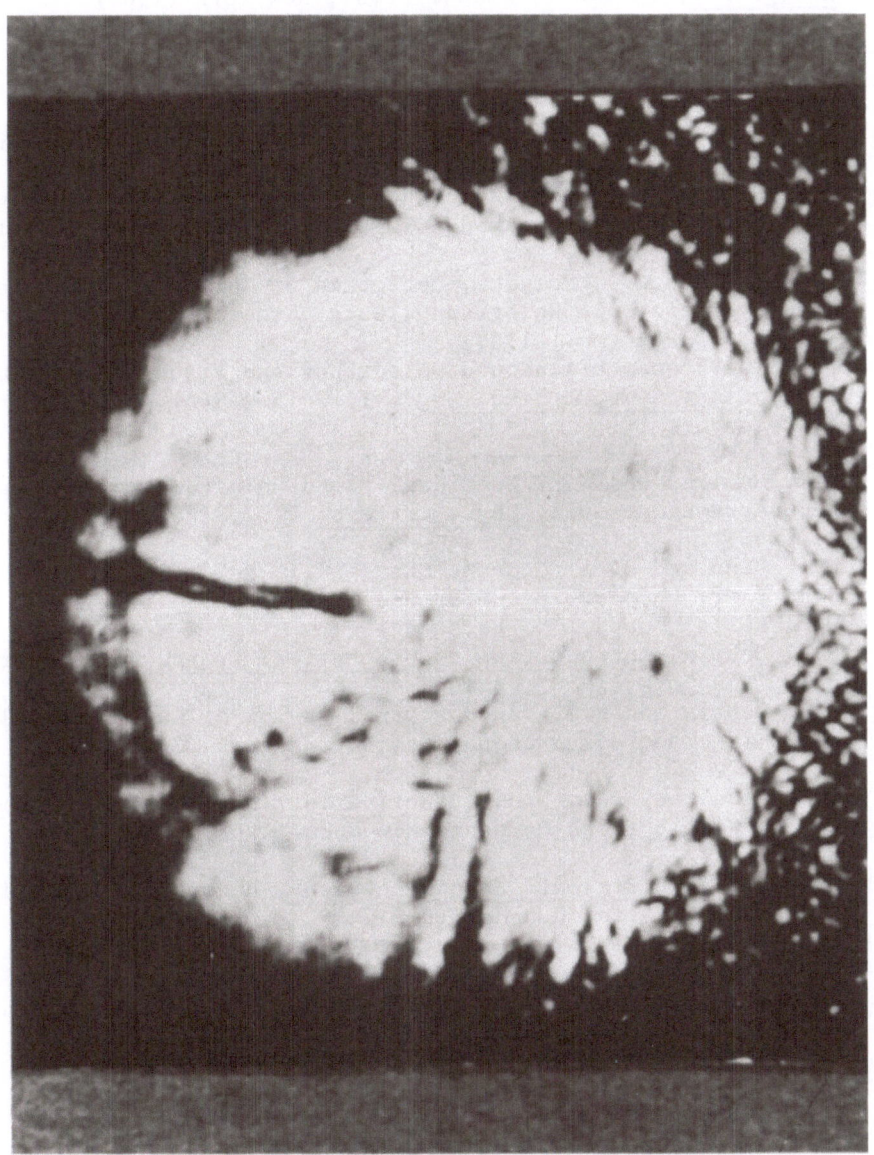

Fig. 4.8 - Steel Cylinder Acoustical.

REFERENCES

(1)   A. F. Metherell, H.M.A. El-Sum and Lewis Larmore, Acoustical Holography, Volume 1, Plenum Press, New York (1969).

(2)   A.F. Metherell and Lewis Larmore, Acoustical Holography, Volume 2, Plenum Press, New York (1970).

(3)   A.F. Metherell, Acoustical Holography, Volume 3, Plenum Press, New York (1971).

(4)   B.B. Brenden, Acoustical Holography as a Tool for Nondestructive Testing, Materials Evaluation, 27(6):140-144 (1969).

(5)   R.B. Smith and B.B. Brenden, Refinements and Variations in liquid Surface and Scanned Holography, Ultrasonics, 7(2):125-126 (April, 1969).

(6)   L. Weiss and E.D. Holyoke, Detection of Tumors in Soft Tissues by Ultrasonic Holography, Surg. Gynecol. and Obstet. 128(5):953-962 (May, 1969).

(77)   F.L. Thurstone, Ultrasound Holography and Visual Reconstruction, Proc. Symp. Biomed. Eng. 1:12-15 (1966), also Chapter 7, Reference 1.

(8)   R.K. Mueller and N.K. Sheridan, Sound Holograms and Optical Reconstruction, Appl. Phys. Letters 9(9):328-329 (November 1, 1966).

(9)   W.A. Penn and J. L. Chovan, Chapter 11, Reference 2.

(10)   J.B. Farr, Chapter 16, Reference 2.

(11)   W.H. Wells, Chapter 8, Reference 2.

(12)   M.B. Dobrin, Introduction to Geophysical Prospecting, McGraw-Hill, New York (1960), p. 64.

(13)   M.B. Dobrin and R.G. Van Nostrand, Review of Current Developments in Exploration Geophysics, Geophysics, 21:142-155 (1956).

(14)   R.C. McMaster, Ed., Nondestructive Testing Handbook, Volume 2, The Ronald Press Company, New York (1963), Section 46, pp. 3-5.

(15)   D.N. White, Ultrasonic Encephalography, Hanson and Edgar Inc., Kingston, Ontario, Canada (1970).

(16)   J.B. Swint, B.G.W. Yee and N.H. Godbold, Chapter 9, Reference 3.

(17)   H. Clements, Chapter 8, Reference 3.

FINAL REMARKS AND FAREWELL!

E. Camatini

Politecnico di Milano, Italy

Scientific Director of the NATO Advanced Study
Institute on Holography

I am not attempting to conclude how much of the original ideas
of Gabor in his basic papers of 1949 and 1951 is now under current
interest and how many deviations and application there have been
from his very ingenious method of the two steps imaging procedure.

Of course much was changed or rather made possible by the advent of the lasers in the early 60's, proposed by Vander Lugt and
realized first in the two beautiful papers by Leith and Upatnieks
in the years 1962 and 1963, which in many ways initiated the interest in holography which since then has been so rapidly growing.

I am going to make only few remarks on the basis of what has
been presented so clearly by the experts.

So, I shall start from hologram interferometry.

Holography has had greater influence on measuring technique,
in its wider sense, than as an imaging tool.

Thus, to be specific, it is not the demonstration holograms,
"the boast holograms" which have fascinated so many, which are
important. The 3 D imaging properties can seldom be utilized,
partly due to the fact that it is difficult to transfer them over
distances. Of course, I am not referring now to a specific class
of cases where the importance of this 3 D property is really
obvious.

What I mean with measuring technique is mainly hologram interferometry.

403

The advantages over classical interferometry are in principle
very well known: diffuse surfaces and their shapes can be examined
even better than highly reflecting ones; components in the optical
system which are far from perfect can be tolerated, on account of
the different character of the double or multiple exposures which
are made on the holograms; and a few other advantages.

Industries in many countries have begun to realize what aid
can be furnished by application of hologram interferometry, and it
has been very stimulating to hear what has also been brought forward on various techniques to measure deformation, stress parameters, etc; in many applications.

Special emphasys has been given to three applications of holo
gram interferometry during this seminar, i.e. comparison of shapes,
use of computer-generated holograms, holographic contouring.

The wavefront reflected from surface may be stored in a holo
gram and subsequently reconstructed to take part in optical interference with any other wavefront.

In application where surface deformation is to be measured,
the two wavefronts, although complex, are very similar, and the
concept of homologous rays can be used.

If two different surfaces are to be compared, however, the
wavefronts scattered from each will in general be too dissimilar
in detail to yield an interference pattern showing the mean difference in shape.

This difficulty can be overcome to some extent by illuminating
the surfaces at a high angle of incidence, when an increasing proportion of the light will be scattered specularly. The nearly
smooth waves so generated can then be compared interferometrically
(comparison of shapes).

In precise comparison of shapes,a computer-generated hologram
can be used to generate the wave from the master shape. The principle use to which this has been applied is the testing of aspheri
cal optical surfaces (use of computer-generated holograms).

Holograms of scattering objects record the phase of the light
which has travelled from the source via the surface of the object.

Without altering hologram or object positions, this phase can
be varied by changing one of the three recording parameters: the
illuminating angle, the wavelength, the refractive index of the
surrounding medium.

A double exposure hologram recorded when such a change is made between exposures will reconstruct interference fringes giving surface depth contours (holographic contouring).

However, much remains to be done to simplify the identification of geometrical parameters into fringe geometry. The literature at present is too incomplete in this respect and lacks greatly advice for experimental procedures of obtaining reliable and rapid identification.

It is a general opinion that very soon the whole field of interferometry will be so much widened by the advent of holographic techniques that the classical interferometers will stand out as important but only alternative techniques.

Spatial filtering methods being accessible for holography but not so accessible for classical interferometry underline the importance of holographic interferometry.

In this field, it should be stressed in particular the aid we have gathered from time average holograms of vibrating objects. In the application of interferometry to vibration studies much progress has been made and some of the techniques are also so simple as to make them attractive also to workers not much experienced in optical techniques.

An extremely important property of hologram interferometry is that it can, with some ease, be made more or less sensitive. It is understood that a few Å of amplitudes are still measurable instead of, as usual, the order of $\lambda/2$.

It is really of importance in several cases of physical or chemical research that holographic technique is also applicable in this domain of optical pathlengths. Still more common, for workshop and other industrial applications, is that the ordinary interferences are too sensitive, and here also fine materials have been introduced to reduce the sensitivity, by very oblique angles, by moiré formed by two wavelength interferences, and even with plain moiré methods, which of course in then pure form are basically other phenomena than holography, but also in combination of the moiré methods with hologram interference methods.

On the whole, workers in the fields of hologram interferometry at this stage of its development show an inventiveness which imposes very much, and much skill is devoted to overcoming those difficulties which are also connected to the new techniques.

It should only be mentioned the speckling effects, often or mostly felt as a disadvantage, but also the new possibilities of

designing speckle interferometers, a field of which is not holography but is coming out as a sort of "spin-off" in work with laser
holography.

Holography records phase and amplitude of a wavefront, which
can be related to the detailed surface structure. In metrology,
one is concerned only with comparison of phases, which relate to
the change in length, so that much of the information in a hologram
is superfluous.

It is also very often difficult to extract from hologram inter
ferograms the motion of a surface resolved in one particular direction (especially the in-plane directions) since the fringe pattern
relates to the generalized deformation.

Both these drawbacks can be overcome by making use of coherent
speckle effects.

Examination of some three-dimensional time-varying objects is
another aspect which is worth to be underlined.

Examination of some three-dimensional time-varying objects
gives us the possibility of reconstruction and examination of a
relatively large volume of the (x,y,z) space. Since now, with the
advent of powerful pulsed lasers with convenient degree of temporal
coherence holograms can be taken up with short exposures and high
repetition frequency of the pulses: holography therefore offers
means for the study of rapid events when information of (x,y,z and
t) is needed. The holograms may afterwards be examined by classical optical methods. Applications are: gas and liquid fluids, size
of particles in the streams, aerosols, events in plasma physics
and in ballistic cases. Holographic microscopy for the same reason, namely unlimited depth of focus within the coherence domain,
makes it possible to observe a real 3 D image in the plane wanted.

According to the basic idea of Gabor, the two stages of holography may be performed with different coherent waves: light waves,
microwaves, sound waves, X-ray waves, etc.

Good progress has been registered in the field of microwave
holography.

Holographic interferometry in the visible part of the spectrum
has a number of advantages over the classical interferometry and
is widely used in investigating optical inhomageneities of transparent media or objects of great dimensions.

Holographic interferometry in the microwave region can be used
for similar purposes in studying optically opaque media or objects

of large dimensions.

Binary, detour-phase, optical holograms have been developed
to represent computer-generated, Fourier transform holograms.
These holograms have binary transmittance. They represent amplitude of the hologram through the size of transparent apertures in
an opaque screen and represent phase through lateral position of
the apertures. Detour-phase holograms are analogous to gratings
with modulated spacings.

Binary, detour-phase representations of microwave data seem
interesting as a relatively simple way to encode phase and amplitu
de data in a tangible hologram. Such representations are an alternative to somewhat non-linear, empirical relationships between the
amplitude transmittance of an optical hologram and the intensity
of microwave fringes. Such non-linearities can occur in oscillo-
scope displays of microwave fringes.

Other empirical relationships have been used in microwave holography; for example the size of an opaque region has been used
to encode transmittance in microwave holograms and optical holograms and optical holograms produced from microwave holograms.

Results with acoustic waves of short wavelengths for the
first step have been promising.

There are great difficulties involved: large aberrations due
to the large ratio λ/λ'; and extra regard to reflexions at all
boundaries between media of different acoustical density, besides
of the electrical transducers and network for the recording of the
acoustical hologram. However the potential uses are so attractive
and important in material testing, in biological and medical measurements and diagnosis, that a fair guess would be that the next
few years will bring considerable progress in this field.

Other fields of application of acoustic holography seem to be
underwater viewing and underground exploration.

Acoustic holography's full utilization must await development
of certain key components for the imaging system and sophistication
of the over-all technology involved.

As a tool for nondistructive testing acoustic holography may
well surpass optical holography. Further development will result
in near real time scanned holography and will increase the sensitivity, efficiency and quality of the liquid surface. Other improvements will be the result of increased understanding of possible variations in scanning, variations in acoustical illumination
and judicious use of acoustical lenses, spatial filters and other

components which are counterparts from the practice of optics.

An activity who has really grown rapidly in the search for
new materials besides the classical photographical ones for the
registration of holograms.

Also, as is well known, work on specific photographic emulsions has been quite considerable, and not only in laboratories of
large manufacturers of emulsions. But for the future development
of methods of data storing the photoresist, the photopolymers etc.
are most likely of great importance, and the progress seem to promise that also the speed and the time constants of these straylight
free photopolymer media will make them useful for an increasing
number of applications.

This progress is likely to go on in parallel with those on
electrooptical devices which are moreover connected with the utilization of holograms for data storing. As is well known, this
sort of electro-optics is now a technical branch of great actuality.

I will conclude the remarks stating the real interest of
holographic techniques to examine spatial and temporal coherence.

In the Besançon Symposium someone said in a paper: <u>A hologram
is an interferogram</u>, and as even students of elementary optics now
should know, coherence can be measured as the visibility of interference fringes. Holographic techniques in fact are a versatile
tool for this, and it can be done very simply. The concept of coherence is not the only one now so directly accessible with holography: a concept earlier thought of as very abstract, namely the
evenescent waves, also has had new experimental reality as the
waves called subwaves.

These realizations of new techniques by means of holography
strengthen one concept. Holography is very central in present day
optics, and it has already proved that optics is an important
branch of Physics in a stage of rapid development. It will continue
to grow with at least the same speed in the next years.

At this point let me as a Professor of a Polytechnic and as
a professional man in the field of engineering to say that in many
countries including Italy, optics is rather neglected in the
curricula for the degrees in engineering. I think that optics is
going through the same experience as the classical mechanics which
in the rather recent past was considered as a science without any
perspective of further development but now is in the stage of a
new escalation.

Lasers and Holography being a glamorous subject for so many

media should help teachers to make optics popular with students.
Elementary pedagogical introductions into hologram optics are
needed, also intended for those many people from industry and other
places which apply holographic techniques.

Very many - and I also know several skilled physicists among
them - have been at the beginning frightened by the formal system
mainly borrowed from concepts current in electrical network and
communication theory, however beautiful and efficient, and for
them it has obscured the simple physical fact inherent to the phenomena.

We therefore welcome all elementary texts, in the field, as
well as in the more elaborate communications a more simple and
unified conceptual system than can be obtained when progress is so
rapid as now. In this way, more and more physicists and technical
people are introduced to live and work with Fourier optics which is
such powerful tool for present day development.

Let me come now to the conclusion.

It is said that Frederick, the Great, made wars first and
after he ordered jurists to justify them. I am not going to follow
the same procedure, in other words am not asking you to find for
me any justification to my initiative of promoting this Seminar
which is next to the end.

I would rather ask you if this Advanced Study Institute has
reached its aim to start a critical conversation between lecturers
and participants and to promote a stimulating meeting to both the
physicist and the engineer. Your frank comments on this aspect
will be highly appreciated as they will be helpful to me in case
I would intend to promote another Seminar on the same subject in
the future, and will permit an evaluation of the results achieved
through this meeting.

In this connection let me recall the text of a poster I saw
in the development department of an engineering company, which
summons that the steps to a successful project are six.

They are: enthusiasm; disillusion; despair; searching for
scapegoats; blaming the innocent; and finally the last step:
honor and medals for those who contributed nothing. In such a case,
honor and medals would come to me, because as a matter of fact the
success of this Seminar, if any, is yours, and yours only.

I don't want to take any more of your time, but please let me
express again the deepest gratitude: to the Scientific Affairs
Division of NATO, from which this Institute was sponsored; to the
FAST, which has been our excellent and very cooperative host; to

Prof. Sona, who assisted me very friendly during the Seminar; to
all lecturers, who brought into this meeting the high contribution
of their science and experience; to all participants, who made vi-
vid and fruitful the discussion with the lecturers.

Last but not least, a special thank to my Secretary, Mrs.
Valeria Sarchi, whose contribution to the organization of the
Seminar has been extremely precious and brought into this Advanced
Study Institute a note of refinement of manners and exquisite lady-
likeness, besides the intelligent cooperation. To all of you I
cordially wish a good return journey back home.

IMAGE SHARPENING BY HOLOGRAPHY*

GEORGE W.STROKE

State University of New York,Stony Brook,N.Y.
and
Harvard University

Greatly sharpened images may be extracted from photographs which have been blurred either by accident or by unsurmountable instrumental imperfections.In simple words, it has recently become truly possible to turn a bad photograph into a good image in a great number of situations, including cases when photographs were blurred by motion, imperfect focus,atmospheric turbulence and by instrumental defects(including aberrations) among other causes.

The principles of optical image deblurring are highly mathematical in nature,and their detail requires considerable development beyond the scope of a brief introduction such as this[1-4].However,the principles may be readily sketched out with the aid of illustrative diagrams,such as that of FIG.1 and with the aid of a small number of basic equations,such as those given below.

Perhaps the most dramatic image deblurring results obtained to date are shown in FIG.2. and FIG.3.Both figures illustrate the application of the powerful image deblurring method first described by Stroke and Zech in 1967,under the name of "holographic Fourier-transform image-deblurring method"[1] and since that time brought to the present stage of perfection by Stroke and Halioua [2-4]. In contrast with many applications of holography [5,6],the holographic image deblurring method requires considerable photographic care as well as theoretical sophistication.By means of a new type of "holographic filter"[2-4] illustrated in FIG.1 the holographic image

*This paper was not presented at the conference.

deblurring arrangement is capable of carrying out the
complex image-deblurring computation with a speed and
data-capacity unattainable with methods using even the
most powerful digital electronic computers[7-9]in their
present state of the art.

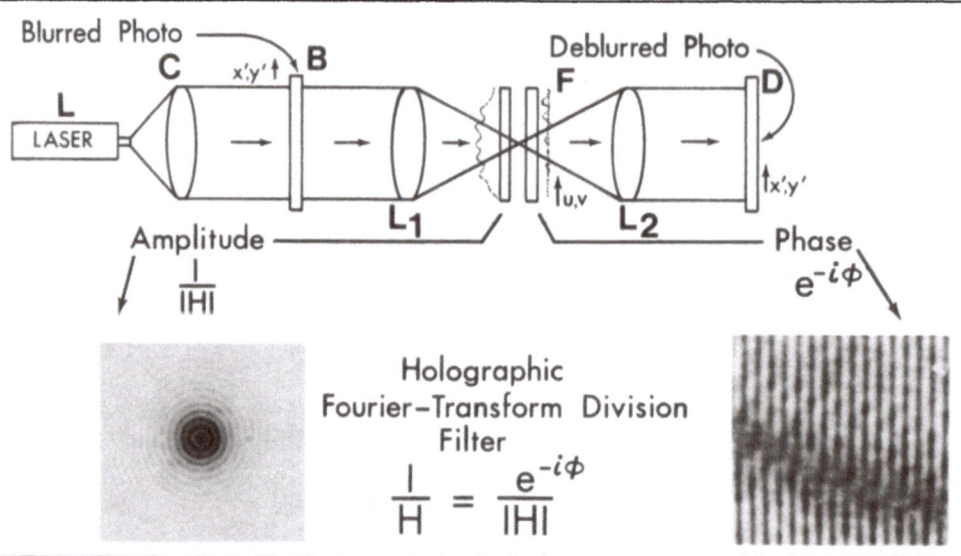

FIG.1. Holographic Image Deblurring Arrangement

A blurred photograph may be deblurred and a sharpened
image extracted from it because the "blurring" in the
original photograph did not irretrievably loose the imaging information,notably in the spatial-frequency regions
required for faithful imaging and for high resolution[6].
An irretrievable loss may have been assumed when simply
looking at an out-of-focus or motion-blurred photograph,
for example[see FIG.2. and FIG.3],or at a photo scanned
with a beam of finite width[FIG.3],as is the case with
electron micrographs recorded with even the most powerful electron microscopes,working in the scanning(reflection or transmission)mode[10,11].In fact,the desired
faithful,sharp image is "encoded" in a decodable convo-
lution-integral form in the blurred photograph.The decoding operation may be readily carried out by means of
the optical analogue holographic computing arrangement,
shown above in FIG.1,where it is presented in the form
first conceived for this purpose by Stroke and his collaborators in 1965,notably for the purpose of deblurring
photographs recorded in incoherent illumination,accor-
ding to the equations given below.

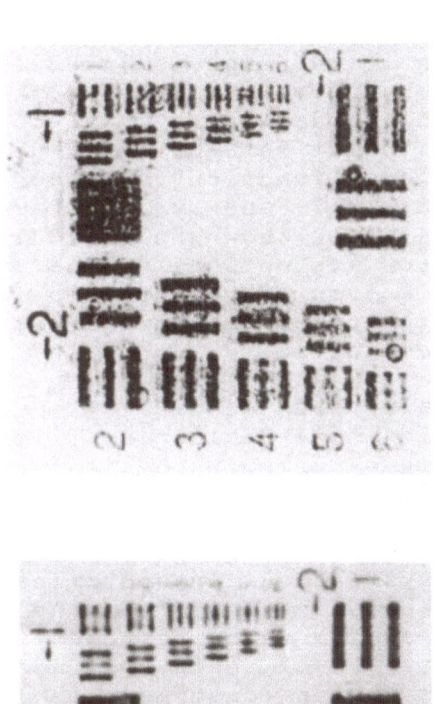

BLURRED PHOTO   **fidelity**   **recognition**   **DEBLURRED IMAGES**

FIG.2. Examples of Holographically-Deblurred Images in the Case of Out-of-Focus Photo-graph, using the Arrangement of FIG.1. The insert h=h(x,y) shows the point spread function of the blurred photograph. Two different holographic deblurring filters were used to extract the "deblurred images" from the "blurred photo", in order to illustrate the capabilities of the method. In the case of 'unknown' objects, where imaging "fidelity" is of primary importance, as for instance in electron microsco-py, emphasis was placed on restoration of low and middle-range spatial frequencies. For 'recognition' of 'known' objects, emphasis on highest spatial frequency resto-ration is in order(even at the expense of some undistracting 'noise'[3]).

In many cases the deblurring may result in a practically perfectly sharp and completely faithful image.In
other cases,the process aims at changing the more or less
blurred photograph into an image which is immediately
interpretable by a human observer.In all cases,the decoding "key" is contained in the defective image of a single
point(commonly called the "point spread function","spread
function" or "impulse-response function",for short,as
illustrated by h=h(x,y) in FIG.2. and by its diameter,
corresponding to 5Å at the scale shown,in FIG.3).It is
the point spread function(or its optically-generated
analogue,in some of the electron microscopy applications)
which is used to produce the powerful Fourier-transform
division filter,illustrated in FIG.1.

Even though mathematically and technically complex,
the role of the holographic image-deblurring filter may
be explained in rather simple terms.In physical terms,
the principle of optical image deblurring may be quite
readily compared to electrical signal filtering methods,
such as those used for example in high-fidelity sound
or TV image reproduction systems. As in acoustical sound
"frequency equalization",where the amplitude and the phase
of the distorted and shifted frequency components are
'restored' by suitable networks of resistors,capacitors
and inductors,we find that the optical image-deblurring
process consists in suitably acting on the spatial frequency components in the blurred photograph by means of
the optical deblurring filter.

These concepts may in fact be summarized with the aid
of a small number of rather simple-looking equations as
follows. The original image intensity in the object is
represented by f(x,y) and that in the blurred photograph
by g(x',y').The intensity in the desired "deblurred"
image is represented by f(x',y'): it is ideally equal
to f(x,y),and in practice approximates the original
object with considerable improvement,in comparison to
the blurred photograph,as illustrated in FIG.2. and FIG.3,
even though certain spatial-frequency components may have
been irretrievably removed from f(x,y) by the blurring
as described below.Mathematically,the blurring process
may be represented by the convolution integral of f(x,y)
with the point-spread function h(x'-x,y'-y) as follows:

$$g(x',y') = \iint f(x,y)h(x'-x,y'-y) \ dx \ dy \qquad [1]$$

Eq.[1] may be written in symbolical form[6] as

$$g = f \otimes h \qquad\qquad [1]$$

Eq.[1] represents the blurred photograph. The convolution
integral of eq.[1] takes on the form of a product in
the Fourier-transform(spatial frequency) domain:

$$G(u,v) = F(u,v)H(u,v) \qquad [2]$$

where G,F and H are the spatial Fourier transforms of
g,f and h, for instance

$$H(u,v) = \iint h(x,y)\exp[2\pi i(ux+vy)dx\,dy \quad [3]$$

in the usual normalized form[6].It may be in order to
recall that the function H(u,v) = F.T.[h(x,y)] according
to eq.[3] is the well-known spatial-frequency transfer
function,usually described as MTF function,for short[6].
It may thus be readily seen from eq.[2] that the 'blurring'
process described by eq.[1] results in an overall attenuation (and shift) in spatial-frequency components,in
the general case,and in the total suppression of some
spatial-frequency components in some cases(see e.g.FIG.3).
The deblurring of the "blurred" photograph is carried
out by performing the division of G by H,according to
the equation

$$\frac{G}{H} = \Gamma \qquad [4]$$

in order to give F,from which the desired function f
is extracted by another Fourier transformation,symboli-
zed by the equation

$$F \longrightarrow f \qquad [5]$$

within the limitations discussed in ref.[1],and briefly
summarized above.

The holographic Fourier-transform division filter
represents the function

$$\frac{1}{H} = \frac{e^{-i\phi}}{|H|} \qquad [6]$$

The filter may be materialized by means of two photographic transparencies, in the form of a "sandwich".The
first transparency has a transmittance equal to the am-
plitude $1/|H|$,and the second is a hologram with a trans-
mittance,in one of its side-band[6]image forming waves
equal to $[e^{-i\phi}]$. An important feature of the method is
that the two components of the holographic filter may be

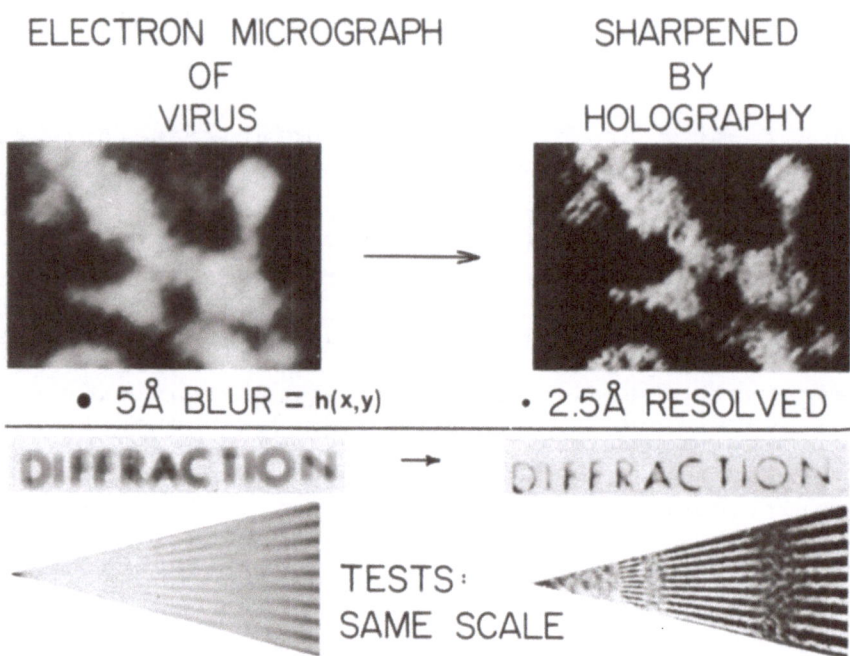

FIG.3.Holographic sharpening of electron micrograph of
     virus and illustration of corresponding imaging
     improvements by optical test photographs at same
     scale.
The original micrograph is that of an fd-virus(bacterio-
phage)recorded with the 5Å scanning transmission electron
microscope by A.V.Crewe and J.Wall as a part of the collaborative work with G.W.Stroke and M.Halioua[2].The
sharpened image was extracted from the blurred photo by
means of the arrangement of FIG.1. and shows the attain-
ment of the 2.5Å "limit"set by diffraction,together with
the restoration of imaging faithfulness,as illustrated
by the restoration of contrasts in the test photos.Note,
for example that the original micrograph,even though only
blurred to twice the diffraction pattern 'width'[the ultimate attainable in the present state of the art]was
obtained under conditions of serious image degradations,
as illustrated by the optical tests.Note,for example,the
shift in the location of the maxima in the blurred test
chart(and corresponding inversion of contrast in the center of the letter "A").Also note 'restoration' by the holographic image deblurring in sharpened images on right.

optically generated from the experimental point-spread
function(or from its optical equivalent,in some of the
electron microscopy applications).

   Details of the production of the two-component holographic filter involve as much art(and technique) as
they involve sophisticated theory and mathematical ex-
pertise[1-4].A full description exceeds the scope of
this presentation. However, by way of example,we may
mention one of the problems which we had to solve in
order to produce successful filters.In order to obtain
the deblurring results shown, the transmission of the
amplitude component of the filter $1/|H|$ had to be made
to be linear over a range of more than $10^4:1$(i.e. 4 on
the logarithmic scale of the H&D curve) with a slope of
1. We may recall that usual photographic materials have
normally only a 'dynamic' range of 30:1,with this slope,
and not $10^4:1$ as required.The dynamic range of $10^4:1$
with a gamma=1 was achieved by a new improvement of our
'masking' method[2,4],of which we gave a first description in ref.[12]. A reproduction of an actual experimental result for an early case is given in FIG.4.

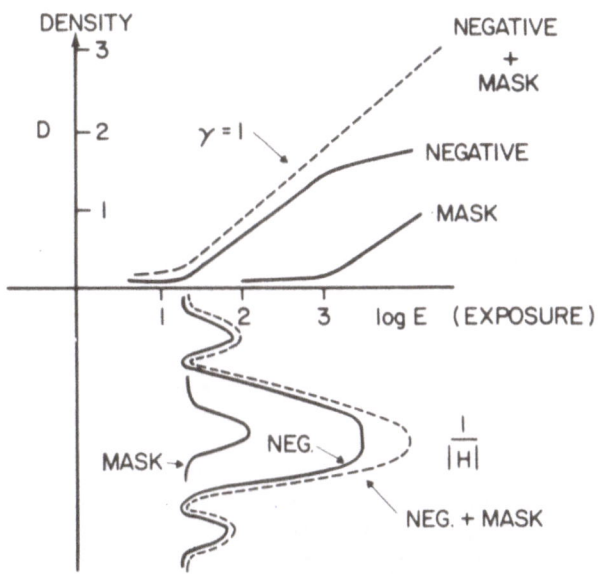

FIG.4. Amplitude component of filter,with 'extended'
       dynamic range[2,12],as required for successful
implementation of holographic deblurring according to
FIG.1. and eq.[4-6].(actual experimental result).

In an earlier version of our deblurring[13]which we used also in deblurring of motion-blurred photos[12] a filter with a 10^6:1 range in the amplitude component was realized in the implementation according to $1/H = H^*/|H|^2$(FIG.5).

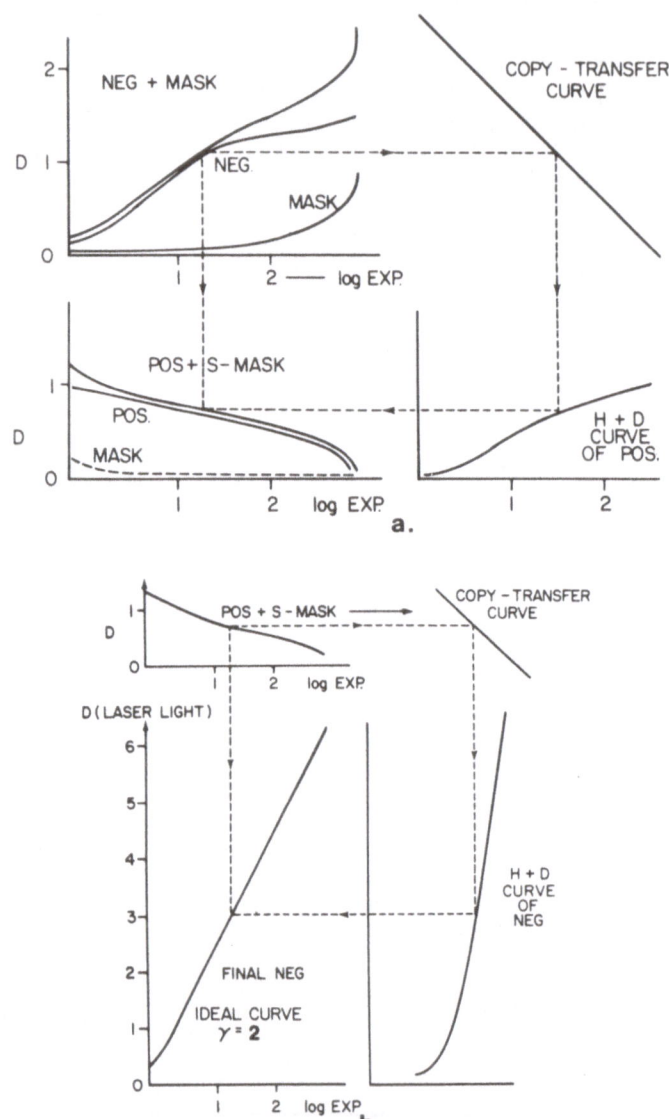

FIG.5.Experimental results showing realization of amplitude component $1/|H|^2$ for filter $1/H = H^*/|H|^2$(ref.1,13]and having linear range of 10^6:1 with gamma=2.

The realization of the phase component of the filter
with transmittance exp[-iØ] is achieved by holography
[1-4,6,10,12,13].It requires great care,of course,but
in general we have found that 'amplitude weighting' is
required only for the purposes of improving imaging
faithfulness(see e.g. FIG.2)and for the realization of
the "Wiener-filter"noise-suppressing refinement,as we
describe also in ref.[1-4]. The arrangement of the
holographic image-deblurring system according to FIG.1
in the author's Electro-Optical Sciences Laboratory at
Stony Brook is shown in FIG.6.

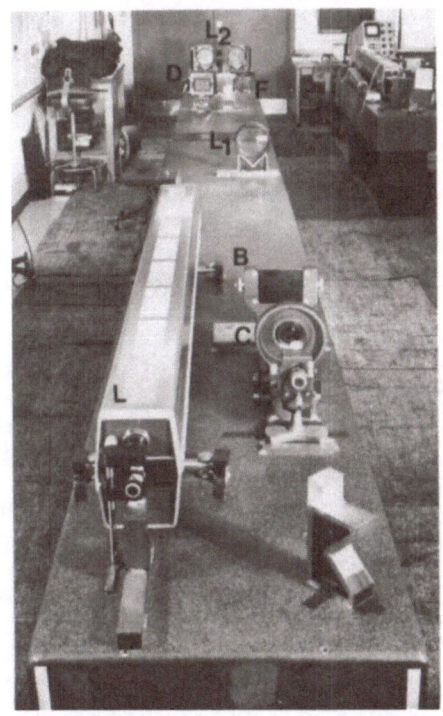

FIG.6. Holographic image-deblurring system according
      to schematic diagram of FIG.1. as arranged in
the author's Electro-Optical Sciences Laboratory at
Stony Brook.L=laser,C=collimator,B=blurred photo,F=ho-
lographic Fourier-transform division filter[eq.6],D=de-
blurred image ,L_1 and L_2 = lenses[see FIG.1]. Two mirrors
on each side of L_2 help to fold the arrangement and to
bring the D plane and F plane into the same plane,next
to each other,for ease of adjustment.

The nature and results of the image deblurring may
be additionally clarified qualitatively with reference
to FIG.1, FIG.2 and FIG.3. The optical tests of FIG.3.
show,as example,the degradation of a radial test-bar
chart when it is badly out of focus by the amount corresponding to the 5Å h(x,y) function shown.This,incidentally
also represents the degree of image degradation(blurring,
artifacts)which would appear in an electron micrograph
of such a test patternfor instance when scanned by means
of a beam having an intensity profile with a width of the
h(x,y) function shown,in a high-resolution scanning electron microscope[2-4,10].In the original test-bar chart
all the radial spokes(bars) in the fan-like chart had
the same intensity,either all black or all white,respec-
tively.There are several features of the blurred photo
which are noteworthy. First of all,we notice that the
overall contrast of the spokes(bars) decreases with increasing closeness of the bars.The "closeness" of the
bars is generally measured in the number of white or
black bars(or bar pairs) per linear inch(or millimiter)
at right angles to the bars.The number of bars per inch
(or millimeter) is called the "spatial frequency". In
the case shown the spatial frequency increases from right
to left.We notice that the overall contrast in the blurred
photo of the radial test-bar chart decreases with increasing spatial frequency.In fact, the blurred photo of the
test-bar chart materializes the MTF function H=F.T.[h]
according to the definition given above[see ref.6].A closer look at the blurred photo of the radial test-bar
chart(i.e.at the MTF function)reveals an even more striking phenomenon.In starting from the righ(lowest spatial
frequencies) we observe not only,as first stated,that the
contrast(ratio of black-white to black) of the bars decreases towards the higher spatial frequencies.In fact,
the contrast drops to zero altogether,somewhere one
third from the right.At that spatial frequency the image
of the bars having that frequency is so badly blurred
that there is no image at all.However,as we proceed towards still higher spatial frequencies,the contrast is
seen to gradually improve again,and this may well seem
to be encouraging.However,if we closely examine the location of the black maxima in that region,compared to the
location in the low-frequency third of the chart on the
right,we notice that the maxima are shifted by half a
period(half an interval between consecutive bars).This
shift is known as a "phase shift" for the bars having a
frequency in that domain(say the spatial frequencies in
the middle third of the chart in this case).What this
shift means,in terms of a more complicated image,is that

regions in the original object which have spatial frequency components in that region will have their <u>intensi-
ty</u> incorrectly represented in the blurred photograph.As
a result regions which should be black may be white,and
vice versa.That this reversal of intensities really happens under such circumstances may also be seen by looking
at the letter "A" in the blurred photograph of the word
"DIFFRACTION" in the same FIG.3: it may be seen that the
center of the letter "A" appears "black" in place where
it should have been "white"! The results of the holographic image-deblurring "restoration" are illustrated
for this case by the sharpened images in FIG.3(right).

It should be emphasized that the image deblurring
results shown may <u>not</u> be obtained by the several methods
of photographic(or electronic) 'contrast enhancement',
such as high-contrast printing,as one may perhaps be temp-
ted to assume when first exposed to such results.Clearly
no "high-contrast printing" alone could conceivably shift
the incorrectly located bars back to their correct posi-
tion(and to equalize the contrast throughout the radial
bar chart).Nor could "high-contrast printing" help in
the case of the reversed contrast in the letter "A".In
fact,high-contrast printing would have made the center
of the letter "A" even more "black",compared to the
rest,rather than restoring it to "white".

By again making reference to the electron micrograph
of FIG.3. and to the corresponding optical "test"photographs,we may further note that these image "artifacts"
were of course characteristic of all such "high-resolution"micrographs heretofore :they resulted in incorrect
structure and image interpretation,without doubt in many
cases.Even though this may not be readily apparent by
simple inspection of the electron micrograph of the virus,it is characterized by the same inversions of contrasts and shifts in spatial frequency components as
those which characterize the test photographs.

The restoration of intensities (i.e.imaging faithfulness) in a blurred photograph thus requires acting on
two parameters,to correct for contrast and location
errors in the spatial-frequency components:one must be
able to appropriately act on the amplitude(i.e.intensity)
and on the phase(i.e.the location of the maxima)of the
different components.

The solution to this problem is provided by the holographic Fourier-transform division filter in the manner

already presented above,but which may be further clarified as follows.The blurred photograph(transparency)is
illuminated by a collimated beam of laser light as shown
in FIG.1. The beam,upon transmission through the transparency is made to pass through the holographic filter
located in the back focal plane of the lens L_1 which
follows the transparency.We recall that a lens,used as
shown is capable of carrying out a Fourier transformation,
for example according to eq.[3]. It may be shown that
when a grating-like transparency,consisting of a regular
sinusoidal "bar" grating is illuminated by a laser beam
under these conditions,then a similar grating-like image
(spectrum) will appear in the "Fourier plane" where the
filter is located[14].In effect,the lens performs "Fourier transformations" on the images when considering
the transformation from electric field amplitude in
the wave upon transmission through the transparency to
the field in the focal plane of the lens which follows
the transparency.

What is remarkable is that the spatial frequency of
the spectrum in the Fourier plane is inversely proportional to that in the grating-like test-bar chart.In
other words,regions of the bar chart with coarse spatial frequencies(widely spaced bars) will produce a spectrum with very high spatial frequency(closely spaced
spectral "lines"),and vice versa. Moreover, no matter
where in the "blurred" transparency a given spatial frequency component section is located,its spectrum will
appear in the <u>same</u> place in the Fourier plane,provided
only that the components in the transparency have the
same orientation.If the bar grating is rotated,its spectrum rotates with it.Thus the spectra of all spatial
frequency(grating-like) components of the transparency
are all <u>superposed</u> in the Fourier plane,for components
of same orientation.It is this fact which permits one
to restore all the incorrectly imaged spatial frequency
components in the blurred photograph simultaneously
with only a single filter in the Fourier plane.In this,
the optical deblurring method has a considerable advantage over methods using for example point by point scanning of the blurred photograph,in a microdensitometer,
in view of digitizing and electronic digital computer
deconvolution. It will become clear now also why the
holographic filter which we use consists of two components,an amplitude and a phase component.

The amplitude component $1/|H|$ for the case of the stron-
gly out of focus photos of FIG.2. and FIG.3. consists of

the ring system shown in FIG.1(it happens to be an Airy
disk,with "gaussian"weighting).We can readily imagine
that this ring system is formed by rotating a grating-
-like spectrum about the optical axis.We note immediately that the amplitude component of the filter is dar-
kest at the center(the region corresponding to the
lowest spatial frequencies in the out-of-focus photographs)and that it becomes increasingly more transparent
away from the center(the regions corresponding to the
highest spatial frequencies,for instance in the radial
test-bar chart).Since the blurred photo of the test-bar
chart is characterized by a very weak contrast(intensity) in the high-frequency regions(left side of the
chart)compared to the low-frequency regions(right side
of the chart),the restoration of correct intensity is
performed by the amplitude component of the filter,which
greatly enhances the high-frequency components as a result of its great attenuation of the low-frequency components by the dark-ring portions in its central region.

The restoration of the phase(correct location of the
maxima of the frequency components in the radial-bar
test chart) is equally straightforward.A greatly enlarged section of the phase component[$e^{-i\emptyset}$] of the filter
is also shown in FIG.1.It is a part of the filter strad-
dling two of the white rings,near the center of the amplitude component.Close inspection of the phase component
of the filter also reveals a grating-like structure:it
is in fact the "carrier"grating of the hologram which
forms the phase component of the filter.An even closer
look reveals,moreover, that the carrier "fringes" in
the holographic grating are displaced in one ring by
exactly half a grating interval relative to the adjacent one[14].It may be shown that this half-interval
displacement in the filter is exactly the displacement
required to compensate for the half-interval displacement of the bars in the central third of the blurred
radial test-bar chart photograph.Similar phase shifts
characterize the corresponding shifts in the spatial
frequency components of the other blurred images shown.

Finally,we may mention that perhaps one of the most
important applications of our method of holographic
image deblurring is that which we recently initiated
in high-resolution electron microscopy,notably in view
of sharpening of electron micrographs of biological
specimens beyond the ultimate that can be achieved with
even the most powerful electron microscopes,in the present state of the art. Thus,most recently,as a part of

our collaborative effort with A.V.Crewe,we have used
our method,as mentioned,to successfully enhance electron micrographs of fd-virus test specimens obtained
under the best realizable conditions with his famous
scanning transmission electron microscope[11],opera-
ting at a magnification in excess of 800,000 and giving resolution of about 5Å,as shown in FIG.3(where
the results are printed with a magnification of 5 million,
using the scale of 1Å=0.5mm).As we discuss in detail in
our ref.[2], the electron micrograph of the virus was
successfully sharpened to a demonstrated resolution of
2.5Å,thus attaining the theoretical "limit of diffraction" ,while simultaneously also correspondingly increasing the S/N ratio compared to that which would
correspond to an original micrograph recorded by scanning
with a beam diameter of these dimensions,had such a
scanning beam not been unrealizable because as yet unsurmountable aberrations of the electron microscope.The
reason why holographic image sharpening may be used to
increase the resolution so considerably in the case of
such electron micrographs[the case of micrographs recorded in partially-coherent illumination is presented
by Stroke and Halioua in ref.3] may not be immediately
obvious. It may indeed be shown[2,3] that the electron
micrographs which result from scanning in the electron
microscope may be very closely approximated by images
which would have been obtained in the microscope by
scanning with an infinitely narrow beam(a physical im-
possibility) had these images been out of focus(according to ordinaryincoherent imaging [e.g.as described
by eq.1] by an amount which would make the spread func-
tion(blur circle)diameter equal to the diameter of the
scanning beam.Accordingly,the image deblurring method
and filter may be made to be ideally suited for the
sharpening of electron micrographs by the considerable
amount shown-much beyond the resolution attainable in
the microscope itself. The results of similar improvements of electron micrographs recorded with 'conventional' transmission electron microscopes,based on the
theory in our ref.[3] will be reported upon completion
of the experimental verifications which we have now
under way.

We may conclude by noting that even further improvements of electron microscopy have become possible as
a result of the successful development of the holographic image-sharpening method in actual practice.By its
ability to compensate for the geometrical aberrations
a posteriori, holographic image sharpening permits one
to use even larger apertures in the recording of the

original micrographs,and thus correspondingly even further increase the ultimate resolutions,as set by diffraction,while at the same time still further increasing the
S/N ratio.

In all cases described,it is essential,of course,that
the blurred transparency used in the filtering have an
electric field E_T transmittance equal(i.e. proportional)
to the intensity $g(x'y')$ of the original micrograph.This
may be achieved[1-4] by copying the original micrograph
("negative"with slope γ_N)onto a transparency("positive"
with slope γ_P) and by respecting the condition $\gamma_N \gamma_P = -2$
for the photographic curves in the linear parts of
the H&D range.

In conclusion,we may recall that holography[5]began
with an attempt to sharpen electron micrographs that were blurred by the spherical aberration of electron objec-
tives.This goal has in fact now been attained,as described,by means of a posteriori image deblurring,as first
proposed by the author in 1957[14].We may recall that
what is theoretically remarkable in our holographic image
deblurring method is that the phase component(complex
function)of the Fourier-transform division filter may be
realized in the form of a hologram.This was first shown
by Stroke and Zech[13]who also first showed that holograms,as originated by Dennis Gabor[5] could be used to
provide a practical solution to the spatial-filtering
(image deblurring)problem according to the principles
first suggested(even though not demonstrated at that
time) for the case of incoherent imaging by Maréchal
and Croce[15].In view of the great pleasure experienced
by the entire scientific community when the 1971 Nobel
Prize for Physics was awarded to Professor Dennis Gabor
for his discovery of holography[5,6,16],and recalling
that his very first hope was that holograms[5]could be
used to improve the imaging performance of electron microscopes,it may be gratifying to note that the results
which we present demonstrate that holography may indeed
be used to approach Gabor's original goal of obtaining
atomic resolutions in electron microscopy,notably with
biological specimens,even though Gabor's original holographic "wave-front reconstruction microscope itself[5]
remains yet to be brought to this level of perfection.

It is a pleasure to acknowledge the collaboration of
Dr.Maurice Halioua in the work described.The theoretical
and experimental results were obtained thanks to support
extended to the author in the form of grants by the

National Science Foundation[Grant GK-27313],the National
Aeronautics and Space Sdministration[Grant NGR-33-015-068]
and the Office of Naval Research[Contract NR-No.0150917].
The author wishes to acknowledge with particular gratitude the continuous fruitful suggestions and encouragement
received from Professor Dennis Gabor. FIG.3 was first
prepared in the form shown for use in Prof.Gabor's
"Nobel Lecture"(10 December 1971).

REFERENCES
1. G.W.STROKE,Optica Acta,16(1969)401
2. G.W.STROKE and M.HALIOUA, Optik(1972)(in print) I.
3. G.W.STROKE and M.HALIOUA, Optik(1972)(in print) II.
4. M.HALIOUA, Ph.D.Thesis,University of Paris(1971)
5. D.GABOR, Nature,161(1948)777
6. G.W.STROKE,An Introduction to Coherent Optics and
    Holography,Second Edition(Academic Press,1969)
7. J.L.HARRIS,J.Opt.Soc.Am,56(1966)569
8. R.NATHAN in Advances in Optical and Electron Microsco-
    py, V.E.Cosslett and R.Barer,eds.(Academic Press,1971)
9. T.C.RINDFLEISCH,J.A.DUNG,H.J.FRIEDEN,W.D.STROMBERG and
    R.M.RUIZ,J.Geophys.Res.76(1971)394
10. G.W.STROKE,M.HALIOUA,A.J.SAFFIR and D.J.EVINS in
    29th Annual Proceedings of the Electron Microscopy
    Society of America, C.J.Arceneaux,ed.(Claitor's
    Publ.Div.,Baton Rouge,Louisiana,1971)
11. A.V.CREWE,Quart.Rev.Biophysics,3(1970)137
12. G.W.STROKE, F.FURRER and D.LAMBERTY,Optics.Commun.1
    (1969)141
13. G.W.STROKE and R.G.ZECH,Physics.Lett.,25A(1967)89
14. G.W.STROKE,"Diffraction Gratings"in Handbuch der
    Physik,S.Flügge ed.,Vol.29(1967)(Springer Verlag)
15. A.MARECHAL et P.CROCE,C.rend.Ac.Sc.(Paris)237(1953)
    607
16. D.GABOR,W.E.KOCK and G.W.STROKE,Science,173(1971)11

INDEX